THE LOGIC OF ESSENTIALISM

The New Synthese Historical Library
Texts and Studies in the History of Philosophy

VOLUME 43

The titles published in this series are listed at the end of this volume.

THE LOGIC OF ESSENTIALISM

An Interpretation of Aristotle's Modal Syllogistic

by

PAUL THOM

The Australian National University,
Department of Philosophy,
Faculty of Arts,
Canberra, ACT, Australia

KLUWER ACADEMIC PUBLISHERS

DORDRECHT / BOSTON / LONDON

A C.I.P. Catalogue record for this book is available from the Library of Congress.

ISBN 0-7923-3987-8

Published by Kluwer Academic Publishers,
P.O. Box 17, 3300 AA Dordrecht, The Netherlands.

Kluwer Academic Publishers incorporates
the publishing programmes of
D. Reidel, Martinus Nijhoff, Dr W. Junk and MTP Press.

Sold and distributed in the U.S.A. and Canada
by Kluwer Academic Publishers,
101 Philip Drive, Norwell, MA 02061, U.S.A.

In all other countries, sold and distributed
by Kluwer Academic Publishers Group,
P.O. Box 322, 3300 AH Dordrecht, The Netherlands.

Printed on acid-free paper

Printed in the Netherlands

PREFACE

Aristotle's modal syllogistic has been an object of study ever since the time of Theophrastus; but these studies (apart from an intense flowering in the Middle Ages) have been somewhat desultory. Remarkably, in the 1990s several new lines of research have appeared, with series of original publications by Fred Johnson, Richard Patterson and Ulrich Nortmann. Johnson presented for the first time a formal semantics adequate to a *de re* reading of the apodeictic syllogistic; this was based on a simple intuition linking the modal syllogistic to Aristotelian metaphysics. Nortmann developed an ingenious *de dicto* analysis. Patterson articulated the links (both theoretical and genetic) between the modal syllogistic and the metaphysics, using an analysis which strictly speaking is neither *de re* nor *de dicto*.

My own studies in this field date from 1976, when my colleague Peter Roeper and I jointly wrote a paper "Aristotle's apodeictic syllogisms" for the XXIInd History of Logic Conference in Krakow. This paper contained the disjunctive reading of particular affirmative apodeictic propositions, which I still favour. Nonetheless, I did not consider that paper's results decisive or comprehensive enough to publish, and my 1981 book *The Syllogism* contained no treatment of the modal syllogism. The paper's ideas lay dormant till 1989, when I read Johnson's and Patterson's initial articles. I began publishing on the topic in 1991. Gradually my thoughts acquired a certain comprehensiveness and systematicity, till in 1993 I was able to take a semester's sabbatical to write up a draft of this book.

The book has developed to its present form under the watchful gaze of a select band of correspondents, prime among them Fred Johnson, but also including Ignacio Angelelli, Friedemann Buddensiek, Storrs McCall, Ulrich Nortmann, Richard Patterson, Peter Roeper, Klaus Schmidt, Robin Smith and Steve Thomason. There have been many others too who have helped improve the work by discussing it, whether in the Philosophy Program of the Research School of Social Sciences at the Australian National University, or the Australasian Society for Ancient Philosophy, or the Ancient Philosophy Colloquium at the University of Texas at Austin. I wish to thank all these for the critical attention they have given my work. Finally, it is a pleasure to record my thanks to the Philosophy Department of the University of Melbourne who gave me the status of Visiting Fellow for a very happy and productive semester in 1993.

P.T.
Canberra
January 1996

TABLE OF CONTENTS

INTRODUCTION

CHAPTER ONE. ELEMENTARY SYNTAX

CHAPTER TWO. AXIOMS

CHAPTER THREE. THESES

CHAPTER FOUR. REJECTIONS

CHAPTER FIVE. FLAWS IN THE FABRIC

CHAPTER SIX. SYSTEM *AL*

CHAPTER SEVEN. SYSTEMS *AQ* AND *AQ'*

CHAPTER EIGHT. SYSTEM *AX'*

CHAPTER NINE. SINGULAR MODAL SYLLOGISTIC

CHAPTER TEN. ESSENTIALISM

REFERENCES AND INDEXES

INTRODUCTION

§0. ENTERING THE REALM OF DARKNESS

Tradition accords to Aristotle the title of first logician, and by so doing recognises his primacy in theorising the central logical notions of form, proof and model. When twentieth-century logicians rediscovered the axiomatic method in logic they found that Aristotle had been first in that too: his logical theory not only dealt with deductive structures, but was itself organised deductively, *more geometrico*. In recent times, with the rise of natural deduction systems in logic, Aristotelian logic has been found to be such a system. The riches of the first logician's thought have perhaps not yet been exhausted.

But the modal syllogistic? This remains one of the most difficult parts of Aristotle's work. Patzig, who along with many recent writers on syllogistic does not delve into it, describes it as "a realm of darkness".[1] The main reasons for the text's obscurity are two. First, Aristotle gives no indication of what semantic interpretation he intends for the forms of proposition with which he deals. Second, when he deals with contingency-propositions he distinguishes several different senses of contingency but gives only scant indications about which sense is intended in particular passages. I shall try to show that imaginative interpretation and careful reading can illuminate the text's obscurities. The most appropriate semantics turns out to be a *de re* approach, adapted to accommodate the laws of conversion and to allow for the differences between Aristotle's treatment of affirmative and negative modals. The senses of *endechesthai* need to be sorted into those that play no essential rôle in the modal logic and those that do. And it turns out that Aristotle's cryptic indications, understood in context, suffice to enable us to tell which of the latter senses is intended in any particular passage.

Becker, having devoted considerable exegetical effort to the text, thought it was inconsistent.[2] It is not my intention in these pages to argue that the modal syllogistic is free of inconsistencies. In the end I have been unable to settle on uncontroversial interpretations of every passage, or to say that the text is always right. But I can say that I have formed a consistent interpretation which makes the bulk of the text not only logically correct but philosophically interesting. This book's main aim is to present and defend that interpretation.

An interpretation always goes beyond its text. However, so long as it remains interpretation, it is not free to go beyond the text in random ways, but enjoys only a constrained freedom. The nature of the constraints varies according to the type of interpretation being undertaken. Ideally, an interpretation of the modal syllogistic would (i) make valid those forms Aristotle says are valid, (ii) make invalid those forms Aristotle says are invalid, (iii) make true, or at least consistent with

---

[1] PATZIG p.86 n.21.

[2] BECKER p.42.

Aristotelian doctrine, those propositions Aristotle says are true, (iv) make sound the methods of validation, and invalidation, used by Aristotle, and (v) establish connections with Aristotelian uses of the concepts of necessity, possibility and contingency outside the *Prior Analytics*. In addition, the interpretation should at least be consistent with plausible hypotheses concerning the text's genesis.

I shall argue that no interpretation of the modal syllogistic can be ideal by these standards, because of the text's mistakes and inconsistencies. All the same, it has seemed to me that Aristotle's system is a great intellectual achievement, containing much of enduring value because of its interlacing of logical and metaphysical theory. If this is so, there is value in presenting an interpretation that falls short of the ideal.

The main lines of my interpretation are as follows.

There are different layers in the modal syllogistic. Fairly clearly distinguishable are a layer dealing with necessity-propositions and one dealing with contingency-propositions (though Aristotle does deal with individual syllogisms containing both these types of proposition). Within the contingency-syllogistic it will be useful to distinguish cases where the subject-term of a contingency-proposition is *ampliated* (*i.e.* extended to cover everything that could possibly fall under the term) from cases where it is not. The logic of possibility-propositions is also covered, but not extensively, cropping up from time to time in the discussion of necessity- or contingency-syllogisms. Finally, propositions that Aristotle calls 'unqualified' assertorics are discussed in one or two places. As far as is feasible, I treat these different layers separately.

The major textual difficulties are as follows. Some aspects of the text are simply hard to make sense of. These include the problem of the two Barbaras. Aristotle wants Barbara LXL (*Every c is b, Every b is necessarily a, so every c is necessarily a*) to be valid while Barbara XLL (*Every c is necessarily b, every b is a, so every c is necessarily a*) is invalid. Similarly with Barbara QXQ (*Every c is b, Every b is contingently a, so every c is contingently a*) and Barbara XQQ (*Every c is contingently b, every b is a, so every c is contingently a*). This suggests that he has in mind a *de re* reading of necessity- and contingency-propositions. But he also holds that Li-conversion (*Necessarily some b is a, so necessarily some a is b*) and Qi-conversion (*Contingently some b is a, so contingently some a is b*) are valid; and this suggests that he reads these propositions *de dicto*. In my opinion, the problem falls short of an inconsistency. It simply challenges the invention of interpreters to come up with a reading which validates Barbara LXL and QXQ, along with Li-conversion and Qi-conversion, but does not validate Barbara XLL or XQQ.

More serious difficulties for interpreters are posed by the presence in the text of passages with which we, as logicians and metaphysicians whose own projects at least overlap with Aristotle's, cannot agree. Here we must distinguish incidental mistakes, which can be rectified without altering the body of theses, from substantial errors. I shall argue that there are two substantial errors.

The first of these occurs in the contingency-syllogistic, when Aristotle claims that no conclusion follows from the premisses *Every b is necessarily a, Every c is contingently a*, or from the premisses *Every b is contingently a, Every c is necessarily a*. Contrary to what Aristotle says, it seems obvious (since the necessary and the contingent are mutually exclusive) that *No b is c* follows.

The second substantial error concerns 'unqualified' assertorics. Aristotle says that the inference *Every c may or may not be b, Every b is a, Therefore every c may be a* is valid if the second premiss is taken without any 'qualification', but is invalid if it is restricted to one particular time. He demonstrates the validity-claim by means of a procedure which has recently been called "upgrading". I argue that the syllogism is valid whether or not the second premiss is 'unqualified', that the procedure of upgrading does not establish validity, and I formulate a restricted version of that procedure which does establish validity.

There are two major inconsistencies between different parts of the modal syllogistic.

I shall argue that there is an inconsistency in Aristotle's necessity-syllogistic. Briefly, it is this. Aristotle rejects the inference *Some c is necessarily not a, Every b is a, so some c is necessarily not b*. On the other hand he accepts as valid the inference *Some c is necessarily not a, Every b is necessarily a, so some c is necessarily not b*. He proves the validity of the latter inference by the process of Ecthesis. My argument will be that this process can also be used to demonstrate the validity of the former inference. (Similar considerations apply to the inferences *Every c is b, Some c is necessarily not a, so some b is necessarily not a* and *Every c is necessarily b, Some c is necessarily not a, so some b is necessarily not a*.)

Secondly, I shall argue that, if we accept Aristotle's statement that some animals are contingently white, then an inconsistency exists between the theory of modal opposition (whereby *Some b is contingently a* is incompatible with *Every b is necessarily a*) and the doctrine that particular contingency-propositions are convertible (*Some b is contingently a* being equivalent to *Some a is contingently b*).

I shall interpret Aristotle's text as containing a deductive formal system, construct a formal semantics for that system, and then examine the extent to which that semantics can be given a metaphysical turn.

In regard to the text, I agree with Becker that the key terms in the modal syllogistic are systematically ambiguous; I do not agree that the recognition of these ambiguities precludes the possibility of a consistent interpretation. Nor do I think that the ambiguity between *de re* and *de dicto* readings is particularly salient, though there are occasions when Aristotle confuses these. I do think that the text contains five ambiguities whose recognition is vital if we are to form an adequate interpretation.

One of these ambiguities was seen by Alexander, when he suggested that the contingency-syllogistic deals with propositions of two distinct forms, which the medievals called unampliated and ampliated (*What* is *b may or may not be a* and *What* may or may not be *a may or may not be b*). I follow this suggestion.

A second important ambiguity is that between the 'qualified' and 'unqualified' readings of assertoric propositions; and my analysis here takes as its starting point a suggestion of Angelelli's, who treats 'unqualified' assertorics as having ampliated subject-terms. I take it that the differences between ampliated and unampliated contingency-statements, and between 'unqualified' and ordinary assertorics, are logical, not merely superficial syntactic, differences. This means that the faithful interpreter should distinguish those logical theses that concern ampliated from those that concern unampliated contingency-forms, and should distinguish those theses that concern 'unqualified' from those that concern ordinary assertorics. Because I make these distinctions, my treatment of the modal syllogistic looks quite different from the interpretation of McCall, for whom all assertoric-forms, and all contingency-forms, are alike.

A third ambiguity is the one pointed out by Patterson between strong and weak modals. Strong modals state relations between natures as such, weak ones refer to whatever happens to be of a given nature, though maybe not described as having that nature. The modal syllogistic is designed to cover modals of both types, though sometimes Aristotle forgets this and makes statements that apply only to strong modals.

A fourth important ambiguity is the one between what may be called "metaphysical" and "non-metaphysical" readings of modal propositions. Metaphysical modals state of the members of the subject-class that their nature requires or forbids or allows that they be members of the predicate-class. Non-metaphysical modals follow from primary ones by conversion. As with the strong/weak distinction, Aristotle sometimes makes statements that apply only to metaphysical modals though his theory is intended to cover both types.

Finally, there is an ambiguity between possibility-propositions stating what a subject can *be* and possibility-propositions stating what the subject can *become*. I call this the static/dynamic ambiguity. Aristotle is sometimes inattentive to this distinction.

I take the modal syllogistic, like the assertoric, to be a formal logical theory, not a heap of *ad hoc* rules. Moreover, I take the approach to be fundamentally extensional, as in the assertoric syllogistic.

In proof theory, I follow the definitive work of Smiley and Corcoran in using a natural deduction analysis of the syllogistic. My presentation differs from theirs in aiming to be readily re-interpreted in axiomatic terms by those who cling to the

Łukasiewicz-Patzig view of the syllogism as an implication rather than an inference. I place considerable emphasis on Aristotle's proofs by Ecthesis; in this way I depart from the admirable work of McCall, who ignores these proofs. Ecthetic proofs may be construed as involving singular modal propositions, and another feature of my approach is that it allows Aristotle's modal syllogistic to be reduced to a logic of such propositions.[3] In this reduction, I differ from Patterson who wishes to understand Aristotle's modal sentences as *sui generis*.

In semantics, I follow the path-breaking work of Johnson and Thomason. Johnson showed us how to understand Aristotle's modal sentences in terms of a structure of sets which makes no appeal to possible worlds.[4] Thomason extended this approach to a range of systems of varying strength. One has to regard this kind of semantical analysis of the modal syllogistic as particularly appropriate if one thinks that Aristotelian metaphysics are in any way implicit in that syllogistic. For, Aristotle's metaphysic envisages a single world. True, in that world there is a fundamental difference between some sets (the Aristotelian Kinds) and others. But this difference is not, to Aristotle's mind, explained by any consideration of what is the case in other worlds. It is a primitive difference which fundamentally structures this world - Nature.

On the question of what Aristotle's motivation might have been in constructing his modal syllogistic, I follow Patterson. The basic idea is that the modal syllogistic formalises the logical features that are common to the strong and the weak modals of Aristotelian metaphysics. Here I take issue with Johnson (who does not in my opinion provide a metaphysical foundation which is sufficiently Aristotelian), and with Łukasiewicz (who wished to cordon off all metaphysical considerations). However, the metaphysical interpretation to be developed here distinguishes between those parts of Aristotelian metaphysics which are implicated in the modal syllogistic and those that are not. A good deal of the metaphysics in the *Categories* is not thus implicated. Nor, significantly, is the theory of potentiality in the *Physics*: the possibilities which the modal syllogistic formalises are not Aristotelian potentialities. They are, with few exceptions, static rather than dynamic possibilities.

Łukasiewicz at the end of his book poses the question whether Aristotle's modal syllogistic "has any useful application to scientific or philosophic problems".[5] His response shows an interesting complexity. As a logician himself, he acknowledges the value of Aristotle's insights into modal propositional logic. But Łukasiewicz, besides being a logician, is an anti-essentialist, and as such he declares Aristotle's

3 This approach was also taken in THOM 1981 §44.

4 JOHNSON 1995, with its elegant decision-procedures for apodeictic syllogisms, appeared too latre for discussion here.

5 ŁUKASIEWICZ §62.

treatment of apodeictic syllogisms "a failure", castigating Aristotle for his "disastrous", Plato-influenced, view that "propositions which ascribe essential properties to objects are ... not only factually, but necessarily true". The disaster to which he alludes is, of course, essentialism, his own view being that "there are no true apodeictic propositions". The logician in the man engages with Aristotle's text; the anti-essentialist cannot do so, and therefore declares the text a disaster. Such is Łukasiewicz's interpretative predicament. But interpretation need not be tied so closely to the interpreter's interests. Aristotle was indeed an essentialist, but his interpreters do not have to be. They will, however, need to find something of value in Aristotle if they are to go to the trouble of interpreting him. My own view is that Aristotle's modal syllogistic remains valuable because it embodies the first attempt to give a formal analysis of the statements that are involved in an essentialist metaphysics - the eternal verities of metaphysical theory as well as statements that apply those truths to the fluctuating contingencies of the everyday world.

CHAPTER ONE. ELEMENTARY SYNTAX

This Chapter contains an exposition of the syntactic basis of Aristotle's modal syllogistic. It defines the variables, functors, modal signs, contradictories and contraries, modal forms (necessity-, possibility-, contingency-, ampliated and unampliated, 'unqualified'), well-formed formulae, theses, and the three Figures, that occur in our representation of Aristotle's system. It catalogues Aristotle's varying linguistic expression of modal forms, and his doctrine of opposition between modal forms. It states the transformation rules of his deductive system - Substitution, Permutation, Cut, *Reductio*, some rules of Modal Ecthesis whose details are a little indeterminate, and a rule of Modal *Reductio* which if stated in its full generality is unsound. (These indeterminacies and unsoundnesses will be remedied in later Chapters.)

§1. ELEMENTARY FORMS

Aristotle uses the letters

$$A, B, \Gamma, \ldots$$

as *term-variables*, intended to represent noun-phrases. It would be nice to follow him exactly, but if we did so our notation would end up looking unnecessarily cluttered; so, for aesthetic reasons, I use instead the letters

$$a, b, c, \ldots$$

The superscripts

$$^{a}, ^{e}, ^{i}, ^{o}$$

will act as *functors* turning pairs of term-variables into propositional forms. They are respectively the universal affirmative, the universal negative, the particular affirmative, and the particular negative, functors. In a string

$$v_1 v_2{}^f$$

v_1 is the *predicate*, v_2 the *subject*. The propositional forms

$$v_1 v_2{}^a,\ v_1 v_2{}^e,\ v_1 v_2{}^i,\ v_1 v_2{}^o$$

are called *assertoric forms*, respectively the universal affirmative, universal negative, particular affirmative and particular negative. The universal affirmative form is intended to be read "v_1 belongs to every v_2" or "Every v_2 is a v_1". The universal negative is intended to be read "v_1 belongs to no v_2" or "No v_2 is a v_1". The particular affirmative is intended to be read "v_1 belongs to some v_2" or "Some v_2 is a v_1". And the particular negative is intended to be read "v_1 belongs to not every

v_2" or "Not every v_2 is a v_1". Members of the pairs $\{ab^a, ab^o\}$, $\{ab^e, ab^i\}$ are said to be *contradictories*. Members of the pair $\{ab^a, ab^e\}$ are said to be *contraries*.

In addition to the above primitive symbols we need the *modal signs*

$$L,\ M,\ Q,\ Q',\ Q'',\ X'$$

for necessity, possibility (one-way possibility), unampliated or ampliated contingency (two-way possibility), and ampliated assertorics. (We will also use the sign X for ordinary assertorics.) One of these, when prefixed to an assertoric form produces a *modal form*.

The modal forms

$$Lv_1v_2{}^a,\ Lv_1v_2{}^e,\ Lv_1v_2{}^i,\ Lv_1v_2{}^o$$

are *necessity-forms* (or apodeictic or L-forms), respectively universal affirmative or negative, and particular affirmative or negative. They are intended to be read respectively "v_1 belongs necessarily to every v_2" or "Every v_2 must be a v_1", "Necessarily v_1 belongs to no v_2" or "No v_2 can be a v_1", "v_1 belongs necessarily to some v_2" or "Some v_2 must be a v_1" , and "Necessarily v_1 does not belong to some v_2" or "Some v_2 cannot be a v_1".

The modal forms

$$Mv_1v_2{}^a,\ Mv_1v_2{}^e,\ Mv_1v_2{}^i,\ Mv_1v_2{}^o$$

are *possibility-forms* (or problematic or M-forms), respectively universal affirmative or negative, and particular affirmative or negative. They are intended to be read respectively "v_1 possibly belongs to every v_2" or "Every v_2 may be a v_1", "Possibly v_1 belongs to no v_2" or "No v_2 may be a v_1", "v_1 possibly belongs to some v_2" or "Some v_2 may be a v_1", and "Possibly v_1 does not belong to some v_2" or "Some v_2 may not be a v_1".

The modal forms

$$Qv_1v_2{}^a,\ Qv_1v_2{}^e,\ Qv_1v_2{}^i,\ Qv_1v_2{}^o$$

are *contingency-forms* (or Q-forms), respectively universal affirmative or negative, and particular affirmative or negative. They are intended to be read respectively "v_1 belongs contingently to every v_2" or "Every v_2 may or may not be a v_1", "Contingently v_1 belongs to no v_2" "Every v_2 may or may not be a v_1", "v_1 belongs contingently to some v_2" or "Some v_2 may or may not be a v_1", and "Contingently v_1 does not belong to some v_2" or "Some v_2 may or may not be a v_1". For the ampliated versions of these forms

$$Q'v_1v_2{}^a, Q'v_1v_2{}^e, Q'v_1v_2{}^i, Q'v_1v_2{}^o$$

$$Q''v_1v_2{}^a, Q''v_1v_2{}^e, Q''v_1v_2{}^i, Q''v_1v_2{}^o$$

replace 'v_2' by 'possible v_2' and 'contingent v_2' respectively.

We need a separate class of ampliated contingency-forms because A13, 32b23-32 notes a syntactic ambiguity in the expression "*kath' hou to B, to A endechesthai*". It appears from that text that contingency-propositions may be either ampliated or unampliated. In the ampliated sense, there is, besides the expressed mode of contingency, an unexpressed mode of possibility affecting the subject-term. Consider universals. In the unampliated sense, the proposition that *a* is contingent for all *b* means that for everything that is *b*, *a* is contingent. In the ampliated sense, it means that for everything that *could* be *b*, *a* is contingent. The text explicitly draws attention to this ambiguity. A further (lexical) ambiguity remains, to which the text does not draw attention. This concerns the meaning of the unexpressed mode of possibility. In referring to what *could* be *b* do we mean "could" in the sense of Q or of M? Aristotle offers no answer, and we defer the question to §29.[1]

Finally we have the ampliated assertoric forms

$$X'v_1v_2{}^a, X'v_1v_2{}^e$$

(universal X'-forms, affirmative and negative). They are intended to be read (respectively) "v_1 belongs to every possible v_2" or "Every possible v_2 is a v_1", "v_1 belongs to no possible v_2" or "No possible v_2 is a v_1" *etc.*

We need a separate class of ampliated assertorics because, as Angelelli notes, the logic of Aristotle's 'unqualified' propositions appears to require that such assertorics be understood as ampliating their subjects to the possible, so that "v_1 belongs without qualification to every v_2" means "v_1 belongs to every possible v_2".[2] As Angelelli puts it in discussing Aristotle's example *All horses may be moving, All moving things are without qualification men, so all horses may be men*:

> What really matters is that horses *can* move even if all moving objects happen to be men .... Then, the premise required ... should be: 'all things that *can* move are men' ....[3]

He continues:

---

[1] The above argument and the notation Q' are used in BRENNER. Independently, they also occur in THOM 1994.

[2] ANGELELLI §9.

[3] ANGELELLI pp.200-201.

> There is no reason why the 'possible' antecedent "that can move" has to be two-sided, although the example in 34b7-18 may suggest it (horses can move). What is needed by [XaQa in Figure 1] is just the one-sided possibility that it is not impossible for horses to move.[4]

Further details on the semantics of unqualified assertorics we defer till §34.

The terms "contradictory" and "contrary" as applied to modal forms will be defined in §2.

Assertoric forms and modal forms together make up the set of *categorical forms*.

Aristotle's linguistic expressions for modal forms vary in three different ways.

(i) Word-order. The linguistic components of a modal form are: MODE-PREDICATE-SUBJECT-QUANTIFIER. This is the order I have adopted for my symbolism, and Aristotle himself sometimes employs it, *e.g.* in expressing the forms Lab^e (A3, 25a29-30, 30-31), Qab^a (A17, 37a20-21) and Qab^e (A17, 37a10-11): we could mimic his word-order in English as "It is $\left\{\begin{matrix}\text{necessary}\\ \text{possible}\end{matrix}\right\}$ that a belongs to b in $\left\{\begin{matrix}\text{every}\\ \text{no}\end{matrix}\right\}$ case". But more frequently Aristotle uses other word-orders. Among the many variations he employs, his favourite seems to be PREDICATE-QUANTIFIER-SUBJECT-MODE (with "belongs" interpolated at different points). He uses this order to express the forms Lab^a (A9, 30a38; A10, 30b29, 31a11-12; A16, 35b38-39, 36a3-4, 18-19; A21, 39b37; A22, 40a13), Lab^e (A10, 30b26-28), Lab^i (A3, 25a33, 34; A9, 30a39-40; A10, 30b28; A11, 31a40-41, 31b29), Lab^o (A10, 30b25-26; A11, 31b36-37; A17, 37a4-5), Qab^a (A14, 32b39-40; A15, 34a35-36; A16, 36a4-5; A20, 39a14-15), Qab^e (A14, 33a9-10), Qab^i (A14, 33a26, 32-33; A16, 36a35), and Qab^o (A14, 33a30; A15, 35b6-7; A20, 39a21-22). We could mimic the word-order: a belongs to $\left\{\begin{matrix}\text{every}\\ \text{no}\\ \text{some}\end{matrix}\right\}$ b $\left\{\begin{matrix}\text{necessarily}\\ \text{possibly}\end{matrix}\right\}$.

(ii) Omission of "belongs". In all the above texts the word "belongs" occurs. But Aristotle frequently omits it. For example, he expresses the forms Lab^a (A22, 40a13-14), Lab^e (A11, 31a36; A15, 34b36-37), Lab^o (A19, 38a24), Qab^a (A14, 32b38, 33a5-6, 23, 29-30; A15, 34a35, 41-b1, 34b4-5, 40; 35a9-10, 18; A16, 36a9; A18, 37b27; A19, 38a19, 38b11-12; A20, 39a16-17, 18), Qab^e (A14, 33a2,

[4] ANGELELLI p.201.

3, 6, 14, 25; A15, 35a7-8, 17; A17, 36b36, 37), Qab^i (A14, 33a24; A20, 39a17, 18-19; A20, 39a34-35; A22, 40a14-15) in ways such as

$$\text{“}a\text{is} \begin{Bmatrix} \text{necessary} \\ \text{possible} \end{Bmatrix} \text{for} \begin{Bmatrix} \text{every} \\ \text{no} \\ \text{some} \end{Bmatrix} b\text{”}.$$

(iii) Use of the language of possibility for L-forms.[5] Aristotle sometimes expresses Le- and Lo-forms in the language of possibility, namely in sentences meaning "*a* is not possible for any *b*" (A10, 30b10, 12, 30b13, 14-15, 16-17, 31a6-7, 8; A11, 31b35-36; A16, 36a12-13; 36a34-35, 37, 36a38; A19, 38a23) or "*a* is not possible for some *b*" (A19, 38a24).

Smith perceptively remarks on a subtle difference between Aristotle's ways of expressing negative contingencies and negative necessities.

> Aristotle often expresses a necessary *e* premise this way (the general form is 'to none is it possible for A to B to belong': *oudein endechetai to A tōi B huparchein*). An almost identical idiom serves for possible *e* premises: 'it is possible for A to no B to belong' (*endechetai to A oudeni tōi B huparchein*). The crucial distinction is the order of 'is possible' (*endechetai*) and 'to no' (*oudeni/mēdeni*): when *oudeni* precedes *endechetai* the meaning is 'necessarily not to any', while otherwise the meaning is 'possibly not to any'.[6]

There is no general agreement about the deep syntax of the modal forms with which Aristotle deals. The status of his modal adverbs is a disputed matter. Are they modifications of the copula, as some commentators maintain; or are they modifications either of the predicate or of the whole proposition (perhaps ambiguously)? Syntax cannot be decided independently of semantics; so we shall for the present adopt a non-committal notation, assigning a definite syntactic structure to these propositions only after we have an overview of their semantic behaviour. Thus for modal forms:

$$Sv_1v_2{}^f$$

we do not, at present, offer anything deeper than the superficial syntactic analysis by which these categories of expression were defined.

For a proposition *p*'s *contradictory* I shall write:

$$\sim p;$$

and for a *contrary* of *p* I shall write:

$$\neg p.$$

---

[5] See WIELAND 1980, p.114.
[6] SMITH p.123.

(A proposition may have more than one contrary.) §2 lists what the contradictories and contraries are, for various forms of proposition.

The *well-formed formulae* of the modal syllogistic consist of a syllogistic sign

$$\overline{\qquad}$$

with a (non-null) sequence of categorical forms above it and one categorical form below it. The forms above the syllogistic sign will be called the wff's *premisses* and the form below it the *conclusion*, without prejudice to the standard usage of these words.

The very nature of Aristotle's syllogistic has been a disputed point. Łukasiewicz and Patzig took the view that a syllogism is a conditional propositional form whose antecedent is the conjunction of the premisses and whose consequent is the conclusion; consequently they viewed the syllogistic as an axiomatic theory based on an unstated system of propositional logic. A succession of commentators have disagreed with this interpretation. Corcoran, Smiley and Lear have proposed instead that a syllogism is a deduction not a propositional form. (Smith[7] goes so far as actually to translate "*sullogismos*" "deduction".) On their view the syllogistic does not rest on any system of propositional logic, but is a natural deduction system in which from a set of basic deductions further deductions are generated by proof-theoretic procedures. One clear advantage of this interpretation is that it generates plausible representations not only of Aristotle's results but also of his methods of establishing them. It is the main aim of the present work to present natural deduction interpretations of the various parts of Aristotle's modal syllogistic. Still, I find it useful to write wffs in such a way that they can be understood as conditional propositions with a variable number of antecedents. For, in addition to the recent studies which generally take a natural deduction standpoint, there is an immensely valuable tradition of scholarly work on the syllogistic (including the work of Łukasiewicz, Patzig and Mignucci) which assumes that the theses of the system are forms of conditional proposition. By retaining this ambiguity we remain in touch with that tradition.

Two-premissed wffs fall into one of three Figures:

TABLE 1.1

1	2	3
$\dfrac{ab \quad bc}{ac}$	$\dfrac{ab \quad ac}{bc}$	$\dfrac{ac \quad bc}{ab}$

The *theses* will be those wffs that are either axioms or else obtainable from the axioms by means of transformation rules. The axioms and transformation rules together comprise the *basis* of the system.

---

[7] SMITH pp.xvi-xvii.

§2 MODAL OPPOSITION

§2.1 *Necessity- and possibility-forms*

We saw in §1 that Aristotle sometimes expresses apodeictic propositions in the language of possibility, formulating negative apodeictics as denials of possibility. This would be natural if he accepted what Bocheński calls the 'classical'[8] theory of equipollences between L- and M-forms, as represented in Figure 2.1.

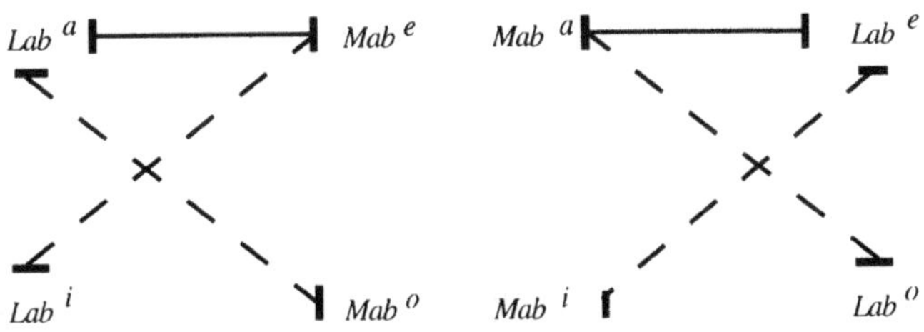

Fig. 2.1

The broken lines join pair of contradictories and the unbroken lines join other incompatible pairs. I shall refer to contradictory pairs by a notation like "$La \mid Mo$", and to incompatible pairs in general by a notation like "$La \parallel Me$".

That Aristotle did accept the contradictory pairs of this theory is indicated by A13, 32a21-29 which states the equivalences $\sim M \equiv L \sim$ and $M \equiv \sim L \sim$:[9]

For 'not possible to belong', 'impossible to belong', and 'necessary not to belong' are either identical or follow from one another; so their opposites, 'possible to belong', 'not impossible to belong', and 'not necessary not to belong', will also either be identical or follow from one another.

These two equivalences can be taken as summary statements of the claims that La- and Mo-forms are contradictories, as are Le- and Mi-forms, Li- and Me-forms, and Lo- and Ma-forms.

It is one thing to state a logical principle, something distinct to use it. Aristotle makes use of the principles of modal equipollence in his proofs by Indirect Reduction. An Indirect Proof of a syllogism

---

[8] BOCHENSKI p.93.

[9] See BECKER pp.11-14 on the authenticity of this passage.

$$\frac{p \quad q}{r}$$

possesses the structure

(1) Suppose p and q

(2) then r

(3) ∵ assume $\sim r$

(4) if $\sim r$ and p then $\sim q$ (or $\neg q$)

(5) but we supposed q.

In this structure there are two salient juxtapositions: the first occurs at steps (2) and (3) when the 'assumption' is juxtaposed with its contradictory, the target conclusion. The second juxtaposition occurs at steps (4) and (5) when the deduced 'impossibility' is juxtaposed with one of the original suppositions, which is its contradictory or contrary. Thus, one way to find out which pairs of forms Aristotle treated as contradictories is to look at these juxtapositions in his Indirect Proofs.

What we find is that he does treat all four 'classical' L-M contradictions as such. The *La* | *Mo* contradiction is used in a (2)-(3) juxtaposition in the Indirect Proof of Bocardo QXM:

... *a* will possibly not belong to some *b*. For if *a* belongs necessarily to all *b* ... (A21, 39b34-36)

The *Le* | *Mi* contradiction is used in a (2)-(3) juxtaposition in his Indirect Proof of Le-conversion:

... it is necessary that *b* belongs to no *a*. For, if it is possible for some, ... (A3, 25a30-31)

The *Li* | *Me* contradiction is used in a (2)-(3) juxtaposition in the Indirect Proof of Celarent X'QM:

... *a* possibly belongs to no *c*. For suppose that is not possible ... (A15, 34b21-22)

(The context makes it clear that the 'assumption' here is Lac^i.) The *Lo* | *Ma* contradiction is used in a (2)-(3) juxtaposition in the Indirect proof of Barbara X'QM:

a possibly belongs to all *c*. For suppose that is not possible ... (A15, 34a35-36)

(The context makes it clear that the 'assumption' here is Lac^o.) Aristotle therefore accepted all the classical M-L contradictions:

La | *Mo*   *Le* | *Mi*   *Li* | *Me*   *Lo* | *Ma*.

That he also accepted the incompatibility *La* ‖ *Me* appears from his Indirect Proof of the partial convertibility of La-forms (for which, see §8.2). Given this, it seems natural also to accept the incompatibility $Le \parallel Ma$. Thus I attribute the full 'classical' doctrine of L-M oppositions to Aristotle.

§2.2 Necessity- and contingency-forms

According to what again may be called the classical doctrine, as specified in McCall,[10] the incompatibilities holding among Q- and L-forms are those shown in Figure 2.2. Did Aristotle subscribe to this classical doctrine? He discusses the opposition of Q- and L-forms in the following passage:

For, not: if *a* is not possible for no *b*, it necessarily belongs to some. For, 'not to be possible for none' is said in two ways: if it belongs necessarily to some, and if it necessarily does not belong to some. For, what necessarily does not belong to some *a* is not truly said to possibly not belong to all, any more than what necessarily belongs to some [is truly said] to possibly belong to all. If then someone claims that because *c* does not possibly belong to all *d*, it necessarily does not belong to some, he is mistaken. For it belongs to all, but because it belongs necessarily to some, it is clear that it is not possible for all. So to 'it possibly belongs to all' both 'it necessarily belongs to some' and 'it necessarily does not belong to some' are opposed [*antikeitai*]. Similarly for 'it is possible for none'. (A17, 37a14-26)

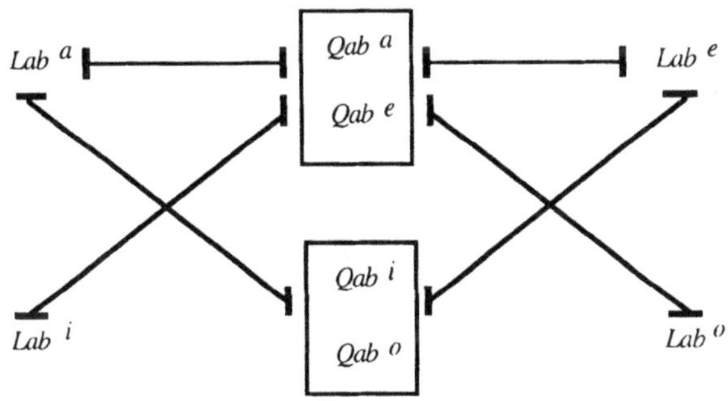

Fig. 2.2

Aristotle here asserts the modal incompatibilities $Qa \parallel Li$, $Qe \parallel Li$, $Qa \parallel Lo$ and $Qe \parallel Lo$. Since, as we shall see shortly, La- and Le-forms imply their Li- and Lo-analogues, this commits him to the incompatibilities $Qa \parallel La$, $Qe \parallel La$, $Qa \parallel Le$ and $Qe \parallel Le$, *i.e.* to the full classical theory so far as universal Q-forms go.

---

[10] McCALL 1963 §23.

This is how the passage is understood by Mignucci.[11] According to him, Aristotle thinks that instead of the relation shown in Figure 2.3.

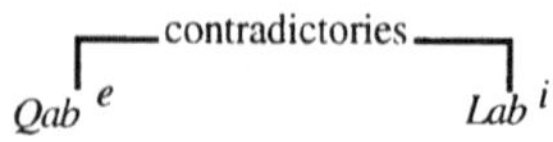

Fig. 2.3

we in fact have that shown in Figure 2.4.

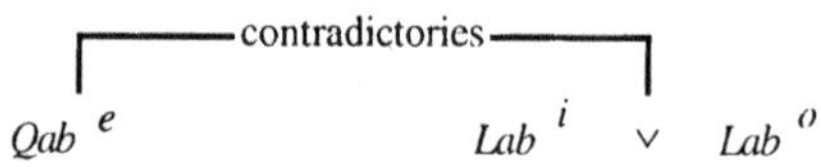

Fig. 2.4

And, of course, if this is so and if Qab^e does not imply either Lab^i or Lab^o separately, then the incompatibilities in question are (as McCall notes)[12] contrarieties.

Still, the passage carries no implication regarding the opposition of *particular* Q-forms. The oppositions of particular Q-forms are logically independent of those for universals, so Aristotle's commitment to the former does not entail commitment to the latter.

One may think that the modal oppositions $Qo \parallel La$, $Qo \parallel Le$, $Qi \parallel La$ and $Qi \parallel Le$ are of a piece with those which Aristotle formulates. Though, while it is plausible that the contradictory of Qab^e is $Lab^i \vee Lab^o$, it is not plausible to suggest that the contradictory of Qab^i is $Lab^a \vee Lab^e$: this leaves out the possibility that some *b*s are necessarily *a* and other *b*s are not possibly *a*. The logical alternatives to Lab^a are shown in Figure 2.5.

---

[11] MIGNUCCI *ad* 37a31.
[12] McCALL 1963 §23.

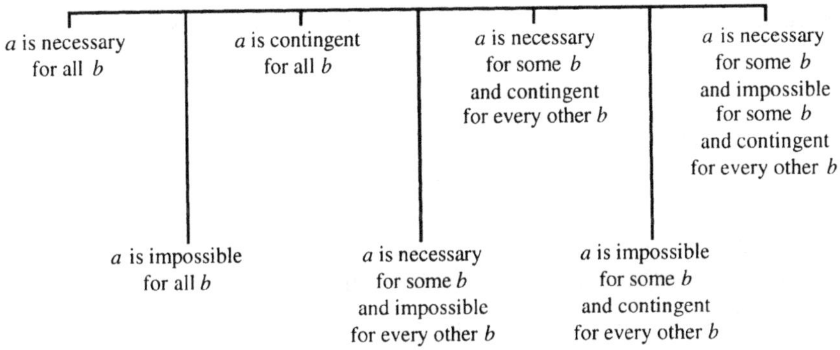

Fig. 2.5

Distinct from the question of what oppositions Aristotle stated, is the question what oppositions he *used* in his reasonings. Here we find that Aristotle did use some but not all of the classical oppositions, including some of those for particular Q-forms. He uses the oppositions $Qa \parallel Lo$, $Qa \parallel Le$, $Qi \parallel Le$, $Qo \parallel La$ and $Qe \parallel Li$ in Indirect Reductions of various syllogisms. This is clear from a consideration of the turning points in the proofs of those syllogisms (steps (4) and (5) when the 'impossible' consequence is juxtaposed with the supposition with which it is incompatible). For Celarent LQX ($Qa \parallel Lo$, $Qa \parallel Le$):

So *b* possibly belongs to not all or to no *c*. But at the start we supposed all. (A16, 36a14-15)

What was supposed "at the start" was that *b* is *contingent* for all *c* (Qbc^a); what has been deduced is that *b* is not *possible* for some, or for any, *c* (Lbc^o or Lbc^e). For Cesare LQX and Camestres QLX ($Qa \parallel Lo$):

... *a* is not possible for some *c*. But it was supposed possible for all. (A19, 38a24-26)

For Ferio LQX ($Qi \parallel Le$):

... *b* is possible for no *c*. But it was supposed possible for some. (A16, 36a38-39)

Bocardo QXM ($Qo \parallel La$):

... *a* will necessarily belong to all *c*, for we proved this before. But it was supposed that it possibly did not belong to some. (A21, 39b37-39)

The inclusion of $Qi \parallel Le$ and $Qo \parallel La$ among the oppositions used by Aristotle suffices to show that he was committed, in practice, to the classical theory for particular Q-forms. We shall see in §12.1 that Aristotle uses the opposition $Qe \parallel Li$ in his proof of Ferio QLM.

There is no case where we must suppose him to be making use of $Qa \parallel Li$. Thus there is one (and only one) of the classical oppositions which he does not actually use. But $Qa \parallel Li$ is equivalent to $Qe \parallel Li$ since affirmative and negative Q-forms are equivalent. And his statements in A17 commit him to this opposition. So I shall operate, for the present, on the hypothesis that Aristotle does endorse the full classical theory of Q-L oppositions, *i.e.* the contraries $Qa \parallel La$, $Qa \parallel Li$, $Qa \parallel Lo$, $Qa \parallel Le$, $Qe \parallel La$, $Qe \parallel Li$, $Qe \parallel Lo$, $Qe \parallel Le$, $Qi \parallel La$, $Qi \parallel Le$, $Qo \parallel La$ and $Qo \parallel Le$. Eventually, we shall find reason to revise this hypothesis in the light of a wider view of his system.

§3. TRANSFORMATION RULES: SUBSTITUTION, PERMUTATION, *REDUCTIO* AND CUT

Because the modal syllogistic is a deductive system, it makes explicit or implicit use of transformation rules to derive new theses from ones that have already been proved. Some of these transformation rules are identical with rules used in the assertoric syllogistic; some are not. In this section I discuss the former.

§3.1 Substitution

Aristotle's usage of variables is remarkably uniform in the modal syllogistic. He standardly uses the ordered triple $\langle A, B, \Gamma \rangle$ in his formulation of syllogistic theses; in Figure 1 these are major-middle-minor, in Figure 2 middle-major-minor, in Figure 3 major-minor-middle. This uniform symbolism contrasts with his practice in the assertoric system where the triples $\langle M, N, \Xi \rangle$ and $\langle \Pi, P, \Sigma \rangle$ are standardly used for Figures 2 and 3.[13] Even so, his usage in the modal syllogistic is not *absolutely* uniform: at A15, 34a19-21 he uses the triple $\langle \Gamma, \Delta, Z \rangle$ for a Figure 1 syllogism, and his process of Direct Reduction sometimes generates Figure 1 syllogisms whose major, middle and minor terms are $\langle B, A, \Gamma \rangle$ (*e.g.* A10, 30b11-13) or $\langle A, \Gamma, B \rangle$ (*e.g.* A11, 31a28-30) or $\langle B, \Gamma, A \rangle$ (*e.g.* A11, 31a32-33) or $\langle \Gamma, A, B \rangle$ (*e.g.* A10, 30b15-17).[14]

---

[13] See THOM 1981 §5.

[14] The remaining possible permutation $\langle \Gamma, B, A \rangle$ is not used for Figure 1 syllogisms. Such a syllogism could be generated by a Direct Reduction only by converting the major premiss of a second figure syllogism in $\langle B, \Gamma, A \rangle$, or by converting the minor of a third figure syllogism in $\langle \Gamma, A, B \rangle$, or by

Variants on accepted syllogisms are thus asserted as theses. Variants are one kind of substitution in a thesis, namely improper substitutions. They are to be distinguished from *proper* substitutions, which identify distinct variables in a thesis. The formula

$$\frac{ab^a \quad ba^a}{aa^a}$$

is a proper substitution in Barbara

$$\frac{ab^a \quad bc^a}{ac^a}$$

obtained by identifying the variables c and a. Nowhere in the modal syllogistic does Aristotle assert a thesis on the basis of identifying distinct variables. Because of this we can attribute to him the tacit use of a rule of Substitution:

U^* If x is a thesis, and y is a variant of x, then y is a thesis.

§3.2 Permutation

Aristotle standardly states the premisses of a syllogism in the order major-minor; but he knew that in modal as in assertoric syllogistic, premiss-order is logically immaterial. Sometimes his proofs assume that the order of the premisses can be varied without affecting validity. This happens regularly with the modal as well as the assertoric varieties of Camestres and Disamis (which is why their mnemonic names include the letter 'm'). It also happens (as we shall see in §9.2 and §11) with Darapti XLL and Bocardo Q'Q'Q'. Clearly then he made tacit use of a rule of Permutation:[15]

M $$\frac{R}{p} \longrightarrow \frac{R'}{p}$$

---

converting both premisses of a figure 1 syllogism in $\langle \mathrm{A}, \mathrm{B}, \Gamma \rangle$; and none of these processes occurs in the modal syllogistic.

[15] For more on Permutation see THOM 1981 §6.

where p is a categorical form, and R, R' are possibly null sequences of categorical forms differing only in the ordering of their elements. The arrow here indicates that if there is a thesis of the left-hand form then there is one of the right-hand form.

§3.3 *Indirect reduction*

The assertoric syllogistic makes use of two rules of Indirect Reduction. Each generates one syllogism from another, by making the contradictory (or contrary) of the latter's conclusion a premiss leading to the contradictory of one of the original premisses, the other premiss remaining constant.

$$C \qquad \frac{p \quad R}{q} \longrightarrow \frac{\sim q \quad R}{\sim p}$$

$$K \qquad \frac{p \quad R}{q} \longrightarrow \frac{\neg q \quad R}{\sim p}$$

where p, q are categorical forms, $\sim p$ is the contradictory of p, $\neg q$ is a contrary of q, and R is a possibly null sequence of categorical forms.

These rules apply too in the modal syllogistic. We saw in the previous section that Aristotle had an account of modal oppositions. There, we used the expression "contrary" to signify all relations of incompatibility other than contradiction. This is not how Aristotle understands logical contrariety. He regards any propositional form as having at most one contrary. But, as we saw in the previous section, there are four equally good candidates for the rôle of contrary of the modal form Qab^a, namely Lab^a, Lab^e, Lab^i and Lab^o.

However, this does not mean we cannot use rule K in modal syllogistic. The validity of that rule does not depend on taking the sign $\neg$ as an operator generating a unique form from a given form. The rule is still valid if $\neg p$ stands for *any* form (other than $\sim p$) which is incompatible with the form p, *i.e.* in our sense for any contrary of p. This is the way I use the sign $\neg$ in this book. On this understanding, we are entitled to use both rule C and rule K in the modal syllogistic. That Aristotle himself does so is clear from the following examples.

Aristotle uses rule C in his Indirect Proof of Le-conversion

> If it is necessary that *a* belongs to no *b*, it is also necessary that *b* belongs to no *a*. For, if it is possible for some, *a* is possible for some *b*. (A3, 25a29-32)

which we may represent

$$Mi-conv\frac{Mba^i}{Mab^i} \xrightarrow{C} \frac{Lab^e}{Lba^e}$$

Aristotle uses rule *K* in his Indirect Proof of Celarent LQX

Let *a* be possible for no *b*, and *b* be possible for all *c*. Necessarily then *a* belongs to no *c*. For let it belong to some or all. It was supposed possible for no *b*. By converting the negative, *b* is possible for no *a*. But *a* was let belong to some or all *c*. So *b* possibly belongs to not all or no *c*. But at the start we supposed all. (A16, 36a8-15)

We may represent this:

$$\begin{array}{l} Le\text{-}conv \\ Ferio\ LXL \end{array} \frac{\frac{Lab^e}{Lba^e} \quad ac^i}{Lbc^o} \xrightarrow{K} \frac{Lab^e \quad Qbc^a}{ac^e}$$

§3.4 *Cut*

In modal as in assertoric syllogistic, several theses can be strung together to generate a new thesis. This happens in the process of Direct Reduction, when premisses that imply given premisses are said also to imply the conclusion from those given premisses. The process is justified by a rule of Cut

T_1

$$\left.\begin{array}{c} \dfrac{p \quad R}{q} \\ \\ \dfrac{q \quad S}{r} \end{array}\right\} \longrightarrow \frac{p \quad R \quad S}{r}$$

where *p, q, r* are categorical forms, and *R, S* are possibly null sequences of categorical forms.

As an example of Aristotle's use of Cut in modal syllogistic we may take his Direct Reduction of Cesare LXL to Celarent LXL:

First let the negative be necessary, and *a* be possible for no *b*, and just belong to all *c*. Then, since the negative converts, *b* is possible for no *a;* but *a* belongs to all *c*, so *b* is not possible for any *c*, for *c* is under *a*. (A10, 30b9-13)

which can be represented

$$\left.\begin{array}{r} Le-conv\ \dfrac{Lab^e}{Lba^e} \\[4ex] Celarent\ LXL\ \dfrac{Lba^e \quad ac^a}{Lbc^e}\end{array}\right\} \xrightarrow{T_1} \dfrac{Lab^e \quad ac^a}{Lbc^e}$$

or, for short[16]

$$\dfrac{{}_{Le-conv}\dfrac{Lab^e}{Lba^e} \quad ac^a}{Lbc^e}\ {}_{Celarent\ LXL}$$

Strictly speaking, the rule of Cut that is used in Aristotle's assertoric syllogistic, is not T_1 but a restricted version which does not permit the generation of incestuous derivations, *i.e.* applications of Cut where $\frac{R}{p}$ shares with $\frac{p \quad S}{q}$ some variable not in p.[17] Incestuous derivations do not occur in the assertoric syllogistic.[18] But there are some in the modal syllogistic:

$$\dfrac{{}_{x}\dfrac{ab^a \quad Lbc^a}{Lac^a} \quad Lbc^a}{Lab^i}\ {}_{Darapti\ LLL}$$

This occurs in Aristotle's disproof of Barbara XLL (30a23-27). (The formula to be disproved is marked x.) The use of the unrestricted rule T_1 in modal syllogistic marks an important difference between it and the assertoric theory.

As a consequence of its permitting the generation of incestuous derivations, the modal syllogistic also includes theses which are proper substitutions in theses. Here is an example, in which two Barbaras combine incestuously

$$\dfrac{{}_{Barbara}\dfrac{ab^a \quad bc^a}{ac^a} \qquad cb^a}{ab^a}\ {}_{Barbara}$$

to generate a thesis

---

[16] THOM 1981 §7.

[17] See THOM 1981 §7.

[18] But there are proper substitutions in the assertoric syllogistic. See THOM 1981 §5 for details, and for a fuller account of proper and improper substitution.

$$\frac{ab^a \quad bc^a \quad cb^a}{ab^a}$$

which is a proper substitution in

$$\frac{ab^a \quad bc^a \quad cd^a}{ad^a}$$

This is another respect in which the modal differs from the assertoric syllogistic. Nonetheless, the fact that the modal syllogistic contains theses which are proper substitutions in theses does not mean that it contains the unrestricted rule of Substitution. For, the proper substitutions it contains are not there by virtue of their impropriety, but by virtue of being the progeny of incestuous Cuts.

§4. TRANSFORMATION RULES: CONTRACTION, ECTHESIS

In addition to the transformation rules he uses in the assertoric syllogistic Aristotle appears to formulate and use some others in his modal system. These new rules are: (1) a rule of Contraction (A9), (2) one or more rules of modal ecthesis (A8), and (3) one or more rules of modal *reductio* (A15). This section will outline rules of the first two types, and examine their validity. The third type will be discussed in the next section.

§4.1 Contraction

The incestuous derivations noted in the previous section occur in the context of Aristotle's attempt (30a23-27) to show that Barbara XLL is invalid. Aristotle argues that if Barbara XLL

$$\frac{ab^a \quad Lbc^a}{Lac^a}$$

were valid then, using Barbara XLL and Darapti LLL, we could reason, using rule T_1:

$$x\,\frac{\dfrac{ab^a \quad Lbc^a}{Lac^a} \quad Lbc^a}{Lab^i}\,\text{Darapti LLL}$$

But, argues Aristotle, the premisses are compatible with the denial of Lab^i, so step x (Barbara XLL) must be rejected.

It might be suggested that this reasoning makes use of a rule of Contraction

Cont $\dfrac{p \quad p \quad R}{q} \longrightarrow \dfrac{p \quad R}{q}$

where *p, q* are categorical forms and *R* is a possibly null sequence of categorical forms.

Two arguments tell against this suggestion. First, Aristotle's reasoning is indifferent to the presence or absence of Contraction. If we assume Contraction then Aristotle is saying that ab^a and Lbc^a are compatible with Lab^i; if we do not, then he saying that ab^a and Lbc^a and Lbc^a are compatible with Lab^i. But each of these alleged compatibilities is as good, or as bad, as the other. We have no good reason then for finding Contraction in this passage. Second, we have a good reason against attributing Contraction to Aristotle. We know[19] that it cannot happen in Aristotle's assertoric syllogistic that there is a pair of theses

$$\frac{p \quad R}{q} \text{ and } \frac{p \quad p \quad R}{q}$$

We also know[20] that if the assertoric syllogistic is extended by generalising its rule of Cut to T_1, then, though there will be such pairs of theses, in all cases *p* will be a variant on aa^a or aa^i. Now, these things would not be true of the modal syllogistic were Contraction a rule there. So the metatheory of the modal would be quite different from the metatheory of the assertoric syllogistic. Just how different it would be becomes clear when we consider one of the utterly basic properties of Aristotle's assertoric syllogistic, and of such extensions as were mentioned a moment ago: the property of interweaving.[21] A wff interweaves iff its premisses can be so ordered that each premiss shares one variable with one of its neighbours and the other variable with the other neighbour (where 'neighbour' is defined so that the conclusion is neighbour to the first and the last premisses, the second premiss is neighbour to the first, the penultimate to the last, and all other premisses have as their neighbours their predecessor and successor). In all these systems, every thesis interweaves.[22] But in a modal syllogistic that included rule T_1 plus Contraction, this would not be so. For, the thesis

$$\frac{ab^a \quad Lbc^a}{Lab^i}$$

[19] See THOM 1981 §50 'False Cause' Theorem, Corollary 3.
[20] THOM 1981 §50 'False Cause' Theorem, Corollary 2.
[21] THOM 1981 §4.
[22] THOM 1981 §46 Theorem.

does not interweave, since the minor term c is not paired with anything. I therefore reject the suggestion that the modal syllogistic uses Contraction.

§4.2 Ecthesis

Ecthesis is a procedure used in the assertoric syllogistic, whereby one derives theses containing particular premisses ab^i or ab^o from corresponding theses containing instead premiss-pairs $\{ad^a, bd^a\}$ or $\{ad^e, bd^a\}$. The validity of the procedure rests on the equivalences:

ab^i is true iff for some d: ad^a, bd^a are true;
ab^o is true iff for some d: ad^e, bd^a are true.
These equivalences underlie the rules of ecthesis:

i-ecthesis $$\frac{R \quad ad^a \quad bd^a}{p} \longrightarrow \frac{R \quad ab^i}{p}$$

o-ecthesis $$\frac{R \quad ad^e \quad bd^a}{p} \longrightarrow \frac{R \quad ab^o}{p}$$

Ecthesis is also used in the modal syllogistic. The following passage describes an ecthetic proof of Baroco LLL.

> ... it is necessary for us to 'exhibit' that to which each does not belong and make a syllogism about this. For it will be necessary about these; and if it is necessary of what is exhibited, then it will be necessary of some of the former (for what is exhibited is just a certain 'that'). Each of these syllogisms occurs in its home figure. (A8, 30a9-14)

The procedure here is, instead of using the particular negative apodeictic premiss Lac^o, to consider a premiss relating to 'that part' of c which is not a. This part cannot be b, says Aristotle, and so some c cannot be b. There are two possible readings of this. On one reading, Lac^o is thought of as replaceable by the pair $\{Lad^e, cd^a\}$; on another reading it is thought of as replaceable by the pair $\{Lad^e, Lcd^a\}$. Therefore there are two possible readings for the ecthetic proofs of Baroco LLL and Bocardo LLL. Either they depend on the rule

Lo-ecthesis A $\dfrac{R \quad Lad^e \quad bd^a}{p} \longrightarrow \dfrac{R \quad Lab^o}{p}$

or on the rule

Lo-ecthesis AA $\dfrac{R \quad Lad^e \quad Lbd^a}{p} \longrightarrow \dfrac{R \quad Lab^o}{p}$

At this stage of our exposition we have no way of determining which rule to include in Aristotle's system. Nor do we have any way of determining which, if any, rules of modal ecthesis for Li-, La-, Le-, Q- or M-forms to include. Both questions will receive extended discussion as we proceed.

§5. TRANSFORMATION RULES: MODAL *REDUCTIO*

Łukasiewicz[23] finds Rules of Extensionality for necessity and possibility in Aristotle. For necessity:

Rule L1 If $\dfrac{p}{q}$ is a thesis, and p is necessary then q is necessary;

So if one puts '*a*' for the premisses and '*b*' for the conclusion, it follows not only that *a* being necessary *b* is necessary, but that if the former is possible the latter is possible. (A15, 34a22-24; *cf.* 34a5-7, 34a29-31)

and for two-premissed theses:

Rule L2 If $\dfrac{p \quad q}{r}$ is a thesis, and p and q are necessary then r is necessary.

Aristotle makes no use of these rules in the modal syllogistic.

---

[23] Łukasiewicz §39. Because he wishes to treat 'M' as a truth-functional operator (albeit in a system with more than two truth-values) he accepts the Rule in the stronger form CCpqCMpMq, which does not require that antecedent and consequent be necessary connections. Aristotle, on the other hand, clearly imposes this requirement of necessary connections.

Aristotle's initial statement of the possibility-rule (as Łukasiewicz hints)[24] gives not a purely logical but an ontological rule that concerns coming-to-be. This is clear from the proof he gives. Aristotle's proof of this rule (as Alexander realised)[25] depends on the principle that what is possible may, without any impossibility resulting, be supposed actual. The proof is that, given that *b* is a necessary consequence of *a*, let us suppose *a* is possible and *b* impossible. Then we may (without any impossibility resulting) suppose *a* while *b* is impossible. This supposition, however, does involve the impossibility that *a* while *not-b*.

Aristotle then says the same rule holds for the logical as for the ontological cases:

Rule M1 If $\frac{p}{q}$ is a thesis and p is possible then q is possible.

One must apply 'impossible' and 'possible' [*dunaton*] not only to coming-to-be but to being true and to belonging[26] and all other senses of possibility. For, it is the same in all cases. (A15, 34a12-15)

No proof is offered, but we can adapt the proof of the ontological rule. He then applies this rule to syllogistic statements:

Rule M2 If $\frac{p \quad q}{r}$ is a thesis and p and q are jointly possible then r is possible.

But '*a* being, *b* is' must not be taken as meaning that when some one thing *a* is, *b* will be (for, some one thing being, nothing is necessary, but there must be at least two as in a syllogism). For, if *c* of *d*, and *d* of *e*, then necessarily *c* of *e*; and if both [*hekateron*] are possible the conclusion too is possible. (A15, 34a16-22)

This rule, which is correct, must be distinguished from the invalid rule:

If $\frac{p \quad q}{r}$ is a thesis and p and q are severally possible then r is possible.

If premisses necessitate a conclusion, it is their joint possibility, not their several possibilities, that necessitate the conclusion's possibility. The medievals were very clear about this. Pseudo-Scotus, for instance, comments:[27]

Now, if both premisses are both possible and compossible, it is necessary for the conclusion to be possible.

---

24 §40.

25 ALEXANDER 177.11.

26 Following SMITH's translation of 34a12.

27 SCOTUS Q. 32 Part I conclusion 2.

Scotus is also very clear here that the possibility in question is *in sensu composito* (*i.e. de dicto*).

For the possibility-case, Aristotle draws a corollary:

> This proved, it is clear that the conclusion from a false[28] but not impossible [*adunaton*] supposition will itself be false but not impossible. For example, if *a* is false but not impossible, and *a* being *b* is, then *b* will be false but not impossible. (A15, 34a25-29)

for which he supplies the proof:

> For, since it has been proved that if *a* being *b* is, then *a* being possible *b* will be possible, supposing *a* to be possible *b* too will be possible. For, if it is impossible, the same thing will be both possible and impossible. (A15, 34a29-33)

In the succeeding passage (A15, 34a34-b2) Aristotle advances an argument having the form:

$$\frac{p \quad bc^a}{r} \longrightarrow \frac{\sim r \quad Qbc^a}{\sim p}$$

His reasoning seems to be this. Suppose $\sim r$ and Qbc^a are true. Then bc^a may be false but it is not impossible (relative to our suppositions). Now r is impossible (given our suppositions). So, by the Corollary, any premisses leading to r as conclusion must themselves be impossible. Now, p and bc^a are such premisses. But, as we said, bc^a is possible. So p must be impossible, and we have $\sim p$. In other words, the suppositions that $\sim r$ and Qbc^a lead to the conclusion that $\sim p$.

Smith's analysis is:[29]

> Aristotle constructs a proof from impossibility by supposing both the contradictory of the desired conclusion and a premise *consistent with* one of the premises, and from this he deduces (as in the nonmodal cases) the contradictory of the other premise.

In other words, the reasoning proceeds in accordance with one or other of

Rule C'

$$\left.\begin{array}{c}\dfrac{\sim r \quad s}{\sim p}\\[2em] NOT\ \dfrac{q}{\sim s}\end{array}\right\} \longrightarrow \frac{p \quad q}{r}$$

---

[28] ROSS *ad* 34a25-27 comments that the reference to falsity here is wrong, and that we should understand not 'false' but 'unwarranted by the data'.

[29] SMITH *ad* 34a34-b2.

Rule K'

$$\left.\begin{array}{r}\dfrac{\sim r\quad s}{\neg p}\\[4ex] NOT\ \dfrac{q}{\sim s}\end{array}\right\}\longrightarrow \dfrac{p\quad q}{r}$$

Two things have to be said about this. First, Rules C' and K' do not follow from *M1-2*. If p and Qbc^a are true bc^a may also be true. If it is true then r will be true. But *M1-2* give us no reason for saying that if p and Qbc^a are true *and* bc^a *is false*, then r is true. The second thing to be said is that Rules C' and K' are invalid, being equivalent to

$$\left.\begin{array}{r}\dfrac{p\quad s}{r}\\[4ex] NOT\ \dfrac{q}{\sim s}\end{array}\right\}\longrightarrow \dfrac{p\quad q}{r}$$

which is not sound, as it would licence the substitution in valid formulae of weakened (but consistent) premisses, *e.g.*[30]

[30] I do not accept TREDENNICK's analysis (*ad* 34b2) of the reasoning as requiring a rule

$$\left.\begin{array}{r}\dfrac{p\quad q}{r}\\[4ex] \dfrac{s}{\sim r}\\[4ex] NOT\ \dfrac{s}{\sim p}\end{array}\right\}\longrightarrow \dfrac{s}{\sim q}$$

which would hold only on the supposition that what is inconsistent with a conjunction must be inconsistent with some conjunct. (This analysis is endorsed by PATTERSON 1995 p.161.)

$$\frac{ab^i}{ab^o} \text{ is not a thesis and } \frac{ab^a \quad bc^a}{ac^a} \text{ is a thesis}$$

$$\text{but } \frac{ab^i \quad bc^a}{ac^a} \text{ is not a thesis.}$$

Thus, the modal syllogistic does indeed use rules of Indirect Reduction not found in the assertoric syllogistic. A detailed argument, apparently involving rules of extensionality for possibility, leads up to the adoption of these rules. But the rules are neither proven nor sound. In §18.6 we shall see whether the theses Aristotle supports with these rules can be given alternative proofs. In §37 we shall formulate a sound restricted version of *K'*.

CHAPTER TWO. AXIOMS

This Chapter presents a version of the assertoric syllogistic as a foundation on which to build various modal systems. It distinguishes five layers in the modal syllogistic, and lists Aristotle's axioms, or if you wish his axiomatic derivations, in each of these layers. We investigate the possibility of basing the M moods on the L moods, and also the converse possibility. Distinguishing ampliated from unampliated contingency-forms, we take the uniform contingency-moods as involving the former, the mixed moods the latter. Finally, we sketch what we can of Aristotle's rudimentary logic of 'unqualified' assertorics.

§6 . X AND L AXIOMS

Aristotle begins his discussion of modal syllogistic by observing that there is a difference between belonging, necessary belonging and contingent belonging (A2, 25a1-2), and that corresponding to this difference "there will be different syllogisms" (A8, 29b29-35). Different the logic of necessity and contingency may be; but that difference does not preclude their being treated together as parts of the one system. Aristotle, however, does not do this. He expounds his modal syllogistic in five groups of chapters with few cross-references. These are: (1) the LLL syllogisms (A8), (2) the L-X syllogisms (A9-11), (3) the QQQ syllogisms (A14, A17, A20), (4) the Q-X syllogisms (A15, A18, A21), (5) the Q-L syllogisms (A16, A19, A22). In making this division, Aristotle did not carve things at the logical joints. He sometimes puts disparate material together, as well as separating matters that should go together. I shall make use of a different division, a division into layers each of which presupposes material from earlier layers. In addition to the assertoric syllogistic, we have (1) the L moods (including L-X moods), (2) the mixed Q moods (Q-X and Q-L moods), (3) the pure Q moods, (4) the M moods (including M-L, M-X, M-Q-X and M-L-X moods), and (5) some moods involving a distinct type of 'unqualified' assertoric form.

§6.1 X axioms

We need to assume some version of assertoric syllogistic,[1] preferably a version not containing Indirect Reductions since there are none of these in the L layer. I adopt the following axioms:

[1] See THOM 1981 §9.

e-conversion	$\dfrac{ab^e}{ba^e}$	*Barbara*	$\dfrac{ab^a \quad bc^a}{ac^a}$
a-conversion (partial)	$\dfrac{ab^a}{ba^i}$	*Celarent*	$\dfrac{ab^e \quad bc^a}{ac^e}$
i-conversion	$\dfrac{ab^i}{ba^i}$	*Darii*	$\dfrac{ab^a \quad bc^i}{ac^i}$
a-repetition[2]	$\dfrac{ab^a}{ab^a}$	*Ferio*	$\dfrac{ab^e \quad bc^i}{ac^o}$

plus rules U^*, M, T_1 and o-ecthesis.

§6.2 L axioms

The L layer, whose basis is presented in chapters A3, A8, can be taken either as resting on L-axioms or as resting on a rather sketchy logic of M-forms. The situation may be compared with the ecthetic proof of e-conversion in A2. That proof can be seen as depending on a system of singular syllogisms which is only partly acknowledged by Aristotle.[3] The logic of M-forms, like that of singular forms, is only sketchily developed by Aristotle. I shall begin by taking the laws of L-conversion as axiomatic. Later, in looking at the M layer, we shall investigate and further develop the logic of possibility in ways which will provide a foundation for the L layer.

§6.2.1 Pure L axioms

Aristotle takes three laws of conversion as axiomatic:

[2] o-repetition is derivable from Felapton by o-ecthesis.

[3] See THOM 1981 §44.

Le-conversion $\dfrac{Lab^e}{Lba^e}$

If it is necessary that *a* belongs to no *b*, it is also necessary that *b* belongs to no *a*. (A3, 25a29-31)

La-conversion (partial) $\dfrac{Lab^a}{Lba^i}$ *Li-conversion* $\dfrac{Lab^i}{Lba^i}$

If *a* belongs necessarily to all or some *b*, it is also necessary that *b* belongs to some *a*. (A3, 25a32-33)

By a double application of Le-conversion it can be shown that Le-forms imply themselves:

$$\dfrac{\dfrac{Lab^e}{Lba^e}\,Le\text{-}conv}{Lab^e}\,Le\text{-}conv$$

Similarly, for Li-forms:

$$\dfrac{\dfrac{Lab^i}{Lba^i}\,Li\text{-}conv}{Lab^i}\,Li\text{-}conv$$

But if we want the analogous principles of Repetition for La- and Lo-forms, we have to add them explicitly:

La-repetition $\dfrac{Lab^a}{Lab^a}$ *Lo-repetition* $\dfrac{Lab^o}{Lab^o}$

For reasons of systematic coherence, I shall include these among the axioms for the L layer, even though Aristotle does not mention them.

In addition to these one-premissed axioms, A8, 29b36-30a14 treats four First Figure LLL syllogisms as axiomatic, describing them as 'perfect'.[4]

[4] PATTERSON 1995 ch.7 contains an excellent discussion of perfection.

Barbara LLL $\dfrac{Lab^a \quad Lbc^a}{Lac^a}$ *Celarent LLL* $\dfrac{Lab^e \quad Lbc^a}{Lac^e}$

Darii LLL $\dfrac{Lab^a \quad Lbc^i}{Lac^i}$ *Ferio LLL* $\dfrac{Lab^e \quad Lbc^i}{Lac^o}$

The remaining LLL syllogisms he reduces to these four using Direct Reduction or (in the case of Baroco LLL and Bocardo LLL) Ecthesis.[5]

§6.2.2 L-X axioms

The L-X axioms are presented in A9-11, where Aristotle assumes the laws of apodeictic conversion. In addition I attribute to him the laws of L-subordination[6]

L-subord $\dfrac{Lab^a}{ab^a}$ $\dfrac{Lab^e}{ab^e}$ $\dfrac{Lab^i}{ab^i}$ $\dfrac{Lab^o}{ab^o}$

Aristotle does not expressly acknowledge these, but he does assert a number of syllogisms from (partly) apodeictic premisses to an assertoric conclusion, among them *Felapton XLX*:

For let *bc* be affirmative and necessary, and let *ac* be negative and not necessary. Then since the affirmative converts, *c* also belongs necessarily to some *b*. So if *a* belongs to no *c*, and *c* to some *b*, *a* does not belong to some *b*. (A10, 31a10-13)

And the proof he gives for this appears to assume L-subord, specifically for Li-forms. For, he reasons:

$$x\,\dfrac{ac^e \quad \dfrac{Lbc^a}{Lcb^i}\,La\text{-}conv}{ab^o}$$

---

[5] Indirect Reduction is not useable in the pure L layer because it requires a form's contradictory to be statable - a condition not met in this layer.

[6] A term derived from McCALL 1963 §13.

and syllogism x has never been proved; but the obvious way to prove it is to reduce it to Ferio by L-subord. Thus Wieland errs in contending that for Aristotle assertoric conclusions can be derived only from assertoric premisses.[7]

In addition to all this, Aristotle relies on four First Figure mixed L-X axioms:

$$\text{Barbara LXL} \quad \frac{Lab^a \quad bc^a}{Lac^a} \qquad\qquad \text{Celarent LXL} \quad \frac{Lab^e \quad bc^a}{Lac^e}$$

If a is taken as belonging or not belonging necessarily to b, and b as just belonging to c, ... a will belong or not belong necessarily to c. For since a belongs or does not belong necessarily to all b, and c is one of the bs,[8] clearly one or other of these will be necessary for c too. (A9, 30a17-23)

$$\text{Darii LXL} \quad \frac{Lab^a \quad bc^i}{Lac^i} \qquad\qquad \text{Ferio LXL} \quad \frac{Lab^e \quad bc^i}{Lac^o}$$

Let a belong necessarily to all b, and b just belongs to some c. It is necessary then that a belongs necessarily to some c for c is under b[9] and it belongs necessarily to all b. (A9, 30a37-b1)

It is the same if the syllogism is negative. (A9, 30b1-2)

Given the perfect LXL syllogisms, the perfect LLL moods follow by L-subordination:

$$\text{Barbara LXL} \frac{Lab^a \qquad \dfrac{Lbc^a}{bc^a}\, L-subord}{Lac^a}$$

So there is no need to include LLL moods as axioms.

Thus, as axioms for the L-layer I take Le-conversion, La-conversion, Li-conversion, La-repetition, Lo-repetition, L-subordination, Barbara LXL, Celarent LXL, Darii LXL, Ferio LXL.

---

7 WIELAND 1966 p.58. For a critique of Wieland see PATTERSON 1995 §4.3.6.

8 SMITH *ad* 30a15-23 scents *ecthesis* here. See §36 for a proof of Barbara LXL which uses an ecthetic procedure.

9 SMITH *ad* 30a40 observes that "'under' in this sort of context usually means either 'within the extension of' or 'a subject of predication of'. In the present case, it has to mean something like 'part of C falls under B', if Aristotle's argument is to work".

§7. Q AXIOMS

A13, 32b32-37 notes that there are two different ways of reasoning to a contingency-predication. Sometimes, such a conclusion is derivable from two contingency-premisses; sometimes, a contingency-predication is derivable from a contingency-proposition together with an assertoric. Thus, included within Aristotle's purview are *uniform* derivations (in which all propositions contain such a mode), and *mixed* ones (in which some propositions contain no such mode).

Prima facie it seems that different meanings are required to validate the uniform and the mixed syllogisms which Aristotle claims as valid. (i) The unampliated sense will not validate the uniform syllogisms. Consider Barbara QQQ. If its premisses are that *a* is contingent for every (actual) *b*, and that *b* is contingent for every (actual) *c*, then nothing will follow from them, because there may be individuals for which *b* is contingent, but which are not actually *b*. (ii) On the other hand, the ampliated sense will not validate the mixed syllogisms. Consider Barbara QXQ. If its premisses are that *a* is contingent for every individual for which *b* is contingent, and that *b* is true of every *c*, then nothing will follow from them, because there may be *b*s for which *b* is not contingent. And if Barbara QXQ's premisses are that *a* is contingent for every individual for which *b* is possible, and that *b* is true of every *c*, then, while it follows that *a* is contingent for every *actual c*, it does not follow that *a* is contingent for every *possible c*.[10]

Thus, the interpreter must either revise Aristotle's account of the admissible meanings of contingency-propositions, or reject some Aristotelian contingency-syllogism, or read the contingency-propositions in the uniform syllogisms differently from those in the mixed syllogisms.

This last alternative was first proposed by Alexander:[11]

> So there are two senses. However, whenever the sentence means the first of these, then the premisses are not both contingent, rather the minor is assertoric and the major contingent (this is a mixed conjunction); but whenever we have the second, then both premisses are contingent.

Alexander takes Aristotle's distinction between ampliated and unampliated senses to be generated from a consideration of two ways in which a contingency-proposition may be the conclusion of a syllogism - either as the conclusion of a QQ syllogism or of a QX syllogism. If he is right, contingency-syllogistic is not a single system but two distinct systems. While one system (A14, A17, A20) concerns ampliated

---

[10] The argument of this paragraph will be subject to an important qualification in §30, in the light of semantic theory to be developed in the meantime.

[11] ALEXANDER 166,5-8. Alexander's strategy is used by STAHL 1977 pp.214-215.

contingency-predications and contains the uniform moods, the other (A15-16, A18-19, A21-22) concerns unampliated contingency-predications and contains the mixed moods.

Becker[12] objects that if Alexander's interpretation were accepted we should expect A14-22 to contain references back to A13's distinction of unampliated and ampliated senses, indicating which sense applies in which case; but we find no such references.

It is true that no later passage unambiguously refers to the distinction between ampliated and unampliated senses. Nonetheless, the objection can be answered. Aristotle's comments on the four perfect Figure 1 QQQ syllogisms can be read in the required way. He says that the perfection of Barbara QQQ and Darii QQQ is clear *"ek tou horismou"*, and that "the demonstration is the same" for the perfection of Ferio QQQ (A14, 32b40, 33a24, 33a27). His comment at 33a3 on Celarent QQQ makes it clear what this *horismos* is:

For the statement that it is possible [*endechesthai*] for *a* not to belong to that of which *b* may be true means (as we saw) that none of those things which can fall under *b* is left out of account. (Jenkinson/Barnes translation)

This is most naturally read as a reference to the introduction of the ampliated sense of contingency-propositions at 32b25-37. Given this, it seems natural to read the adjacent references to "the definition" as references to the same passage (though Alexander himself, strangely, reads Aristotle's comment on the perfection of Barbara QQQ as referring to the definition of "belonging to all" given at 24b29).

These passages are to be contrasted with others containing the expression *"kata ton diorismon"* (A14, 33b23; A15, 33b27; A17, 37a27-28). Aristotle uses the expression *"ek tou horismou"* or related expressions in the justification of perfect syllogisms with two contingency-premisses, and never in justifying a syllogism with only one contingency-premiss. The definition referred to is best taken to be that of ampliated universal contingency-propositions in our text; for, this definition, and not the accompanying definition of unampliated propositions, is linked in that text to syllogisms with two contingency-premisses. By contrast, Aristotle uses the expression *"kata ton diorismon"* and its relatives when referring to the definition of contingency as two-sided possibility.

Alexander's suggestion for solving the problem of the ampliation of contingency-propositions can therefore be defended against Becker's objection. We shall adopt it for the present. Alternative strategies will be considered later.[13]

---

[12] BECKER pp.32-34.

[13] The argument of this section is based on THOM 1994.

§7.1 Pure Q axioms

The reducibility of imperfect to perfect contingency-syllogisms requires that a number of one-premissed derivations be valid, namely the partial conversion of contingency a-forms, the simple conversion of contingency i-forms, and the 'complementary'[14] conversion of all contingency-predications (whereby contingency a- and e-forms are mutually equivalent, as are contingency i- and o-forms). The partial convertibility of a-forms and the full convertibility of i-forms are said at 25a39-40 to hold in all senses of "possible". Thus we have:

Qa-conversion $\dfrac{Qab^a}{Qba^i}$ *Qi-conversion* $\dfrac{Qab^i}{Qba^i}$

If *a* is possible for all or some *b*, *b* too will be possible for some *a*. For if for none, neither *a* for *b*. For this was proved before. (A3, 25a40-b3)

There is a thought here that the convertibility of affirmative possibility-forms is derivable from that of universal negative necessity-forms. This may be so for one-sided possibility; but it is not true for contingency, because there is no contradiction $Qi \mid Le$. For Aristotle, it seems, these laws of conversion are axiomatic.

Characteristic of two-way possibility are the axioms:

Q-conversion $\dfrac{Qab^a}{Qab^e}$ $\dfrac{Qab^e}{Qab^a}$ $\dfrac{Qab^i}{Qab^o}$ $\dfrac{Qab^o}{Qab^i}$

It follows that all possibility-premisses are mutually convertible. I do not mean affirmatives into negatives, but that all having an affirmative form [convert] by opposition. For example, 'possibly belongs' to 'possibly does not belong', 'possible for all' to 'possible for none' or 'not all', and 'for some' to 'not some'. The same way with others. For since the possible is not necessary, and the not necessary may not belong, it is clear that if it is possible for *a* to belong to *b*, it is also possible for it not to belong. And if for all it is possible to belong, also for all it is possible not to belong. Similarly for particular affirmatives, for the proof is the same. (A13, 32a29-b1)

Negative Q-forms do not according to Aristotle convert in the same way as negative assertorics. While Qe-forms are not convertible, Qo-forms are taken as convertible.

---

[14] ROSS *ad* 32a21-29 introduced this term. BROGAN prefers the term "non-reversed change" (p.58). PATTERSON 1995 p.125 uses the term "qualitative conversion".

Qo-conversion $\dfrac{Qab^o}{Qba^o}$

... the particular [negative Q-] premiss converts. (A3, 25b17-18)

This is derivable from Qi-conversion and Q-conversion:

$$\dfrac{\dfrac{\dfrac{Qab^o}{Qab^i}\,Q\text{-}conv}{Qba^i}\,Qi\text{-}conv}{Qba^o}\,Q\text{-}conv$$

By a double application of Qi-conversion it can be shown that Qi-forms imply themselves:

$$\dfrac{Qab^i}{Qab^i}$$

Similarly, for Qo-forms:

$$\dfrac{Qab^o}{Qab^o}$$

The analogous principles of Repetition for Qa- and Qe-forms follow from Q-conversion:

$$\dfrac{\dfrac{Qab^a}{Qab^e}\,Q\text{-}conv}{Qab^a}\,Q\text{-}conv \qquad \dfrac{\dfrac{Qab^e}{Qab^a}\,Q\text{-}conv}{Qab^e}\,Q\text{-}conv$$

A further result, noticed by Wieland,[15] is $\dfrac{Qab^e}{Qba^o}$, which is provable from Qa-conv and Q-conv.

---

[15] WIELAND 1980 p.112.

§7.2 *Q-X axioms*

The Q-X theses are presented in A15 ,18 ,21. Aristotle considers some two-premissed derivations to be perfect. Other derivations involving contingency-propositions are perfected by being reduced to these perfect moods by a variety of methods.

Aristotle takes QXQ analogues of the four perfect assertoric syllogisms as axiomatic:

Barbara QXQ $\dfrac{Qab^{a} \quad bc^{a}}{Qac^{a}}$

Let *a* be possible for all *b*, and suppose that *b* belongs to all *c*. Since c is under *b*, and *a* is possible for all *b*, it is clear that it is also possible for *c*. A perfect syllogism is produced. (A15, 33b33-36)

Celarent QXQ $\dfrac{Qab^{e} \quad bc^{a}}{Qac^{e}}$

Similarly, the *ab* premiss being negative and the *bc* premiss affirmative, if the first is taken as a possible premiss and the second as assertoric, there will be a perfect syllogism that it is possible for *a* to belong to no *c*. (A15, 33b36-40)

Darii QXQ $\dfrac{Qab^{a} \quad bc^{i}}{Qac^{i}}$

Ferio QXQ $\dfrac{Qab^{e} \quad bc^{i}}{Qac^{o}}$

If one interval is universal, one particular, when the premiss relating to the major extreme is taken as universal and possible, be it affirmative or negative, the particular being assertoric, there will be a perfect syllogism just as when the terms are universal. (A15, 35a30-35)

§7.3 *Q-L axioms*

The Q-L theses are presented in A16,19,22. Four First Figure QLQ syllogisms are taken as axiomatic.

$$\text{Barbara QLQ} \quad \frac{Qab^a \quad Lbc^a}{Qac^a}$$

Suppose that it is possible that *a* belongs to all *b* and that *b* belongs necessarily to all *c*. There will be a syllogism that *a* possibly belongs to all *c*. (A16, 36a2-5)

$$\text{Celarent QLQ} \quad \frac{Qab^e \quad Lbc^a}{Qac^e}$$

Again, let the affirmative premiss be necessary, and let *a* possibly belong to no *b* and *b* necessarily belong to all *c*. The syllogism will be perfect, not that it does not belong but that it possibly does not belong. (A16, 36a17-21)

$$\text{Darii QLQ}^{16} \quad \frac{Qab^a \quad Lbc^i}{Qac^i} \qquad \text{Ferio QLQ} \quad \frac{Qab^e \quad Lbc^i}{Qac^o}$$

When the premisses signify [respectively] necessary and possible belonging, there will be a syllogism when the terms are related as previously. And it will be perfect when the minor extreme is necessary. (A16, 35b23-26)

Aristotle does not notice that the perfect QLQ syllogisms follow from the perfect QXQ moods by L-subordination:

$$\text{Barbara QXQ} \frac{Qab^a \quad \dfrac{Lbc^a}{bc^a} L-subord}{Qac^a}$$

In sum, for the mixed Q layer I take as axiomatic:
Qa-conversion, Qi-conversion,
Q-conversion,
Barbara QXQ, Celarent QXQ, Darii QXQ, Ferio QXQ.

16 WAITZ *ad* 36a39 points out that Aristotle omits to list this thesis explicitly. It is, however, implicitly included in his summary remarks (quoted).

§8. Q', M AND X' AXIOMS

§8.1 Q'-axioms

The pure Q theses are presented in parts of A3 and in A13, A14,17,20. I take this layer to involve ampliated contingency-propositions.

The laws of conversion are the same as for unampliated Q-forms.

Q'a-conversion $\dfrac{Q'ab^a}{Q'ba^i}$ *Q'i-conversion* $\dfrac{Q'ab^a}{Q'ba^i}$

Q'-conversion $\dfrac{Q'ab^a}{Q'ab^e}$ $\dfrac{Q'ab^e}{Q'ab^a}$

$\dfrac{Q'ab^i}{Q'ab^o}$ $\dfrac{Q'ab^o}{Q'ab^i}$

plus four perfect Q'Q'Q' syllogisms:

Barbara Q'Q'Q' $\dfrac{Q'ab^a \quad Q'bc^a}{Q'ac^a}$

When *a* is possible for all *b*, and *b* for all *c*, there will be a perfect syllogism that *a* possibly belongs to all *c*. This is clear from the definition [*ek tou horismou*]. For, 'possibly belonging to all' we explained in this way. (A14, 32b38-33a1)

Celarent Q'Q'Q' $\dfrac{Q'ab^e \quad Q'bc^a}{Q'ac^e}$

Similarly, if *a* is possible for no *b*, and *b* for all *c*, then *a* is possible for no *c*. For, 'What is possibly [*endechetai*] *b* is possibly not *a*' leaves out nothing that is possibly [*endechomenon*] under *b*. (A14, 33a1-5)

Darii Q'Q'Q $\dfrac{Q'ab^a \quad Q'bc^i}{Q'ac^i}$

When one premiss is universal, one particular, the premiss relating to the major extreme being universal, there will be a perfect syllogism. If *a* is possible for all *b*, and *b* for some *c*, *a* is possible for some *c*. This is clear from the definition [*ek tou horismou*] of 'possible'. (A14, 33a21-25)

Ferio Q'Q'Q' $\dfrac{Q'ab^e \quad Q'bc^i}{Q'ac^o}$

Again, if *a* is possible for no *b*, and *b* possibly belongs to some *c*, it is necessary that *a* possibly does not belong to some *c*. Same proof. (A14, 33a25-27)

Q'a-conversion is provable:

$$\dfrac{\dfrac{Q'ab^a}{Q'ab^e}\,Q'\text{-}conv}{Q'ab^a}\,Q'\text{-}conv$$

§8.2 M axioms

Aristotle accepts the laws of conversion for Mi-forms and the partial convertibility of Ma-forms. These are said at 25a39-40 to hold in all senses of "possible":

Mi-conversion $\dfrac{Mba^i}{Mab^i}$ *Ma-conversion* $\dfrac{Mba^a}{Mab^i}$

If [*b*] is possible for some [*a*], *a* is possible for some *b*.... If *a* is possible for all or some *b*, *b* too will be possible for some *a*. For if for none, neither *a* for *b*. For this was proved before. (A3, 25a31-32, 40-b3)[17]

---

[17] One might think that Aristotle intends the following ecthetic proof of Mi-conversion:

$$\dfrac{Mbc^a \quad Mac^a}{Mab^i} \longrightarrow \dfrac{Mba^i}{Mab^i}$$

This assumes the validity of Darapti MMM, which I shall argue in §23 is invalid.

As we saw in §7.1, there is a thought here that the convertibility of affirmative possibility-forms is derivable from that of universal negative necessity-forms. Aristotle also entertained the converse thought, that the convertibility of universal negative necessity-forms could be derived from that of particular affirmative possibility-forms.

$$Mi-conv\frac{Mba^i}{Mab^i}\xrightarrow{C}\frac{Lab^e}{Lba^e}$$

If it is necessary that a belongs to no b, it is also necessary that b belongs to no a. For, if it is possible for some, a is possible for some b. (A3, 25a29-32)

It is possible too to construe his remarks about La- and Li-conversion:

If a belongs necessarily to all or some b, it is also necessary that b belongs to some a. For if it is not necessary, neither will a belong necessarily to some b. (A3, 25a32-34)

as a derivation from Me-conversion:

$$Me-conv\frac{Mba^e}{Mab^e}\xrightarrow{K}\frac{Lab^a}{Lba^i}$$

which would then be axiomatic. This conversion is stated explicitly in terms of one-sided possibility:

Me-conversion $\dfrac{Mab^e}{Mba^e}$

All that is said to be possible by virtue of necessarily belonging or not[18] necessarily not belonging, *e.g.* if it said 'man possibly is not horse' or 'white belongs to no cloak' ... converts in the same way. (A3, 25b4-13)

He gives examples but no proof.

His intention regarding which conversions are primary is unclear. Alexander draws attention[19] to the threat of circular proof here. In this section we pursue the thought that there are axiomatic M-formulae, including Ma-, Mi- and Me-conversion.

As with assertoric forms, we should assume axioms of Repetition for M-forms:

---

[18] There is a textual difficulty here. On the question whether to include both 'not's see MIGNUCCI *ad* 25b5.

[19] ALEXANDER p.36, 7 ff.

Ma-repetition $\dfrac{Mab^a}{Mab^a}$ *Mo-repetition* $\dfrac{Mab^o}{Mab^o}$

And, corresponding to L-subordination we have

M-subord $\dfrac{ab^a}{Mab^a}$ $\dfrac{ab^e}{Mab^e}$ $\dfrac{ab^i}{Mab^i}$ $\dfrac{ab^o}{Mab^o}$

The second of these theses is used by Aristotle in the following passage:

It is clear that a syllogism 'it possibly [*endechesthai*] does not belong' is produced since there is one that 'it does not belong'. (A16, 36a15-17)

Aristotle does not propose any axiomatic two-premissed M-syllogisms, unless the following is meant as such:

Barbara MXM $\dfrac{Mab^a \quad bc^a}{Mac^a}$

For, if *b* belongs to all *c*, and *a* is possible for all *b*, then *a* is possible for all *c*. (A15, 34b4-5)

It seems that Aristotle toyed alternatively with the idea of basing an L-system on an M-system and that of basing an M-system on an L-system. We shall see in §12 that all of the M formulae we have noticed, with the possible exception of the last,[20] are provable from L-theses. We shall not further pursue the idea of a system with M-axioms, instead deriving M-theses from a base of L-axioms in §23.

§8.3 'Unqualified' assertoric axioms

Aristotle considers what he calls 'unqualified' propositions in parts of A15, 18, 21. Only universal propositions of this type get mentioned, never particulars. Their logic is but scantily indicated by Aristotle. But the indications we have appear to confirm Angelelli's suggestion that their logic is that of ampliated assertorics. We shall follow Angelelli's suggestion, at least for the purposes of logical theory. (When we come to metaphysical theory in Chapter Ten, we shall have to modify Angelelli's analysis.)

---

[20] Which I discuss in §20.1.

Aristotle assumes that unqualified assertorics imply ordinary assertorics. For instance, in his proof of Celarent X'QM at A15, 34b19-27:

... it is necessary that *a* possibly belongs to no *c*. If it is not possible, *b* being stated to belong [*sc*. without qualification] to *c* .... (A15, 34b19-27)

he assumes that Lab^i is incompatible with $X'ab^e$, but this can be so only if $X'ab^e$ implies ab^e. So we may attribute to him

X'-subord $\frac{X'ab^a}{ab^a}\quad\frac{X'ab^e}{ab^e}$

He also takes it that X'e-forms convert:

X'e conv $\frac{X'ab^e}{X'ba^e}$

Let *a* be taken to belong to no *b* but to be possible for all *c*. Converting the negative, *b* belongs to no *a*. (A18, 37b24-29)

Syllogisms involving 'unqualified' forms can, thinks Aristotle, all be proved; none need be taken as axiomatic.

CHAPTER THREE. THESES

In this Chapter we detail all formulae that Aristotle asserts as theses. These can be organised in the five layers we distinguished earlier. They include (1a) deductions of L-conclusions from pure L-premisses, (1b) deductions of L-conclusions from mixed L-X premisses, (1c) deductions of X-conclusions from mixed L-X premisses, (2a) deductions of Q-conclusions from mixed Q-X premisses, (2b) deductions of Q-conclusions from mixed Q-L premisses, (3) deductions of Q'-conclusions from pure Q'-premisses, (4a) deductions of M-conclusions from X-premisses, (4b) deductions of M-conclusions from L-premisses, (4c) deductions of M-conclusions from mixed L-Q premisses, (4d) deductions of M-conclusions from mixed X-Q premisses, (4e) deductions of M-conclusions from mixed X-M premisses, (5) deductions of M-conclusions from mixed X'-Q premisses.

We also list important theses which Aristotle does not assert but which nonetheless can readily be deduced from his axioms by his rules. We note two possible readings of Aristotle's ecthetic proofs for Baroco LLL and Bocardo LLL.

§9. NECESSITY MOODS

§9.1 Pure L theses

Aristotle himself does not do so, but we deduce the laws of modal subalternation from M-axioms:

La-subalt

$$Me-conv\,\frac{Me-conv\,\dfrac{Mab^e}{Mba^e}}{Mab^e}\;\xrightarrow[Me\,\|\,La]{K}\;\frac{Lab^a}{Lab^i}$$

Le-subalt

$$Ma-rep\,\frac{Mab^a}{Mab^a}\;\xrightarrow[Ma\,\|\,Le]{K}\;\frac{Lab^e}{Lab^o}$$

(Under the arrow we mention the contrariety relied on.) From Le-conversion and Le-subalternation we can deduce

Le-conv (partial)

$$\frac{\dfrac{Lab^e}{Lba^e}\,Le-conv}{Lba^o}\,Le-sub\,alt$$

Aristotle's proofs by Direct Reduction or Ecthesis in the assertoric syllogistic are summarised in Figure 9.1.

A8, 29b36-30a2 says that any assertoric syllogism can be transformed into an LLL syllogism. This means that for each proof represented in Figure 9.1 there is a proof of the corresponding LLL syllogism. The argument for this theorem has two parts. Firstly, 30a2-14 points out that the laws of L-conversion are perfectly analogous to those for assertoric conversion, allowing for Direct Reduction of all but Baroco and Bocardo LLL; secondly, it points out that Baroco and Bocardo LLL are provable by Ecthesis.

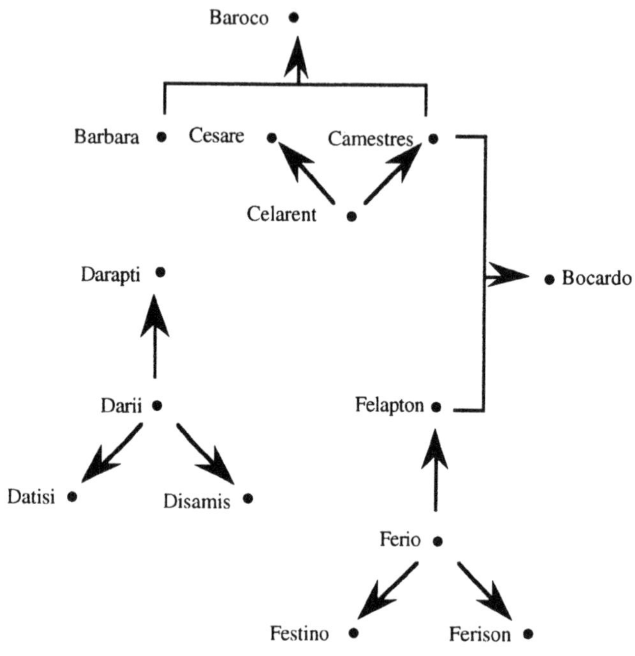

Fig. 9.1

The required Direct Reductions are:

Cesare LLL

$$\text{Celarent } LLL\frac{{}_{Le-conv}\dfrac{Lab^e}{Lba^e} \quad Lac^a}{Lbc^e}$$

Camestres LLL

$$\text{Celarent LLL}\;\dfrac{Lab^a \qquad \dfrac{Lac^e}{Lca^e}\,\text{Le-conv}}{\text{Le-conv}\;\dfrac{Lcb^e}{Lbc^e}}$$

Festino LLL

$$\dfrac{\text{Le-conv}\;\dfrac{Lab^e}{Lba^e} \qquad Lac^i}{Lbc^o}\;\text{Ferio LLL}$$

Darapti LLL

$$\text{Darii LLL}\;\dfrac{Lac^a \qquad \dfrac{Lbc^a}{Lcb^i}\,\text{La-conv}}{Lab^i}$$

Felapton LLL

$$\text{Ferio LLL}\;\dfrac{Lac^e \qquad \dfrac{Lbc^a}{Lcb^i}\,\text{La-conv}}{Lab^o}$$

Disamis LLL

$$\text{Darii LLL}\;\dfrac{\text{Li-conv}\;\dfrac{Lac^i}{Lca^i} \qquad Lbc^a}{\dfrac{Lba^i}{Lab^i}\,\text{Li-conv}}$$

Datisi LLL

$$\text{Darii LLL}\;\dfrac{Lac^a \qquad \dfrac{Lbc^i}{Lcb^i}\,\text{Li-conv}}{Lab^i}$$

Ferison LLL

$$\text{Ferio LLL}\;\dfrac{Lac^e \qquad \dfrac{Lbc^i}{Lcb^i}\,\text{Li-conv}}{Lab^o}$$

As pointed out in §4.2, there are two possible readings for the ecthetic proofs of Baroco LLL and Bocardo LLL - either via the deductions

$$\textit{Felapton LXL}\,\dfrac{\textit{Camestres LLL}\,\dfrac{Lab^a \quad Lad^e}{Lbd^e} \qquad cd^a}{Lbc^o} \xrightarrow{\textit{Lo-ecthesis}\ \mathrm{A}} \dfrac{Lab^a \quad Lac^o}{Lbc^o}$$

$$\textit{Felapton LLL}\,\dfrac{Lad^e \qquad \textit{Barbara LXL}\,\dfrac{cd^a \quad Lbc^a}{Lbd^a}}{Lab^o} \xrightarrow{\textit{Lo-ecthesis}\ \mathrm{A}} \dfrac{Lac^o \quad Lbc^a}{Lab^o}$$

or via the deductions

$$\textit{Felapton LLL}\,\dfrac{\textit{Camestres LLL}\,\dfrac{Lab^a \quad Lad^e}{Lbd^e} \qquad Lcd^a}{Lbc^o} \xrightarrow{\textit{Lo-ecthesis}\ \mathrm{AA}} \dfrac{Lab^a \quad Lac^o}{Lbc^o}$$

$$\textit{Felapton LLL}\,\dfrac{Lad^e \qquad \textit{Barbara LLL}\,\dfrac{Lcd^a \quad Lbc^a}{Lbd^a}}{Lab^o} \xrightarrow{\textit{Lo-ecthesis}\ \mathrm{AA}} \dfrac{Lac^o \quad Lbc^a}{Lab^o}$$

On the first interpretation, the ecthetic proofs use L-X theses along with pure LLL theses; on the second interpretation, they appeal only to LLL theses.[1]

Smith asks[2] why Aristotle does not derive Bocardo LLL from Barbara QLM, itself a consequence of the axiom Barbara QLQ. The answer is, Because Bocardo LLL does not follow from Barbara QLM. The derivation

$$\textit{Barbara QLM}\,\dfrac{Qab^a \quad Lbc^a}{Mac^a} \longrightarrow \dfrac{Lac^o \quad Lbc^a}{Lab^o}$$

would be valid only if there were a contradiction $Qa \mid Lo$, which there is not. Still, as we saw in §8.2, Aristotle asserts Barbara MXM; and this implies acceptance of Barbara MLM, whence Bocardo LLL follows:

$$\textit{Barbara MXM}\,\dfrac{Mab^a \qquad \textit{La - subord}\,\dfrac{Lbc^a}{bc^a}}{Mac^a} \xrightarrow{C} \dfrac{Lac^o \quad Lbc^a}{Lab^o}\,\textit{Bocardo LLL}$$

[1] When we come to consider modal syllogistic with singular terms in §38.3, we shall find other ways of representing these ecthetic proofs.

§9.2 L-X theses

From the axioms of L-subordination, together with LL-theses, the following 1-premissed LX-theses are deducible (though Aristotle himself does not perform these deductions):

$$\dfrac{La-subalt\,\dfrac{Lab^a}{Lab^i}}{ab^i}\,Li-subord \qquad \dfrac{Le-subalt\,\dfrac{Lab^e}{Lab^o}}{ab^o}\,Lo-subord$$

$$\dfrac{La-conv\,\dfrac{Lab^a}{Lba^i}}{ba^i}\,Li-subord \qquad \dfrac{Le-conv\,\dfrac{Lab^e}{Lba^e}}{ba^e}\,Le-subord$$

$$\dfrac{Le-conv\,\dfrac{Lab^e}{Lba^o}}{ba^o}\,Lo-subord \qquad \dfrac{Li-conv\,\dfrac{Lab^i}{Lba^i}}{ba^i}\,Li-subord$$

From the basis of Barbara, Celarent, Darii and Ferio LXL, Aristotle derives the following theses:

Cesare LXL

$$\dfrac{Le-conv\,\dfrac{Lab^e}{Lba^e} \quad ac^a}{Lbc^e}\,Celarent\ LXL$$

First let the negative be necessary, and *a* be possible for no *b*, and just belong to all *c*. Then, since the negative converts, *b* is possible for no *a*; but *a* belongs to all *c*, so *b* is not possible for any *c*, for *c* is under *a*. (A10, 30b9-13)

---

2 SMITH *ad* 30a3-14.

Camestres XLL

$$\dfrac{\dfrac{ab^{a} \qquad \dfrac{Lac^{e}}{Lca^{e}}\,{\scriptstyle Le-conv}}{Lcb^{e}}\,{\scriptstyle Celarent\ LXL}}{Lbc^{e}}\,{\scriptstyle Le-conv}$$

If a is not possible for any c, neither is c possible for any a; but a belongs to all b, so c is not possible for any b; for the First Figure is clearly produced. So b cannot belong to any c, for it converts similarly. (A10, 30b13-18)

Festino LXL

$$\dfrac{\dfrac{Lab^{e}}{Lba^{e}}\,{\scriptstyle Le-conv} \qquad ac^{i}}{Lbc^{o}}\,{\scriptstyle Ferio\ LXL}$$

First let the negative be universal and necessary, and let a not possibly belong to any b but belong to some c. Since the negative converts, b does not possibly belong to any a; but a belongs to some c. So necessarily b does not belong to some c. (A10, 31a5-10)

Darapti LXL

$$\dfrac{Lac^{a} \qquad \dfrac{bc^{a}}{cb^{i}}\,{\scriptstyle a-conv}}{Lab^{i}}\,{\scriptstyle Darii\ LXL}$$

First let both premisses be affirmative, and let both a and b belong to all c, and let ac be necessary. Since b belongs to all c, c will belong to some b, by conversion of the universal to a particular. So, if a belongs necessarily to all c, and c to some b, a will belong necessarily to some b, for b is under c. The First Figure is produced. (A11, 31a24-30)

Darapti XLL

$$\dfrac{\dfrac{\dfrac{ac^{a}}{ca^{i}}\,{\scriptstyle a-conv} \qquad Lbc^{a}}{Lba^{i}}\,{\scriptstyle Darii\ LXL}}{Lab^{i}}\,{\scriptstyle Li-conv}$$

The proof will be similar if bc is necessary. For by conversion c [belongs] to some a, so if b necessarily belongs to all c, it also belongs to some a of necessity. (A11, 31a31-33)

Angelelli notes here[3] that "the text does not perform, curiously, the expected final conversion", which we supply.

Felapton LXL

$$\text{Ferio LXL}\,\frac{Lac^e \quad \dfrac{bc^a}{cb^i}\,a-conv}{Lab^o}$$

Now let ac be negative and bc affirmative, with the negative necessary. Since by conversion c [belongs] to some b, and a to no b of necessity, a does not belong of necessity to some a. For b is under c. (A11, 31a33-37)

Disamis XLL

$$\frac{\dfrac{\dfrac{ac^i}{ca^i}\,i-conv \quad Lbc^a}{Lba^i}\,\text{Darii LXL}}{Lab^i}\,Li-conv$$

Datisi LXL

$$\text{Darii LXL}\,\frac{Lac^a \quad \dfrac{bc^i}{cb^i}\,i-conv}{Lab^i}$$

If one is universal, one particular, both being affirmative, when the universal is necessary the conclusion too will be necessary. The proof is the same as before, for the affirmative converts. If then b necessarily belongs to all c, a is under c, and b necessarily belongs to some a. But if b [belongs] to some a then a belongs necessarily to some b. For it converts. Similarly if ac is the necessary universal. For b is under c. (A11, 31b12-20)

Ferison LXL

$$\text{Ferio LXL}\,\frac{Lac^e \quad \dfrac{bc^i}{cb^i}\,i-conv}{Lab^o}$$

For if a is possible for no c, and b belongs to some c, a necessarily does not belong to some b. (A11, 31b35-37)

---

[3] ANGELELLI p.186.

For it is possible that animal belongs to no white. Man then belongs to no white, but not necessarily. (A10, 30b31-37)[4]

Baroco LXX

$$\frac{Lab^a \quad ac^o}{bc^o}$$

Let the universal be affirmative and necessary, and let the affirmative be stated in relation to *b*. Then if *a* belongs necessarily to all *b*, and does not belong to some *c*, clearly *b* does not belong to some *c* .... (A10, 31a10-13)

Felapton XLX

$$\text{Ferio } XLX\,\frac{ac^e \quad \dfrac{Lbc^a}{Lcb^i}\,La\text{-}conv}{ab^o}$$

For let *bc* be affirmative and necessary, and let *ac* be negative and not necessary. Then since the affirmative converts, *c* also belongs necessarily to some *b*. So if *a* belongs to no *c*, and *c* to some *b*, *a* does not belong to some *b*. (A10, 31a10-13)

He offers no proof of these moods; neither does he say they are perfect. We can only assume that they reduce to assertoric Camestres, Baroco and Ferio by L-subord. The proofs of these moods would then go:

$$\frac{La-subord\,\dfrac{Lab^a}{ab^a} \quad ac^e}{bc^e}\,Camestres \qquad \frac{La-subord\,\dfrac{Lab^a}{ab^a} \quad ac^o}{bc^o}\,Baroco$$

$$\text{Ferio}\,\frac{ac^e \quad \dfrac{\dfrac{Lbc^a}{Lcb^i}\,La\text{-}conv}{cb^i}\,Li\text{-}subord}{ab^o}$$

But if these proofs are acceptable then any premisses yielding an L-conclusion will also yield an X-conclusion.

Finally we should note the 3-premissed thesis

4 This passage is noteworthy for its careful distinction between *necessitas consequentis* and *necessitas consequentiae*.

$$\dfrac{\dfrac{Lab^a}{ab^a}\,La-subord \quad ac^e}{bc^e}\,Camestres \qquad \dfrac{\dfrac{Lab^a}{ab^a}\,La-subord \quad ac^o}{bc^o}\,Baroco$$

$$\dfrac{ac^e \quad \dfrac{\dfrac{Lbc^a}{Lcb^i}\,La-conv}{cb^i}\,Li-subord}{ab^o}\,Ferio$$

But if these proofs are acceptable then any premisses yielding an L-conclusion will also yield an X-conclusion.

Finally we should note the 3-premissed thesis

$$\dfrac{ab^a \quad Lbc^a \quad Lbc^a}{Lab^i}$$

whose proof was given in §4.1.

§10. UNAMPLIATED CONTINGENCY MOODS

The contingency-syllogistic deals with the logic of derivations involving propositions that contain an expressed mode of contingency. There are no pure Q-theses: the theses with pure contingency-premisses involve Q'-premisses. And there are no Q-X theses in the Second Figure. So we move immediately to the Third Figure.

§10.1 Q-X theses

Aristotle proves the following theses:

Darapti QXQ

$$\dfrac{Qac^a \quad \dfrac{bc^a}{cb^i}\,a-conv}{Qab^i}\,Darii\ QXQ$$

Felapton QXQ

$$\dfrac{Qac^e \quad \dfrac{bc^a}{cb^i}\,a-conv}{Qab^o}\,Ferio\ QXQ$$

Similarly if *bc* is assertoric and *ac* problematic, or if *ac* is negative and *bc* affirmative, whichever one is assertoric; in each case the conclusion will be problematic. For the First Figure is produced again, and it has been shown that in this [Figure] if either premiss signifies possibility, the conclusion will be problematic. (A21, 39b16-22)

The last sentence of this passage seems to derive the validity of Felapton XQQ from that of Ferio XQQ. But the appearance is misleading. Aristotle's language here does not distinguish between Q and M conclusions, or between X and X' premisses. In fact, as we shall see in §13.3, Felapton X'QM is deducible from Ferio X'QM; and, as we shall see in §15.2, Aristotle is committed to rejecting Ferio XQQ and XQM.

Datisi QXQ

$$\text{Darii } QXQ\ \dfrac{Qac^a \quad \dfrac{bc^i}{cb^i}\ i-conv}{Qab^i}$$

Disamis XQQ

$$\text{Darii } QXQ\ \dfrac{\dfrac{Qbc^a \quad \dfrac{ac^i}{ca^i}\ i-conv}{Qba^i}}{Qab^i}\ Qi-conv$$

Ferison QXQ

$$\text{Ferio } QXQ\ \dfrac{Qac^e \quad \dfrac{bc^i}{cb^i}\ i-conv}{Qab^o}$$

If one premiss is universal, one particular, both being affirmative, or the negative universal and the affirmative particular, there will be the same moods of syllogisms, for all are completed through the First Figure. (A21, 39b26-30)

Overlooked by Aristotle is

Darapti XQQ[5]

$$\dfrac{\dfrac{\dfrac{ac^a}{ca^i}\,a\text{-}conv \quad Qbc^a}{Qba^i}\,\text{Darii } QXQ}{Qab^i}\,Qi\text{-}conv$$

Mignucci argues for the validity of Bocardo QXQ.[6] But, intended or not by Aristotle, this mood cannot be proved by Aristotelian methods. The derivation

$$\text{Barbara } LXL\,\frac{bc^a \quad Lab^a}{Lac^a} \longrightarrow \frac{bc^a \quad Qac^o}{Qab^o}$$

is not justified by rule C, because Qab^o is not contradictory to Lab^o. Nor is it justified by rule K, because Qab^o is not contrary to Lab^o. Mignucci's derivation

$$\left.\begin{array}{l} \text{Barbara } LXL\,\dfrac{\dfrac{Lab^a bc^a}{Lac^a}}{\sim Qac^o} \\[2em] \text{Celarent } LXL\,\dfrac{\dfrac{Lab^e bc^a}{Lac^e}}{\sim Qac^o} \end{array}\right\} \xrightarrow{R} \frac{\sim Qab^o \quad bc^a}{\sim Qac^o} \xrightarrow{C} \frac{Qac^o \quad bc^a}{Qab^o}$$

makes the assumption, in using rule R, that if Qab^o is false then either Lab^a or Lab^e is true; but this is incorrect, since the falsity of Qab^o might be due to a's being necessary for some bs and impossible for others. In fact Bocardo QXQ is invalid, as we shall see in §28.4.

Angelelli's derivation[7]

---

[5] See McCALL 1963 §32.
[6] MIGNUCCI *ad* 39b39.
[7] ANGELELLI p.186.

$$\frac{\frac{\frac{\frac{\frac{Qac^o}{Qac^i}\,Q\text{-}conv}{Qca^i}\,Qi\text{-}conv \quad bc^a}{Qba^i}\,Darii\ XQQ}{Qab^i}\,Qi\text{-}conv}{Qab^o}\,Q\text{-}conv$$

assumes the validity of Darii XQQ. We shall see in §15.2.1 that Aristotle is committed to rejecting this mood, though (as we shall see in §12.1 and §13.1) he is committed to asserting Darii XQM and Darii X'QM.

§10.2 Q-L theses

Aristotle establishes the following syllogisms:

§10.2.1 First Figure

Celarent LQX

$$Ferio\ LXL\ \frac{Le\text{-}conv\ \frac{Lab^e}{Lba^e} \quad ac^i}{Lbc^o} \quad \xrightarrow[Lo\,\|\,Qa]{K} \quad \frac{Lab^e \quad Qbc^a}{ac^e}$$

Let *a* be not possible for any *b*, and *b* be possible for all *c*. Necessarily then *a* belongs to no *c*. For let it belong to some or all. It was supposed not to be possible for any *b*. By converting the negative, *b* is not possible for any *a*. But *a* was let belong to some or all *c*. So *b* does not possibly belong to any or some *c*. But at the start we supposed all. (A16, 36a8-15)

Alexander[8] derives Celarent LQL from Disamis MQM:[9]

$$\frac{Mac^i \quad Qbc^a}{Mab^i} \xrightarrow{C} \frac{Lab^e \quad Qbc^a}{Lac^e}$$

This proof could not be accepted by Aristotle, who at 36b3-7 rejects all L-Q moods with Le-conclusion, and thus by implication Celarent LQL. We outline Aristotle's position in §15.3.1.

---

[8] ALEXANDER 249,19-25.

[9] Or is it Disamis MMM? - this mood is invalid - or Disamis QQQ? - this mood not only is not valid, but does not even imply Celarent LQL.

Ferio LQX

$$\text{Celarent } LXL\dfrac{\text{Le-conv}\dfrac{Lab^e}{Lba^e} \quad ac^a}{Lbc^e} \quad \xrightarrow[Le \,\|\, Qi]{K} \quad \dfrac{Lab^e \quad Qbc^i}{ac^o}$$

For example, if *a* does not possibly belongs to any *b*, and *b* possibly belongs to some *c*, necessarily *a* does not belong to some *c*. For if it belongs to all, and it is not possible for any *b*, then *b* does not possibly belong to any *a*. So if *a* belongs to all *c*, *b* is not possible for any *c*. But it was supposed possible for some. (A16, 36a34-39)

It should be mentioned here that Barbara LQX and Darii LQX are not provable in the above manner. If such proofs existed, they would be:[10]

$$\text{Baroco } LXL\dfrac{Lab^a \quad ac^o}{Lbc^o} \xrightarrow[Lo \,\|\, Qa]{K} \dfrac{Lab^a \quad Qbc^a}{ac^a}$$

$$\text{Camestres } LXL\dfrac{Lab^a \quad ac^e}{Lbc^e} \xrightarrow[Le \,\|\, Qi]{K} \dfrac{Lab^a \quad Qbc^i}{ac^i}$$

But Baroco LXL and Camestres LXL, on which these 'proofs' rest, are explicitly rejected by Aristotle. (See §14.2.)

§10.2.2 Second Figure

Cesare LQX

$$\text{Ferio } LXL\dfrac{Lab^e \quad bc^i}{Lac^o} \xrightarrow[Lo \,\|\, Qa]{K} \dfrac{Lab^e \quad Qac^a}{bc^e}$$

Camestres QLX[11]

$$\text{Ferio } LXL\dfrac{Lac^e \quad i-\text{conv}\dfrac{bc^i}{cb^i}}{Lab^o} \quad \xrightarrow[Lo \,\|\, Qa]{K} \quad \dfrac{Lac^e \quad Qab^a}{bc^e}$$

---

[10] WIELAND 1975, p.88, errs in stating that these 'proofs' rest on the oppositions $Qe \,\|\, Li$ (his $MBec \longrightarrow \neg NBiC$) and $Qo \,\|\, La$ (his $MBoc \longrightarrow \neg NBaC$).

[11] BECKER (Tafel III) proves this mood by Direct Reduction to Celarent LQX:

Let a necessarily belong to no b, but be possible for all c. Converting the negative, b belongs to no a but a is possible for all c.... it is clear that b belongs to no c. For let it belong. Then if a is possible for no b, and b belongs to some c, a is not possible some c. But it was supposed possible for all. The proof will be of the same sort if the negation is proposed in relation to c. (A19, 38a16-18, 21-26)

LeQeXe-2

$$\dfrac{Le-conv\ \dfrac{Lab^e}{Lba^e} \quad \dfrac{Qac^e}{Qac^a}\ Q-conv}{bc^e}\ Celarent\ LQX$$

QeLeXe-2

$$Celarent\ LQX\ \dfrac{Q-conv\ \dfrac{Qab^e}{Qab^a} \quad \dfrac{Lac^e}{Lca^e}\ Le-conv}{\dfrac{cb^e}{bc^e}\ e-conv}$$

If the premisses are of similar form, when they are negative a syllogism will always be produced by contingency-conversion as in previous cases. For, let it be stated that a necessarily does not belong to b, and possibly does not belong to c. Converting the premisses, b belongs to no a but a is possible for all c. The First Figure is produced. Similarly if the negative is stated in relation to c. (A19, 38b6-13)

$$Celarent\ LQX\ \dfrac{Qab^a \quad \dfrac{Lac^e}{Lca^e}\ Le{-}conv}{\dfrac{cb^e}{bc^e}\ e{-}conv}$$

Camestres QLX can also be proved by Indirect Reduction to Darii QXM (for whose validity see §12).

$$Darii\ QXM\ \dfrac{Qab^a \quad bc^i}{Mac^i} \xrightarrow{C} \dfrac{Qab^a \quad Lac^e}{bc^e}$$

Similarly provable are Cesare QLX, Festino QLX and Baroco QLX:

$$Ferio\ QXM\ \dfrac{Qab^e \quad bc^i}{Mac^o} \xrightarrow{C} \dfrac{Qab^e \quad Lac^a}{bc^e}$$

$$Celarent\ QXM\ \dfrac{Qab^e \quad bc^a}{Mac^e} \xrightarrow{C} \dfrac{Qab^e \quad Lac^i}{bc^o}$$

$$Barbara\ QXM\ \dfrac{Qab^a \quad bc^a}{Mac^a} \xrightarrow{C} \dfrac{Qab^a \quad Lac^o}{bc^o}$$

Festino LQX[12]

$$Le-conv\,\frac{\dfrac{Lab^e}{Lba^e}\quad Qac^i}{bc^o}\,Ferio\;LQX$$

When the negative is universal and necessary, there will always be a ... syllogism that 'it does not belong'. The proof is by conversion. (A19, 38b25-27)

LeQoXo-2

$$Le-conv\,\frac{\dfrac{Lab^e}{Lba^e}\quad \dfrac{Qac^o}{Qac^i}\,Q-conv}{bc^o}\,Ferio\;LQX$$

When both are negative, and that indicating non-belonging is both universal and necessary, there will be nothing necessary from these very statements, but there will be a syllogism by converting the possible premiss, as in previous cases. (A19, 38b31-35)

Aristotle overlooks

Camestres LQX[13]

$$Ferison\;QXM\,\frac{Qac^e\quad bc^i}{Mab^o}\xrightarrow{C}\frac{Lab^a\quad Qac^e}{bc^e}$$

(A parallel proof exists for Cesare LQX.) We shall see in §15.3.2 that Aristotle regards Camestres LQX as invalid; but in §19.1 we shall question his reasoning, and ultimately we will take this mood as valid.

---

12 PATTERSON 1995 p.195 gives an alternative proof:

$$Celarent\;LXL\,\frac{Lab^e\quad bc^a}{Lac^e}\;\xrightarrow[Le\,\|\,Qi]{K}\;\frac{Lab^e\quad Qac^i}{bc^o}$$

13 For Ferison QXM see §12.3.

§10.2.3 Third Figure

Darapti QLQ

$$\mathit{Darii\ QLQ}\,\frac{Qac^{a} \qquad \dfrac{Lbc^{a}}{Lcb^{i}}\,La-conv}{Qab^{i}}$$

The proof is the same if *bc* is stated as necessary, *ac* as possible. (A22, 40a16-18)

Felapton QLQ

$$\mathit{Ferio\ QLQ}\,\frac{Qac^{e} \qquad \dfrac{Lbc^{a}}{Lcb^{i}}\,La-conv}{Qab^{o}}$$

Again, let one be affirmative, one negative, the affirmative necessary, and let *a* possibly belong to no *c*, and *b* belong necessarily to all. There will be the First Figure again. For the negative premiss signifies possibility. It is clear then that the conclusion will be possible. (A22, 40a18-23)

Felapton LQX

$$\mathit{Ferio\ LQX}\,\frac{Lac^{e} \qquad \dfrac{Qbc^{a}}{Qcb^{i}}\,Qa-conv}{ab^{o}}$$

If the negative premiss is necessary, the conclusion will be ... 'it does not belong'. For, let *a* necessarily not belong to *c*, and *b* be possible for all. Converting the affirmative *bc*, there will be the First Figure, and the negative premiss is necessary. The premisses being thus, it follows that *a* ... does not belong to some *c*. Necessarily then, *a* does not belong to some *b*. (A22, 40a25-32)

Disamis LQQ

$$\dfrac{\mathit{Darii\ \ QLQ}\,\dfrac{Li-conv\,\dfrac{Lac^{i}}{Lca^{i}} \qquad Qbc^{a}}{Qba^{i}}}{Qab^{i}}\,Qi-conv$$

Datisi QLQ

$$\text{Darii } QLQ\ \dfrac{Li-conv\ \dfrac{Lbc^i}{Lcb^i} \quad Qac^a}{Qab^i}$$

Similarly if one term is related universally, the other in part, to the middle. Both being affirmative, there will be a syllogism that 'it is possible'... (A22, 40a39-b1)

Ferison QLQ

$$\text{Ferio } QLQ\ \dfrac{Qac^e \quad \dfrac{Lbc^i}{Lcb^i}\ Li-conv}{Qab^o}$$

And when a negative and an affirmative are posited, the affirmative being necessary. (A22, 40b2-3)

Ferison LQX

$$\text{Ferio } LQX\ \dfrac{Lac^e \quad \dfrac{Qbc^i}{Qcb^i}\ Qi-conv}{ab^o}$$

When the negative is necessary, the conclusion will be 'does not belong'. For the proof will be of the same kind, the terms being either universal or not universal. (A22, 40b3-6)

Also overlooked by Aristotle are

Darapti LQQ

$$\dfrac{La-subord\ \dfrac{Lac^a}{ac^a} \quad Qbc^a}{Qab^i}\ \text{Darapti } XQQ$$

Baroco QLX

$$\text{Barbara } QXM\ \dfrac{Qab^a \quad bc^a}{Mac^a} \xrightarrow{C} \dfrac{Qab^a \quad Lac^o}{bc^o}$$

For Ferio QXM and Barbara QXM see §12.

§11. AMPLIATED CONTINGENCY MOODS

There are no ecthetic proofs of Q'Q'Q' syllogisms. Aristotle's proofs by Direct Reduction are represented in Figure 11.1. They are:

Q'aQ'eQ'a-1

$$\text{Barbara } Q'Q'Q' \frac{Q'ab^a \quad \dfrac{Q'bc^e}{Q'bc^a}\, Q'\text{-}conv}{Q'ac^a}$$

When a is possible for all b and b is possible for no c, no syllogism is produced from the stated premisses, but if bc is converted according to the possible the same one is produced as before. For, since b possibly belongs to no c, it possibly belongs to all, as was said before. So if b [belongs] to all c and a to all b, the same syllogism is produced again. (A14, 33a5-12)[14]

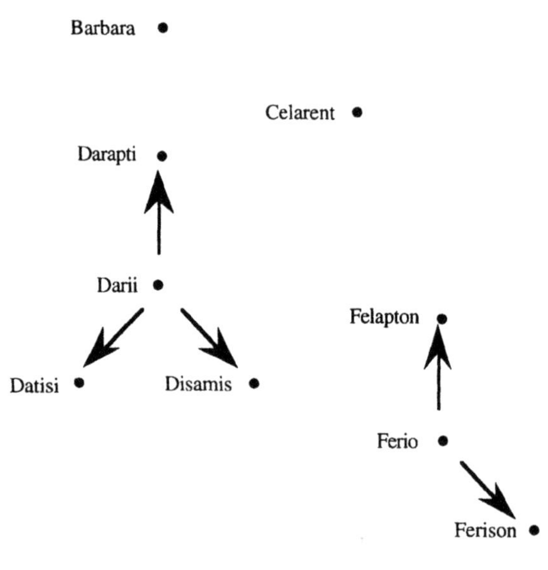

Fig. 11.1

Q'eQ'eQ'a-1

$$\text{Barbara } Q'Q'Q' \frac{Q'\text{-}conv\ \dfrac{Q'ab^e}{Q'ab^a} \quad \dfrac{Q'bc^e}{Q'bc^a}\, Q'\text{-}conv}{Q'ac^a}$$

14 TRICOT *ad loc.* mistakenly describes this as 'non-concluant'. Aristotle makes it clear that the syllogism is valid but imperfect.

Similarly if both premisses state a negative possibility. I mean for example if *a* is possible for no *b* and *b* for no c. For, no syllogism is produced from the stated premisses, but by converting we have the same again as before. (A14, 33a12-17)

Q'aQ'oQ'i-1

$$\text{Darii } Q'Q'Q' \frac{Q'ab^{a} \quad \dfrac{Q'bc^{o}}{Q'bc^{i}}\, Q'\text{-}conv}{Q'ac^{i}}$$

If the particular premiss is taken as negative, the universal as affirmative, the order being the same, (for example if *a* is possible for all *b* and *b* possibly does not belong to some *c*), no clear syllogism is produced through the stated premisses, but by converting the particular, and laying it down that *b* possibly belongs to some *c*, there will be the same conclusion as before. (A14, 33a27-34)

Darapti Q'Q'Q'

$$\text{Darii } Q'Q'Q' \frac{Q'ac^{a} \quad \dfrac{Q'bc^{a}}{Q'cb^{i}}\, Q'a\text{-}conv}{Q'ab^{i}}$$

First, suppose them possible, and let both *a* and *b* possibly belong to all *c*. Since then the affirmative partially converts, and *b* is possible for all *c*, *c* will be possible also for some *b*. So, if *a* is possible for all *c*, and *c* for some *b*, necessarily *a* is also possible for some *b*. For the First Figure is produced. (A20, 39a14-19)

Felapton Q'Q'Q'

$$\text{Ferio } Q'Q'Q' \frac{Q'ac^{e} \quad \dfrac{Q'bc^{a}}{Q'cb^{i}}\, Q'a\text{-}conv}{Q'ab^{o}}$$

And if *a* possibly belongs to no *c*, and *b* to all *c*, necessarily *a* possibly does not belong to some *b*. For there is the First Figure again by conversion. (A20, 39a19-23)

Q'eQ'eQ'i-3

$$\text{Darapti } Q'Q'Q' \frac{Q'\text{-}conv\, \dfrac{Q'ac^{e}}{Q'ac^{a}} \quad \dfrac{Q'bc^{e}}{Q'bc^{a}}\, Q'\text{-}conv}{Q'ac^{i}}$$

For if *a* and *b* possibly do not belong to *c*, then by replacing the possibles, there will be the First Figure again by conversion. (A20, 39a26-28)

Datisi Q'Q'Q'

$$\text{Darii } Q'Q'Q' \; \dfrac{Q'ac^{a} \quad \dfrac{Q'bc^{i}}{Q'cb^{i}}\, Q'i\text{-}conv}{Q'ab^{i}}$$

Disamis Q'Q'Q'

$$\dfrac{\text{Darii } Q'Q'Q' \; \dfrac{Q'bc^{a} \quad \dfrac{Q'ac^{i}}{Q'ca^{i}}\, Q'i\text{-}conv}{Q'ba^{i}}}{Q'ab^{i}}\, Q'i-conv$$

For let *a* possibly belong to all *c*, *b* to some *c*. There will be the First Figure again by converting the particular premiss. For if *a* is possible for all *c*, and *c* for some *b*, *a* [is possible] for some *b*. Similarly if *bc* is the universal. (A20, 39a31-36)

Ferison Q'Q'Q'

$$\text{Ferio } Q'Q'Q' \; \dfrac{Q'ac^{e} \quad \dfrac{Q'bc^{i}}{Q'cb^{i}}\, Q'i\text{-}conv}{Q'ab^{o}}$$

Bocardo Q'Q'Q'

$$Q'\text{-}conv\;\dfrac{Q'i\text{-}conv\;\dfrac{\dfrac{Q'i\text{-}conv\;\dfrac{Q'\text{-}conv\;\dfrac{Q'ac^{o}}{Q'ac^{i}}}{Q'ca^{i}} \quad Q'bc^{a}}{Q'ba^{i}}\,\text{Darii } Q'Q'Q'}{Q'ab^{i}}}{Q'ab^{o}}$$

Similarly again if *ac* is negative, *bc* affirmative. For there will be the First Figure again through conversion.[15] (A20, 39a36-38)

[15] Aristotle does not make it clear that Bocardo as well as Ferison is covered by this. In the case of Bocardo, the conversion required is *complementary* conversion.

Q'eQ'oQ'i-3

$$\text{Datisi } Q'Q'Q' \frac{Q'\text{-conv}\dfrac{Q'ac^e}{Q'ac^a} \quad \dfrac{Q'bc^o}{Q'bc^i}Q'\text{-conv}}{Q'ab^i}$$

Q'oQ'eQ'i-3

$$\text{Disamis } Q'Q'Q' \frac{Q'\text{-conv}\dfrac{Q'ac^o}{Q'ac^i} \quad \dfrac{Q'bc^e}{Q'bc^a}Q'\text{-conv}}{Q'ab^i}$$

If both are stated in the negative, one universal and one particular, there will be no syllogism from them as stated, but there will by conversion, as shown above. (A20, 39a38-39b2)

§12. POSSIBILITY MOODS

Aristotle does not do so, but given the modal oppositions noted in §2.1, and the laws of L-subordination we can deduce

X/M-subord

$$L\text{-subord}\frac{Lab^o}{ab^o} \xrightarrow{C} \frac{ab^a}{Mab^a} \qquad L\text{-subord}\frac{Lab^i}{ab^i} \xrightarrow{C} \frac{ab^e}{Mab^e}$$

$$L\text{-subord}\frac{Lab^e}{ab^e} \xrightarrow{C} \frac{ab^i}{Mab^i} \qquad L\text{-subord}\frac{Lab^a}{ab^a} \xrightarrow{C} \frac{ab^o}{Mab^o}$$

Whence

$$X/M\text{-subord}\frac{L\text{-subord}\dfrac{Lab^a}{ab^a}}{Mab^a}$$

and similarly

$$\frac{Lab^e}{Mab^e} \qquad \frac{Lab^i}{Mab^i} \qquad \frac{Lab^o}{Mab^o}$$

Thus wherever there is an X- or L-conclusion an M-conclusion also follows.

Analogously, we can prove the laws of

Q/M-subord

$$Lo-rep\frac{Lab^o}{Lab^o}\xrightarrow[Lo\,\|\,Qa]{K}\frac{Qab^a}{Mab^a}\qquad Li-rep\frac{Lab^i}{Lab^i}\xrightarrow[Li\,\|\,Qe]{K}\frac{Qab^e}{Mab^e}$$

$$Le-rep\frac{Lab^e}{Lab^e}\xrightarrow[Le\,\|\,Qi]{K}\frac{Qab^i}{Mab^i}\qquad La-rep\frac{Lab^a}{Lab^a}\xrightarrow[La\,\|\,Qo]{K}\frac{Qab^o}{Mab^o}$$

TABLE 12.1

Figure 1	Barbara QXM, QLM,
	Celarent QXM, QLM,
	Darii QXM, QLM,
	Ferio QXM, QLM
Figure 3	Darapti QXM, QLM, XQM, LQM,
	Felapton QXM, QLM,
	Datisi QXM, QLM,
	Disamis XQM, LQM,
	Ferison QXM, QLM

given the oppositions $Qa \,\|\, Lo$, $Qe \,\|\, Li$, $Qi \,\|\, Le$, $Qo \,\|\, La$ respectively. (We shall later find reason to question the second and fourth of these.) Given Q/M-subordination, wherever we have a thesis with a Q-conclusion we have the corresponding M-conclusion. Thus, we have the theses in Table 12.1.

Aristotle proves the following theses by Indirect Reduction.

§12.1 First Figure

Barbara LQM[16]

$$\textit{Baroco LLL}\frac{Lab^a \quad Lac^o}{Lbc^o} \xrightarrow[Lo \parallel Qa]{K} \frac{Lab^a \quad Qbc^a}{Mac^a}$$

For, let *a* be necessary for all *b*, and *b* be possible for all *c*. There will be an imperfect syllogism that *a* possibly belongs to all *c*. That it is imperfect is clear from the proof. For it will be proved in the same way as before. (A16, 35b38-36a2)

Darii LQM[17]

$$\textit{Camestres LLL}\frac{Lab^a \quad Lac^e}{Lbc^e} \xrightarrow[Le \parallel Qi]{K} \frac{Lab^a \quad Qbc^i}{Mac^i}$$

When ... the universal *ab* in affirmatives [is necessary], there will be no syllogism that 'it belongs'. (A16, 36a39-b2)

(But, we must understand, there will be a syllogism that 'it possibly belongs'.)

Prima facie Barbara LQM seems to be susceptible to a counter-example Aristotle deploys against Barbara XQM. Consider the terms ⟨man, moving, horse⟩. If all that moves are men, then all that moves are necessary men. (This follows by Barbara LXL, given that all men are necessary men.) But movement is contingent for all horses. So the premisses of Barbara LQM are true. Yet, man is not possible for any horse; so the conclusion is false. Similar considerations will damage the rest of the LQM moods. The solution to this puzzle lies in the fact that if necessity-propositions can change truth-value, so can contingency-propositions. If

16 This proof is conjectural since Aristotle does not supply one; McCALL 1963 (Table 12) alludes to it. The proof's logic is unimpeachable, but this interpretation has difficulty in supplying a reference for Aristotle's "it will be proved in the same way as before". For there have not been any proofs quite like this before the cited passage, though there is one soon after it (in the proof of Celarent LQX). ROSS *ad* 36a1-2 says that the proof will be analogous to that of Barbara X'QM, and will depend on Bocardo LXX:

$$\textit{Bocardo LXX}\frac{Lac^o \quad bc^a}{ab^o} \longrightarrow \frac{\dfrac{Lab^a}{ab^a}La-sub \qquad Qbc^a}{Mac^a}$$

17 These are McCALL's 1963 derivations. ROSS *ad* 36a1-2 reduces Barbara LQM, not to Baroco LLL, but to Bocardo LXX by the method used at A15, 34a34-b6. (But ROSS *ad* 36b2 reduces Darii LQM to Camestres LLL, like McCall.)

it can become true that all movers are necessarily men, then it can become false that all horses contingently move. If the temporary truth of *All movers are men* implies that of *Necessarily all movers are men* by Barbara LXL, then it also implies that of *Necessarily some horses are not movers* by Camestres LLL:

$$\frac{\text{Necessarily no horses are men} \qquad \text{Necessarily all movers are men}}{\text{Necessarily some horses are not movers}}$$

And if *Necessarily some horses are not movers* is sometimes true then *Contingently all horses are movers* is sometimes false.

Consequential on Barbara LQM is

LaQeMa-1

$$\text{Barbara } LQM\,\frac{Lab^a \quad \dfrac{Qbc^e}{Qbc^a}\,Q-conv}{Mac^a}$$

If the negative is posited in relation to the minor extreme, when it signifies possibility, there will be a syllogism by conversion, as before. (A16, 36a25-27)

Analogous to Darii LQM is

Ferio QLM

$$\text{Datisi } LLL\,\frac{Lac^a \quad Lbc^i}{Lab^i}\ \xrightarrow[Li\,\|\,Qe]{K}\ \frac{Qab^e \quad Lbc^i}{Mac^o}$$

When the particular affirmative is necessary, in negative syllogisms e.g., *bc*..., the syllogism will not be to a conclusion that 'it belongs'. The proof is the same as before. (A16, 36a39-b2)

Ross[18] takes the text to mean that though there is no assertoric conclusion there is a problematic one. This reading is consistent with the statement at 35b29-30 and 36b21-23 that in negative Figure 1 QL syllogisms if the affirmative premiss is necessary there will be a possibility-conclusion. And the manner of proof suggested is analogous with the one we used for Barbara LQM, the difference being that we use the opposition $Qe \,\|\, Li$ rather than $Qa \,\|\, Lo$.

All the other First Figure QLM, and QXM, moods are provable too, as mentioned above.

Aristotle proves one syllogism by means of X/M-subordination:

[18] ROSS *ad* 36b1-2.

Celarent LQM

$$\textit{X / M - subord}\;\dfrac{\textit{Celarent LQX}\;\dfrac{Lab^e \qquad Qbc^a}{ac^e}}{Mac^e}$$

It is clear that a syllogism 'it possibly does not belong' is produced since there is one that 'it does not belong'. (A16, 36a15-17)

Provable by analogous reasoning is

Ferio LQM

$$\dfrac{\textit{Ferio LQX}\;\dfrac{Lab^e \quad Qbc^i}{ac^o}}{Mac^o}\;\textit{X / M - subord}$$

Thus all four First Figure LQM syllogisms are provable.

Overlooked by Aristotle is

Darii XQM

$$\textit{Ferison LQX}\;\dfrac{Lac^e \quad Qbc^i}{ab^o}\;\xrightarrow{\;C\;}\;\dfrac{ab^a \quad Qbc^i}{Mac^i}$$

So Aristotle is committed to asserting Darii XQM; yet, as we shall see in §15.2, he is also committed to rejecting it.

§12.2 Second Figure

Aristotle states at A17, 36b31-34 that Cesare LQM and Camestres QLM are valid. (He indicates one-way possibility for the conclusions by saying that one has to understand "possible [*endechomenon*]" there "as before" (36b33-34).

Cesare LQM

$$\dfrac{\textit{Le - conv}\;\dfrac{Lab^e}{Lba^e} \quad Qac^a}{Mbc^e}\;\textit{Celarent LQM}$$

Camestres QLM

$$\mathit{Me-conv}\,\dfrac{\mathit{Celarent\ LQM}\,\dfrac{Qab^a \quad \dfrac{Lac^e}{Lca^e}\,\mathit{Le-conv}}{Mcb^e}}{Mbc^e}$$

Let a necessarily belong to no b, but be possible for all c. Converting the negative, b belongs to no a but a is possible for all c. The syllogism that b possibly belongs to no c is produced through the First Figure .... The proof will be of the same sort if the negation is proposed in relation to c. (A19, 38a16-18, 21-26)

Further Second Figure moods are

LeQeMe-2

$$\dfrac{\mathit{Le-conv}\,\dfrac{Lab^e}{Lba^e} \quad \dfrac{Qac^e}{Qac^a}\,\mathit{Q-conv}}{Mbc^e}\,\mathit{Celarent\ LQM}$$

QeLeMe-2

$$\dfrac{\mathit{Celarent\ LQM}\,\dfrac{\mathit{Q-conv}\,\dfrac{Qab^e}{Qab^a} \quad \dfrac{Lac^e}{Lca^e}\,\mathit{Le-conv}}{Mcb^e}}{Mbc^e}\,\mathit{Me-conv}$$

If the premisses are of similar form, when they are negative a syllogism will always be produced by conversion of the possible as in previous cases. For, let it be stated that a necessarily does not belong to b, and possibly does not belong to c. Converting the premisses, b belongs to no a but a is possible for all c. The First Figure is produced. Similarly if the negative is stated in relation to c. (A19, 38b6-13)

Festino LQM

$$\dfrac{\mathit{Le-conv}\,\dfrac{Lab^e}{Lba^e} \quad Qac^i}{Mbc^o}\,\mathit{Ferio\ LQM}$$

When the negative is universal and necessary, there will always be a syllogism that 'it is possible'.... the proof is by conversion. (A19, 38b25-27)

LeQoMo-2

$$\dfrac{\mathit{Le-conv}\ \dfrac{Lab^e}{Lba^e} \qquad \dfrac{Qac^o}{Qac^i}\ \mathit{Q-conv}}{Mbc^o}\ \mathit{Ferio\ LQM}$$

When both are negative, and that signifying non-belonging is both universal and necessary, there will be nothing necessary from these very statements, but there will be a syllogism by converting the possible premiss, as in previous cases. (A19, 38b31-35)

Overlooked by Aristotle is the following Indirect Reduction to the First Figure:

Cesare QLM

$$\mathit{Ferio\ QLM}\ \dfrac{Qab^e \quad Lbc^i}{Mac^o} \xrightarrow{\ C\ } \dfrac{Qab^e \quad Lac^a}{Mbc^e}$$

(Compare Aristotle's statement at 36b31-34 that Cesare QLM is invalid.)

Alexander[19] proposes the syllogism LaQaMe-2

$$\mathit{Darii\ LLL}\ \dfrac{Lab^a \quad Lbc^i}{Lac^i} \xrightarrow[Li \parallel Qa]{\ K\ } \dfrac{Lab^a \quad Qac^a}{Mbc^e}$$

This proof is analogous to that of Ferio QLM above, except that instead of relying on Datisi LLL and the opposition $Li \parallel Qe$ Alexander relies on Darii LLL and the incompatibility $Li \parallel Qa$.

§12.3 Third Figure

Darapti LQM

$$\mathit{Darii\ LQM}\ \dfrac{Lac^a \qquad \dfrac{Qbc^a}{Qcb^i}\ \mathit{Qa-conv}}{Mab^i}$$

Let *a* necessarily belong to all *c*, and let *b* be possible for all. Then since *a* is necessary for all *c*, and *c* is possible for some *b*, *a* will be possible (not: will belong) for some *b*. For, this results in the First Figure. (A22, 40a13-16)

That the mood intended is an LQM, not an LQQ, is clear from the proof given. That Darii LQQ is not valid is argued in §15.3.1.

---

[19] ALEXANDER 240.4-11.

Felapton LQM

$$\text{Ferio } LQM\,\frac{Lac^{e} \quad \dfrac{Qbc^{a}}{Qcb^{i}}\,Qa-conv}{Mab^{o}}$$

If the negative premiss is necessary, the conclusion will be 'it possibly does not belong to some' .... For, let *a* necessarily not belong to *b*, and *b* be possible for all. Converting the affirmative *bc*, there will be the First Figure, and the negative premiss is necessary. The premisses being thus, it follows that *a* possibly does not belong ... to some *c*. (A22, 40a25-31; *cf.* 40a8-9)

Again, the proof makes it clear that an M, not a Q, conclusion is intended. That Ferio LQQ is invalid is argued in §15.3.1.

LaQeMi-3

$$\text{Darapti } LQM\,\frac{Lac^{a} \quad \dfrac{Qbc^{e}}{Qbc^{a}}\,Q-conv}{Mab^{i}}$$

When the negative is stated in relation to the minor extreme, if it is possible, there will be a syllogism by transforming the premisses as in previous cases. (A22, 40a33-35)

Datisi LQM

$$\text{Darii } LQM\,\frac{Lac^{a} \quad \dfrac{Qbc^{i}}{Qcb^{i}}\,Qi-conv}{Mab^{i}}$$

Disamis QLM

$$\text{Darii } LQM\,\frac{Lbc^{a} \quad \dfrac{Qac^{i}}{Qca^{i}}\,Qi-conv}{\dfrac{Mba^{i}}{Mab^{i}}\,Mi-conv}$$

Similarly if one term is related universally, the other in part, to the middle. Both being affirmative, there will be a syllogism that 'it is possible'... (A22, 40a39-b1)

As part of his discussion of particular Figure 3 syllogisms, Aristotle briefly acknowledges the validity of

Bocardo QLM[20]

$$\textit{Barbara LLL}\,\frac{Lab^a \quad Lbc^a}{Lac^a} \xrightarrow[La \parallel Qo]{K} \frac{Qac^o \quad Lbc^a}{Mab^o}$$

And when a negative and an affirmative are posited, the affirmative being necessary. (A22, 40b2-3)

Bocardo QXM

$$\textit{Barbara LXL}\,\frac{Lab^a \quad bc^a}{Lac^a} \xrightarrow[La \parallel Qo]{K} \frac{Qac^o \quad bc^a}{Mab^o}$$

If the affirmative is universal, and the negative particular, the proof will be *per impossibile*. For let *b* belong to all *c*, and let *a* possibly not belong to some *c*. Necessarily, *a* will possibly not belong to some *b*. For if *a* belongs necessarily to all *b*, and *b* is proposed to belong to all *c*, *a* will necessarily belong to all *c*, for we proved this before. But it was supposed that it possibly did not belong to some. (A21, 39b31-39)

These proofs both rely on the opposition $La \parallel Qo$. We shall find reason later to question the validity of these two Bocardos. In the meantime, let us note that Bocardo MXM can be proved using the contradiction $La \mid Mo$ in the place of the opposition $La \parallel Qo$:[21]

---

[20] ROSS *ad* 40b4-6 claims that Bocardo LQX is meant here and that this mood is validated by Indirect Reduction, presumably thus:

$$\textit{Barbara X'QM}\,\frac{X'ab^a \quad Qbc^a}{Mac^a} \xrightarrow{C} \frac{\dfrac{Lac^o \quad Qbc^a}{\sim X'ab^a}}{ab^o}\,x$$

Only Barbara X'QM can be the basis for such an Indirect Reduction (Barbara XQM and Barbara XQQ being invalid). But step *x* here is invalid. If ab^a is not true in an unqualified sense, it does not follow that ab^o is currently true. The premisses LoQa-3 do indeed imply $\sim X'ac^a$; but Aristotle seems not to have considered the logic of such negated 'unqualifieds'.

[21] MIGNUCCI *ad* 39b39 claims (i) that Bocardo QXQ is valid and (ii) that Bocardo XQM is valid. (i) We saw in §10.1 that his proof of Bocardo QXQ is fallacious. (ii) His purported proof of Bocardo XQM:

$$\textit{Baroco LXL}\,\frac{Lab^a \quad ac^o}{Lbc^o} \xrightarrow{K} \frac{Qbc^a \quad ac^o}{Mab^o}$$

assumes the validity of Baroco LXL, which Aristotle rejects at A10, 31a10-15 (not 32a4, as McCALL's Table 1 says). (Mignucci maintains, against that text, that Baroco LXL is valid. We discuss his argument in §17.) In fact, Bocardo XQM cannot be valid, as we shall see in §15.2.3.

$$\text{Barbara LXL}\ \frac{Lab^a \quad bc^a}{Lac^a} \xrightarrow{C} \frac{Mac^o \quad bc^a}{Mab^o}$$

Aristotle overlooks

Disamis QXM

$$\text{Festino LQX}\ \frac{Qac^i \quad Lab^e}{bc^o} \xrightarrow{C} \frac{Qac^i \quad bc^a}{Mab^i}$$

This mood, along with four others, is also provable by reasoning analogous to that used for Bocardo QXM:

Disamis QXM

$$\text{Celarent LXL}\ \frac{bc^a \quad Lab^e}{Lac^e} \xrightarrow[Le \,\|\, Qi]{K} \frac{bc^a \quad Qac^i}{Mab^i}$$

Disamis XQM

$$\text{Ferio LXL}\ \frac{\text{Le-conv}\ \dfrac{Lab^e}{Lba^e} \quad ac^i}{Lbc^o} \xrightarrow[Lo \,\|\, Qa]{K} \frac{ac^i \quad Qbc^a}{Mab^i}$$

Datisi QXM

$$\text{Ferio LXL}\ \frac{bc^i \quad Lab^e}{Lac^o} \xrightarrow[Lo \,\|\, Qa]{K} \frac{bc^i \quad Qac^a}{Mab^i}$$

Datisi XQM

$$\text{Celarent LXL}\ \frac{\text{Le-conv}\ \dfrac{Lab^e}{Lba^e} \quad ac^a}{Lbc^e} \xrightarrow[Le \,\|\, Qi]{K} \frac{ac^a \quad Qbc^i}{Mab^i}$$

Ferison QXM

$$\text{Darii LXL}\ \frac{bc^i \quad Lab^a}{Lac^i} \xrightarrow[Li \,\|\, Qe]{K} \frac{bc^i \quad Qac^e}{Mab^o}$$

From Disamis XQM and Disamis QXM, respectively, we can derive

Disamis LQM

$$\dfrac{{}_{L/X-subord}\dfrac{Lac^i}{ac^i}\quad Qbc^a}{Mab^i}\,{\scriptstyle Disamis\ XQM}$$

Darii XQM

$$\dfrac{\dfrac{{}_{Qi-conv}\dfrac{Qbc^i}{Qcb^i}\quad ab^a}{Mca^i}\,{\scriptstyle Disamis\ QXM}}{Mac^i}\,{\scriptstyle Mi-conv}$$

And from Datisi XQM we can derive

XaQoMi-3

$${}_{Datisi\ XQM}\dfrac{ac^a\quad \dfrac{Qbc^o}{Qbc^i}\,{\scriptstyle Q-conv}}{Mab^i}$$

Indirectly reducible to First Figure LQM moods are:

Disamis LQM

$${}_{Celarent\ LQM}\dfrac{Lab^e\quad Qbc^a}{Mac^e}\xrightarrow{\ C\ }\dfrac{Lac^i\quad Qbc^a}{Mab^i}$$

Ferison LQM

$${}_{Darii\ LQM}\dfrac{Lab^a\quad Qbc^i}{Mac^i}\xrightarrow{\ C\ }\dfrac{Lac^e\quad Qbc^i}{Mab^o}$$

Bocardo LQM

$${}_{Barbara\ LQM}\dfrac{Lab^a\quad Qbc^a}{Mac^a}\xrightarrow{\ C\ }\dfrac{Lac^o\quad Qbc^a}{Mab^o}$$

§13. 'UNQUALIFIED' MOODS

Following Angelelli's analysis, as in §8, we treat 'unqualified' forms as ampliated assertorics.

§13.1 First Figure

Barbara X'QM $\dfrac{X'ab^a \quad Qbc^a}{Mac^a}$

Let *a* belong to all *b*, and *b* be possible for all *c*. Necessarily then, *a* possibly belongs to all *c*. Let it be not possible, but let *b* be stated as belonging to all *c*. This is false but not impossible [*adunaton*]. If then *a* is not possible for all *c*, and *b* belongs to all *c*, then *a* is not possible for all *b*. A Third Figure syllogism is produced. But we supposed that it possibly belonged to all. Necessarily then, *a* is possible for all *c*. For, though we supposed the false and not impossible [*adunaton*], the conclusion is impossible [*adunaton*]. (A15, 34a34-b1)

For the 'unqualified' nature of the major premiss, see A15, 34b7-9; for the one-way possibility of the conclusion, see ROSS *ad* A15, 33b25-33.

We saw in §5 that Aristotle's proof here appears to rely on

Rule K'

$$\left.\begin{array}{l} \dfrac{\sim r \quad s}{\neg p} \\[3em] NOT\, \dfrac{q}{\sim s} \end{array}\right\} \longrightarrow \dfrac{p \quad q}{r}$$

If we take the text at face-value, the proof is:

$$\left.\begin{array}{l} \text{Bocardo } LXM\, \dfrac{Lac^o \quad bc^a}{Mab^o} \\[3em] NOT\, \dfrac{Qbc^a}{bc^o} \end{array}\right\} \xrightarrow{K'} \dfrac{X'ab^a \quad Qbc^a}{Mac^a}$$

Understood thus, it assumes that Mab^o is incompatible with $X'ab^a$, and that Bocardo LXM is valid,[22] which it must be given Barbara LXL, and L- and M-subordination:

$$\textit{Barbara LXL}\frac{Lab^a \quad bc^a}{\textit{La - subord}\frac{Lac^a}{\textit{X / M - subord}\frac{ac^a}{Mac^a}}} \xrightarrow{C} \frac{Lac^o \quad bc^a}{Mab^o}$$

In favour of this interpretation is that it follows the text closely. Against it is the fact that when Aristotle proves the assumed "Third Figure syllogism" at A15, 34b2-6, his proof:

$$\textit{Barbara MXM}\frac{Mab^a \quad bc^a}{Mac^a} \xrightarrow{C} \frac{Lac^o \quad bc^a}{Lab^o}$$

establishes not Bocardo LXM but Bocardo LXL.[23]

On a second interpretation, the syllogism on which the proof depends is Bocardo LXL.[24] The proof then assumes that Lab^o is incompatible with $X'ab^a$. In favour of this interpretation are the facts that (i) this is the syllogism which is later proved at A15, 34b2-6, and (ii) Bocardo LXL is the analogue of Disamis LXL, which plays the analogous rôle in Aristotle's proof of Celarent X'QM. But as against this, (i) Aristotle explicitly rejects Bocardo LXL, and (ii) on this reading the text should not say "then *a* is not possible for all *b*" but "then *a* is not possible for some *b*".

There is a third interpretation, according to which the "Third Figure syllogism" is Bocardo LXX,[25] it being assumed that ab^o is incompatible with $X'ab^a$. Textual emendations are required by this interpretation. In §34 we shall find reason to prefer the second or third interpretation to the first.

---

[22] PATTERSON 1995 p.160 takes the third figure syllogism to be Bocardo LXL; but TRICOT *ad* 34b2-4 takes it to be Felapton LXL:

$$\textit{Felapton LXL}\frac{Lac^e \quad bc^a}{Lab^o} \longrightarrow \frac{X'ab^a \quad Qbc^a}{Mac^a}$$

Aristotle does indeed accept Felapton LXL. But the trouble with Tricot's reading is that the inference depends on there being a contradiction $Le \mid Ma$, which there is not.

[23] PATTERSON 1995 p.164 mistakenly takes this to be a "second proof" of Barbara X'QM.

[24] MARSILIUS, Q.43 conclusion 5, takes the syllogism to be 'Brocardo' LXL.

[25] ROSS *ad* 34a25-b2 (followed by McCALL 1963 §30, Table 12) takes the third figure syllogism to be Bocardo LXX. This we may assume to be valid by Aristotelian standards.

Smith detects an equivocation in the proof:[26]

But in fact his argument contains an equivocation. The assumption 'it is not possible for A to belong to every C', if taken in the sense of possibility 'according to the determination' is equivalent to '*Either* A of necessity belongs to some C *or* A of necessity does not belong to some C'. In some places, Aristotle takes note of this (see, for instance, 37a15-20). Here, however, he treats it as equivalent to the second disjunct alone: this ... is to use 'possible' in the broader sense of 'not necessarily not'. As Ross points out, Aristotle should say here, as he does in the case of *eio*, that the only conclusion which follows from these premises is one of possibility in that broader sense.

True it is that Aristotle's "If then a is not possible for all c" has to be understood as Lac^o; however, this is precisely because the conclusion of the target syllogism is its contradictory Mac^a. No fallacy is thereby committed, even though elsewhere in the reasoning "possible" is used in the sense of contingency.

$$\textit{Celarent X'QM} \qquad \frac{X'ab^e \quad Qbc^a}{Mac^e}$$

Let ab be universal and negative, and let a be stated as belonging to no b, and b possibly belong to c. This being stated, it is necessary that a possibly belongs to no c. If it is not possible, b being stated to belong to c, [we proceed] just as above. It is necessary that a belongs to some b. For a third figure syllogism is produced. Now this is impossible [*adunaton*]. Consequently a is possible for no c. For this being supposed false, the conclusion is impossible [*adunaton*]. (A15, 34b19-27)

(For the 'unqualified' nature of the major premiss, see A15, 34b7-9; for the one-way possibility of the conclusion, see A15, 33b28-33, 4b27-31.) The proof

$$\left.\begin{array}{r} \textit{Disamis LXL}\ \dfrac{Lac^i \quad bc^a}{Lab^i} \\[3em] \textit{NOT}\ \dfrac{Qbc^a}{bc^o} \end{array}\right\} \xrightarrow{K'} \frac{X'ab^e \quad Qbc^a}{Mac^e}$$

assumes that Lab^i is incompatible with $X'ab^e$, and that Disamis LXL is valid.[27]

[26] SMITH *ad* 34a34-b2.
[27] McCALL 1963 p.90 takes the 'third figure syllogism' here to be Disamis LXX.

Darii X'QM

$$\left.\begin{array}{r} \textit{Ferison LXL}\;\dfrac{Lac^e \quad bc^i}{Lab^o} \\[2em] \textit{NOT}\;\dfrac{Qbc^i}{bc^e} \end{array}\right\} \xrightarrow{K'} \dfrac{X'ab^a \quad Qbc^i}{Mac^i}$$

Ferio X'QM[28]

$$\left.\begin{array}{r} \textit{Datisi LXL}\;\dfrac{Lac^a \quad bc^i}{Lab^i} \\[2em] \textit{NOT}\;\dfrac{Qbc^i}{bc^e} \end{array}\right\} \xrightarrow{K'} \dfrac{X'ab^e \quad Qbc^i}{Mac^o}$$

X'aQoMi-1

$$\textit{Darii X'QM}\;\dfrac{X'ab^a \quad \dfrac{Qbc^o}{Qbc^i}\,Q-conv}{Mac^i}$$

X'eQoMo-1

$$\textit{Ferio X'QM}\;\dfrac{X'ab^e \quad \dfrac{Qbc^o}{Qbc^i}\,Q-conv}{Mac^o}$$

If the intervals are stated one as universal and one as particular, then ... when that relating to the major extreme is universal and assertoric not problematic, and the other is particular and problematic, then whether both are put in the affirmative or both negative, or if one is affirmative and one negative, there will always be an imperfect syllogism. The proof will be *per impossibile* or by conversion of the possible, as in previous cases. For, if *a* belongs or does not belong to all *b*, and *b* possibly does not belong to some *c*, converting *bc* according to possibility, a syllogism is produced. (A15, 35a30-31,35-b2,5-8)

The Aristotelicity of these proofs is of course conjectural. I have followed the second interpretation above. If you favour the first or the third, then substitute 'M' or 'X' for 'L' in the conclusion of the left-hand syllogism.

---

28 McCALL 1963 claims these moods derive from Ferison LXX and Datisi LXX.

The Aristotelicity of these proofs is of course conjectural. I have followed the second interpretation above. If you favour the first or the third, then substitute 'M' or 'X' for 'L' in the conclusion of the left-hand syllogism.

X'eQeMa-1

$$\text{Celarent } X'QM \frac{X'ab^e \quad \dfrac{Qbc^e}{Qbc^a}\, Q-conv}{Mac^a}$$

Take *a* as belonging to no *b*, and *b* as possible for no *c*. Through these, nothing is necessary. But if *b* is taken as possible for all *c* (as is true), with the *ab* premiss as it is, again there will be the same syllogism. (A15, 35a16-20)

X'aQeMa-1

$$\text{Barbara } X'QM \frac{X'ab^a \quad \dfrac{Qbc^e}{Qbc^a}\, Q-conv}{Mac^a}$$

For, let *a* belong to all *b*, and *b* be possible for no *c*. With the terms thus, there will be no syllogism. But if *bc* is converted and *b* is taken as possible for all *c*, a syllogism will be produced just as before. For, the terms are taken in the same way. (A15, 35a6-11)

§13.2 Second Figure

Aristotle seems to state at A17, 36b29-31 that Cesare XQQ and Camestres QXQ are valid but Cesare QXQ and Camestres XQQ are not. The appearance is not quite accurate. It becomes clear in A18 that the valid moods are in fact Cesare X'QM and Camestres QX'M since reduction is to Celarent with contingency-minor, *i.e.*, to Celarent X'QM.

Cesare X'QM

$$\frac{X'e-conv\, \dfrac{X'ab^e}{X'ba^e} \quad Qac^a}{Mbc^e}\, \text{Celarent } X'QM$$

Camestres QX'M

$$\dfrac{\text{Celarent } X'QM \dfrac{Qab^a \quad \dfrac{X'ac^e}{X'ca^e}\, X'e-conv}{Mcb^e}}{Mbc^e}\, Me-conv$$

Let a be taken to belong to no b but to be possible for all c. Converting the negative, b belongs to no a. But a is possible for all c. So a syllogism is produced in the First Figure that b is possible for no c. Similarly if the negation is posited in relation to c. (A18, 37b24-29)

It is noteworthy that we need to use the convertibility of 'unqualified' e-assertorics as well as that of Me-forms.

Further Second Figure moods with mixed X' and Q premisses are:

X'eQeMe-2

$$\text{Cesare } X'QM\ \frac{X'ab^e \quad \dfrac{Qac^e}{Qac^a}\ Q-conv}{Mbc^e}$$

QeX'eMe-2

$$\frac{Q-conv\ \dfrac{Qab^e}{Qab^a} \quad X'ac^e}{Mbc^e}\ \text{Camestres } QX'M$$

If both are negative, signifying belonging and possibility respectively, from them as stated nothing follows necessarily. But converting the problematic premiss produces a syllogism that b possibly belongs to no c, just as in previous cases. For there will be the First Figure again. (A18, 37b29-35)

Festino X'QM

$$\frac{X'e-conv\ \dfrac{X'ab^e}{X'ba^e} \quad Qac^i}{Mbc^o}\ \text{Ferio } X'QM$$

But when it [*sc*. the assertoric] is negative, there will be [*sc*. a syllogism] by conversion, just as in previous cases. (A18, 38a3-4)

X'eQoMo-2

$$\text{Festino } X'QM\ \frac{X'ab^e \quad \dfrac{Qac^o}{Qac^i}\ Q-conv}{Mbc^o}$$

Again, if both intervals are stated in the negative, the universal being non-assertoric, there will be nothing necessary from these very premisses, but converting the possible just as in previous cases there will be a syllogism. (A18, 38a4-7)

§13.3 Third Figure

Aristotle accepts the following Third Figure X'QM syllogisms:

Darapti X'QM

$$\text{Darii } X'QM\,\frac{X'ac^{a} \quad \dfrac{Qbc^{a}}{Qcb^{i}}\,Qa-conv}{Mab^{i}}$$

For first let them be affirmative and let *a* belong to all *c*, and *b* possibly belong to all *c*. Converting *bc*, there will be the First Figure, and the conclusion that *a* possibly belongs to some *b*. (A21, 39b10-14)

Felapton X'QM

$$\text{Ferio } X'QM\,\frac{X'ac^{e} \quad \dfrac{Qbc^{a}}{Qcb^{i}}\,Qa-conv}{Mab^{o}}$$

Similarly if ... *ac* is negative and *bc* affirmative, whichever one is assertoric; in each case the conclusion will be problematic. For the First Figure is produced again, and it has been shown that in this [Figure] if either premiss signifies possibility, the conclusion will be problematic. (A21, 39b16-22)

X'aQeMi-3

$$\text{Darapti } X'QM\,\frac{X'ac^{a} \quad \dfrac{Qbc^{e}}{Qbc^{a}}\,Q-conv}{Mab^{i}}$$

X'eQeMi-3

$$\text{Felapton } X'QM\,\frac{X'ac^{e} \quad \dfrac{Qbc^{e}}{Qbc^{a}}\,Q-conv}{Mab^{o}}$$

If the negation is placed in relation to the minor extreme, or if both premisses are taken negatively, there will not be a syllogism from the proposed premisses, but by conversion there will be, just as in previous cases. (A21, 39b22-25)

Datisi X'QM

$$\text{Darii } X'QM \frac{X'ac^a \quad \dfrac{Qbc^i}{Qcb^i}\, Qi-conv}{Mab^i}$$

Ferison X'QM

$$\text{Ferio } X'QM \frac{X'ac^e \quad \dfrac{Qbc^i}{Qcb^i}\, Qi-conv}{Mab^o}$$

If one premiss is universal, one particular, both being affirmative, or the negative being universal and the affirmative particular, there will be the same moods of syllogisms, for all are completed through the First Figure. (A21, 39b26-30)

It should be noted that, when Aristotle at A21, 39b26-31 proves a Disamis with assertoric minor plus problematic major and conclusion, the minor must be understood as being 'unqualified'; for, the syllogism is reduced directly to Darii X'QM:

$$\text{Darii } X'QM \frac{X'ab^a \quad \dfrac{Qcb^i}{Qbc^i}\, Qi-conv}{\dfrac{Mac^i}{Mca^i}\, Mi-conv}$$

Similarly for Datisi X'QM.

CHAPTER FOUR. REJECTIONS

Here we detail Aristotle's rejections of formulae by counter-example and by deduction.

Aristotle's own presentation of his counter-examples is often skeletal. We try to flesh the examples out, making explicit their assumptions, and specifying exactly what would be required to verify the propositions Aristotle takes as true and to falsify those he takes as false. It turns out that the great majority of the counter-examples can be made sense of, against a background of assumptions that are either plausibly true or plausibly Aristotelian. The ones which cannot be made sense of in this way are signalled for closer scrutiny in Chapter Five.

Aristotle's presentation of a deductive structure among non-theses is both skimpy and inelegant. We make some additions and some simplifications here. We also look at those three-premissed formulae, to which Hintikka drew attention, and which are relied on by Aristotle in the rejection of certain First and Second Figure L-X moods.

§14. NECESSITY MOODS

Aristotle's usual method of rejecting a formula is to construct a counter-example, a possible case in which the premisses are true and the conclusion false. This procedure rests on intuitions regarding what is and what is not a possible case. From time to time, he rejects a formula on the ground that its assertion would imply the acceptance of a formula already rejected. That procedure rests not on intuitions of possibility, but on an assumed deductive structure holding among rejected formulae. In this section, we shall look at Aristotle's counter-examples to L-formulae, and at his rudimentary efforts to axiomatise the rejections of such expressions.

We saw in §6.2 and §9.1 that Aristotle accepts the 1-premissed LL- and LX-theses in the moods listed in Table 14.1.

TABLE 14.1

direct	indirect
aa	
ai	ai
ee	ee
eo	eo
ii	ii
oo	

Of the remaining 1-premissed LL- and LX-moods, Aristotle explicitly rejects only

Lo-conversion $\dfrac{L(\text{man})(\text{animal})^{o}}{L(\text{animal})(\text{man})^{o}}$

The particular negative does not convert, for the same reason as indicated before. (A3, 25a35-36)

I take it that the text refers us to the counter-example given at A2, 25a25-26 to o-conversion. It presupposes that, while it is true that some animals are necessarily not men, it is false that some men are necessarily not animals. According to Aristotelian metaphysics, the presupposition is satisfied provided that both human and non-human animals exist.

The rejection of Lo-conversion implies that of La-conversion, since the assertion of the latter implies that of the former:

$$\frac{Lab^a}{Lba^a} \xrightarrow{C} \frac{Lba^o}{Lab^o}$$

Aristotle does not consider the corresponding LX-moods:

$$\frac{Lab^a}{ba^a} \qquad\qquad \frac{Lab^o}{ba^o}$$

but their invalidity can be shown using the same terms he deploys against Lo-conversion.

Aristotle does not discuss any individual cases of rejected moods with two L-premisses, because he thinks such combinations yield no L-conclusion when the corresponding pair of assertorics yield no X-conclusion. As Angelelli observes,[1] Aristotle apparently accepts

Rule L If $\frac{Lp \quad Lq}{Lr}$ is valid then $\frac{p \quad q}{r}$ is valid, where p,q are assertoric forms.

He gives no proof of this rule.

Short of proving rule *L*, Aristotle needs to show that no L-conclusion follows syllogistically from any of the premiss-pairs in Table 14.2. The bracketed pairs are secondary: their rejection follows from that of stronger pairs elsewhere in the table. Aristotle needs to show that no conclusion, universal or particular, follows from any of the unbracketed pairs. Thus it would be sufficient for his purposes to show that no particular conclusion follows from any of them, *i.e.* to demonstrate invalidity in the cases listed in Table 14.3.

This is what he needs to do; but in fact he doesn't provide the required demonstrations.

[1] ANGELELLI p.181.

TABLE 14.2

Figure 1	LaLe, (LaLo), (LiLe), (LiLo), LeLe, (LeLo), (LoLe), (LoLo), LiLa, (LiLi), LoLa, (LoLi)
Figure 2	LaLa, (LaLi), (LiLa), (LiLi), LeLe, (LeLo), (LoLe), (LoLo), LiLe, (LiLo), LoLa, (LoLi)
Figure 3	LaLe, (LaLo), (LiLo), LeLe, (LeLo), (LoLe), (LoLo), LiLi, LoLi

TABLE 14.3

Figure 1	LaLeLi-1, LaLeLo-1, LeLeLi-1, LeLeLo-1, LiLaLi-1, LiLaLo-1, LoLaLi-1, LoLaLo-1
Figure 2	LaLaLi-2, LaLaLo-2, LeLeLi-2, LeLeLo-2, LiLeLi-2, LiLeLo-2, LoLaLi-2, LoLaLo-2
Figure 3	LaLeLi-3, LaLeLo-3, LeLeLi-3, LeLeLo-3, LiLiLi-3, LiLiLo-3, LoLiLi-3, LoLiLo-3.

§14.1 Rejected L-X moods: First Figure

Aristotle does not say so, but he clearly holds the following analogues of rule *L* to be correct:

$$\text{If } \frac{Lp \quad q}{Lr} \text{ is valid then } \frac{p \quad q}{r} \text{ is valid}$$

$$\text{If } \frac{p \quad Lq}{Lr} \text{ is valid then } \frac{p \quad q}{r} \text{ is valid}$$

This is clear from the fact that in his discussion of LXL and XLL moods he does not bother to show that an LX or XL premiss-pair yields no L-conclusion in those cases where the corresponding XX premiss-pair yields no X-conclusion. In fact these two rules are corollaries of rule *L* since if either $\frac{Lp \quad q}{Lr}$ or $\frac{p \quad Lq}{Lr}$ is valid, so is $\frac{Lp \quad Lq}{Lr}$.

We saw in §6.2 and §9.2 that Aristotle asserts the 2-premissed LXL and XLL theses in Table 14.4.

TABLE 14.4

Figure 1	Barbara LXL, Celarent LXL, Darii LXL, Ferio LXL
Figure 2	Cesare LXL,Camestres XLL, Festino LXL
Figure 3	Darapti LXL, Darapti XLL, Felapton LXL, Datisi LXL, Disamis XLL, Ferison LXL

Of the remainder he explicitly rejects all those whose XXX analogues are valid.

Barbara XLL
Celarent XLL

$$\frac{(\text{moving})(\text{animal})^{a,e} \quad L(\text{animal})(\text{man})^{a}}{L(\text{moving})(\text{man})^{a,e}}$$

If *ab* is not necessary, and *bc* is necessary, the conclusion will not be necessary.... (A9, 30a23-25,28-33)[2]

The proof rests on the possibility that the propositions above the syllogistic sign are true and that below it false. Aristotle assumes in these examples (i) that $L(\text{animal})(\text{man})^a$ is true, (ii) that even if $(\text{moving})(\text{animal})^a$ is true $L(\text{moving})(\text{man})^a$ may be false, and (iii) that even if $(\text{moving})(\text{animal})^e$ is true $L(\text{moving})(\text{man})^e$ may be false. He does not need to state that $L(\text{moving})(\text{man})^a$ and $L(\text{moving})(\text{man})^e$ actually are false under the stated suppositions, only that they may be; but I take it that for Aristotle those propositions are false no matter what. Aristotle does, here and in many other places, take it that $L(\text{animal})(\text{man})^a$ is true.

Aristotle has a second means of rejecting Barbara XLL and Celarent XLL where he reasons:

$$x\,\frac{ab^a \quad Lbc^a}{\dfrac{Lac^a \quad Lbc^a}{Lab^i}\,\text{Darapti } LLL}$$

If it were [*sc.* if the conclusion were necessary] we could conclude that *a* belongs necessarily to some *b* through the first and the third figure. But this is false. For *b* can be such that *a* belongs to none of it. (A9, 30a25-28)

---

2 BRANDT argues that this and other similar passages deal not with inferences having apodeictic conclusions but rather with inferences whose conclusions are negated apodeictics, so that we should translate "the conclusion will be non-necessary" rather than "the conclusion will not be necessary". If he were right, Aristotle would here be asserting the validity of an inference ("*a* is not necessary for all *b*, *b* is necessary for all *c*, so *a* is not necessary for all *c*"). There are two things wrong with this view: the inference's validity could not be demonstrated by Aristotle's example, and it is invalid anyway, being equivalent to an invalid figure 3 inference:

$$\frac{\sim Lab^a \quad Lbc^a}{\sim Lac^a} \overset{C}{\longleftrightarrow} \frac{Lac^a \quad Lbc^a}{Lab^a}$$

This assumes that Hintikka's formula

$$\text{(Hintikka 1)}\quad \frac{ab^a \quad Lbc^a \quad Lbc^a}{Lab^i}$$

(discussed in §4.1) is a *non sequitur.* Thus what we have here is a reduction of the rejection of Barbara XLL to the rejection of Hintikka 1. Similarly, he relies on the implicit rejection of

$$\text{(Hintikka 2)}\quad \frac{ab^e \quad Lbc^a \quad Lbc^a}{Lab^o}$$

to which Celarent XLL is reduced at A9, 30a32-33. These Hintikka formulae are thus used by Aristotle as axioms of rejection from which other rejections are derived.

$$\begin{array}{l}\textit{Darii XLL}\\ \textit{Ferio XLL}\end{array}\quad \frac{(\text{moving})(\text{animal})^{a,e} \quad L(\text{animal})(\text{white})^i}{L(\text{moving})(\text{white})^{i,o}}$$

If the particular is necessary, the conclusion will not be necessary .... Terms: movement - animal - white. (A9, 30b2-6)

Here the assumptions are (i) that $L(\text{animal})(\text{white})^i$ is true, (ii) that even if $(\text{moving})(\text{animal})^a$ is true $L(\text{moving})(\text{white})^i$ may be false, and (iii) that even if $(\text{moving})(\text{animal})^e$ is true $L(\text{moving})(\text{white})^o$ may be false. I take it that for Aristotle the existence of a white man is sufficient to satisfy these conditions.[3]

---

[3] VAN RIJEN p.196 argues that Darii XLL is valid, using an ecthetic proof:

$$\dfrac{\dfrac{\dfrac{\dfrac{ab^a \quad \dfrac{Lbd^a}{bd^a}\,\scriptstyle La-subord}{ad^a}\,\scriptstyle Barbara}{da^i}\,\scriptstyle a-conv \qquad Lcd^a}{\dfrac{Lca^i}{Lac^i}\,\scriptstyle Li-conv}\,\scriptstyle Darii\ LXL \quad \xrightarrow{\ Li-ecthesis\ AA\ } \quad \frac{ab^a \quad Lbc^i}{Lac^i}$$

The assumed rule of Li-ecthesis *AA*:

$$\frac{R \quad Lad^a \quad Lbd^a}{p} \longrightarrow \frac{R \quad Lab^i}{p}$$

§14.2 Rejected L-X moods: Second Figure

Having accepted Cesare and Festino LXL, Aristotle rejects

$$\textit{Camestres LXL}\quad \frac{L(\text{animal})(\text{man})^a\ (\text{animal})(\text{white})^e}{L(\text{man})(\text{white})^e}$$

It can be proved by terms that the conclusion will not be necessary without qualification, but necessary these things being so. *E.g.* let *a* be 'animal', *b* 'man', *c* 'white'.... (A10, 30b31-40)

Presumably, the premisses of this argument would be true and the conclusion false if no animal were white but some man was not necessarily not white. I take it that this condition could according to Aristotle be met. Even if no animal, and hence no man, were white, man would not be *necessarily* non-white according to him.[4]

At A10, 30b20-24 Camestres LXL is rejected on the ground of the previous rejection of Celarent XLL:

$$\textit{Camestres LXL}\frac{Lab^a\quad ac^e}{Lbc^e}\xrightarrow{e-conv,\,M,\,T_1}\frac{ca^e\quad Lab^a}{Lcb^e}\textit{Celarent XLL}$$

(Since the right-hand formula is rejected, the left-hand one must be rejected too.)

A third method for rejecting Camestres LXL is given at A10, 30b24-31. Assuming the rejection of

$$(\text{Hintikka 3})\ \frac{Lab^a\quad ac^e\quad Lab^a}{Lca^o}$$

---

will be discussed in §32.2. VAN RIJEN pp.196-197 offers a second proof of Darii XLL:

$$\begin{array}{ll} \textit{Me-conv} & \dfrac{Mac^e}{} \\ \textit{Celarent MXM} & \dfrac{Mca^e\quad ab^a}{Mcb^e} \\ \textit{Me-conv} & \dfrac{}{Mbc^e} \end{array}\quad\xrightarrow{C}\quad\frac{ab^a\quad Lbc^i}{Lac^i}$$

Celarent MXM, on which this proof relies, will be discussed in §23.

[4] ANGELELLI p.181 takes the present assertion that no animal is white to be inconsistent with Aristotle's assertion (in refuting Darii XLL, Ferio XLL, Ferison XLL and Bocardo XLL) that some white is necessarily an animal; and, believing that a choice is forced on him, he opts for the falsity of the present assertion. In fact, however, no choice is forced. It is true that some white is necessarily an animal provided a white animal exists; and it is true that no animal is white provided no white animal exists.

on the basis of reasoning analogous to that used earlier with the Hintikka rejections of Barbara XLL and Celarent XLL. We shall later discuss whether this reasoning is sound.

Baroco LXL
$$\frac{L(\text{animal})(\text{man})^{a} \quad (\text{animal})(\text{white})^{o}}{L(\text{man})(\text{white})^{o}}$$

Again, let the affirmative be universal and necessary, and let the affirmative be [stated] in relation to *b*. If then *a* belongs necessarily to all *b*, and does not belong to some *c*, then clearly *b* does not belong to some *c* but not necessarily. For the same terms serve in the proof as for the universal syllogisms. (A10, 31a10-15)

The premisses of this argument are true and the conclusion false if some white thing is not an animal but there is no white thing which cannot be a man. It is difficult to see how these conditions could be met: the falsity of the conclusion seems to require that every white thing can be a man, while the truth of the premisses requires that there be an inanimate white thing. The conditions could be met only if some white thing were both inanimate and not necessarily not a man. What could this be? The matter from which a man develops? We will return to this topic in §22.4.

Among Second Figure XLL moods, we have seen that Aristotle accepts Camestres. He rejects

Cesare XLL
Festino XLL
$$\frac{(\text{animal})(\text{white})^{e} \quad L(\text{animal})(\text{man})^{a,i}}{L(\text{white})(\text{man})^{e,o}}$$

In the second figure ... if the affirmative [premiss is necessary] the conclusion will not be necessary. (A10, 30b7-9)

The above counter-example, while not Aristotle's, is adapted from his counter-example to Camestres LXL. Even supposing no animals are white it does not follow that some, or all, men *cannot* be white. Aristotle also rejects

Baroco XLL
$$\frac{(\text{animal})(\text{man})^{a} \quad L(\text{animal})(\text{white})^{o}}{L(\text{man})(\text{white})^{o}}$$

But if the negative is necessary and stated as a particular, there will not be a necessary conclusion. For the proof will be through the same terms. (A10, 31a15-17)

This example is troublesome. The problem is not, as van Rijen supposes, that the minor premiss 'can never be true';[5] presumably it is true if some snow exists. Rather, it seems that any white thing (such as snow) that verified the minor premiss would also verify the conclusion. We shall return to this mood in §20.1.

§14.3 Rejected L-X moods: Third Figure

Having accepted Darapti, Felapton, Datisi and Ferison LXL, Aristotle rejects

Felapton XLL $$\frac{(\text{good})(\text{horse})^e \; L(\text{animal})(\text{horse})^a}{L(\text{good})(\text{animal})^o}$$

It is clear through terms. For let *a* be 'good', *b* 'animal', *c* 'horse'. Then good possibly belongs to no horse, and animal necessarily belongs to all. But it is not necessary that some animal is not good, since it is possible for all to be good. (A11, 31b4-8)

Aristotle seems to find this example unconvincing, and at 31b8-10 he adds two alternative examples:

$$\frac{(\text{awake})(\text{horse})^e \; L(\text{animal})(\text{horse})^a}{L(\text{awake})(\text{animal})^o}$$

$$\frac{(\text{asleep})(\text{horse})^e \; L(\text{animal})(\text{horse})^a}{L(\text{asleep})(\text{animal})^o}$$

Perhaps the 'good horse' example is unsatisfactory because, no matter what the class of horses was like, some of its members would be good *compared to other horses*. The alternative examples work, given that a horse's state of sleep or waking is not necessary to it (being variable over time).

Aristotle claims at A11, 31b1-4 that Felapton XLL can be rejected because Ferio XLL was rejected. But his reasoning is faulty here. We have

$$\textit{Ferio XLL} \xrightarrow{La-conv,\, T_1} \textit{Felapton XLL}$$

---

[5] VAN RIJEN p.212. Van Rijen p.195-6 suggests that "Another arrangement of terms to the effect that the major becomes 'all men are white', the minor 'some animals are necessarily not white' and the conclusion 'some animals are necessarily not men' would give a more convincing counter example provided that the necessity operator is given a *de re* reading." But this is no help. If this conclusion is false, then one of the premisses is false. For, if men are the only animals, then it cannot be that some animal is necessarily not white, given that all men are white.

but not

$$\textit{Felapton XLL} \xrightarrow{La-conv,\, T_1} \textit{Ferio XLL}$$

so it is incorrect to say that the rejection of Ferio XLL implies that of Felapton XLL; it is the other way around.

$$\textit{Datisi XLL} \quad \frac{(\text{awake})(\text{animal})^a \ \ L(\text{biped})(\text{animal})^i}{L(\text{awake})(\text{biped})^i}$$

It is clear through terms. For let a be 'state of waking', b 'biped', c 'animal'. Then b necessarily belongs to some c, and a is possible for c, and a is not necessary for b. For it is not necessary that some biped is awake or sleeps. (A11, 31b27-33)

If every animal is awake and some animal (a man) is necessarily biped, does it follow that some biped is necessarily awake? Aristotle says No. But he forgets that, since Li-forms are convertible, some biped will be necessarily awake if something awake is necessarily biped. And, something awake *is* necessarily biped given that all men are both awake and necessarily biped (Darapti XLL). So it seems that this counter-example is unsuccessful by Aristotelian standards. Aristotle ought to say that if these premisses are true so is the conclusion. However he is right to say that the *form* of argument is invalid. This can be shown by constructing an example from the terms 'awake', 'animal', 'approaching'. Supposing every animal to be awake and some animal to be necessarily approaching (because something approaching, *viz.* an approaching animal, is necessarily an animal), it does not follow that something approaching is necessarily awake. Nothing is necessarily either awake or approaching.

A11, 31b20-27 reduces the rejection of Datisi XLL to that of Darii XLL:

$$\textit{Datisi XLL} \xrightarrow{Li-conv,\, T_1} \textit{Darii XLL}$$

Aristotle again uses the terms 'awake', 'animal', 'biped' to refute

$$\textit{Disamis LXL} \quad \frac{L(\text{biped})(\text{animal})^i \quad (\text{awake})(\text{animal})^a}{L(\text{biped})(\text{awake})^i}$$

The proof will be similar through the same terms if ac is particular and necessary. (A11, 31b31-33)

Again these terms do not work. And again we get a better example by putting 'approaching' in place of 'biped'.

Bocardo LXL $\dfrac{L(\text{biped})(\text{animal})^{o} \quad (\text{moving})(\text{animal})^{a}}{L(\text{biped})(\text{moving})^{o}}$

When the negative, being particular, is necessary: 'biped'-'moving'-'animal', middle 'animal'. (A11, 32a4-5)

Van Rijen comments that the conclusion here 'seems to be perfectly all right'.[6] Patterson says "Aristotle has probably slipped".[7] I cannot find an Aristotelian way to make the example work. It seems that any way of making the major premiss true (*e.g.* by reference to some goat) would also make the conclusion true. We shall return to this mood in §20.1.

Finally, in Figure 3, having accepted Darapti and Disamis XLL, Aristotle rejects

Ferison XLL $\dfrac{(\text{waking})(\text{white})^{e} \quad L(\text{animal})(\text{white})^{i}}{L(\text{waking})(\text{animal})^{o}}$

Bocardo XLL $\dfrac{(\text{waking})(\text{white})^{o} \quad L(\text{animal})(\text{white})^{i}}{L(\text{waking})(\text{animal})^{o}}$

Terms when the universal affirmative is necessary: 'waking'-'animal'-'man', with 'man' middle. When the particular affirmative is necessary: 'waking'-'animal'-'white'. For animal necessarily belongs to something white, and waking is possible for none, but it is not necessary that waking does not belong to some animal. (A11, 31b40-32a4)

The counter-example works, given that, no matter how things turn out, no animal is necessarily not awake. For given this, we may suppose the premisses true and the conclusion false.

Summing up the counter-examples in the L layer, we have found Aristotelian ways to flesh all of them out, with the exception of those to Datisi XLL, Disamis LXL, Baroco XLL and Bocardo LXL. His counter-examples to the former pair overlook the convertibility of Li-forms; alternative examples can be found. We have not found counter-examples to replace the latter pair, but will return to them.

It is possible to reduce all of Aristotle's two-premissed rejections to the formulae in Table 14.5.

---

[6] VAN RIJEN p.198.

[7] PATTERSON 1995 P.85.

TABLE 14.5

Figure 1	LaLeLi-1, LaLeLo-1, LeLeLi-1, LeLeLo-1, LiLaLi-1, LiLaLo-1, LoLaLi-1, LoLaLo-1, Barbara XLL, Celarent XLL, Darii XLL
Figure 2	LaLaLi-2, LaLaLo-2, LeLeLi-2, LeLeLo-2, LiLeLi-2, LiLeLo-2, LoLaLi-2, LoLaLo-2, Baroco LXL, Baroco XLL
Figure 3	LaLeLi-3, LaLeLo-3, LeLeLi-3, LeLeLo-3, LiLiLi-3, LiLiLo-3, LoLiLi-3, LoLiLo-3, Felapton XLL, Bocardo XLL, Bocardo LXL.

Given rule *L* we can omit all the LLL moods here. Barbara XLL can be replaced by Hintikka 1, and Celarent XLL by Hintikka 2.

§15. UNAMPLIATED CONTINGENCY MOODS

We now pass to Aristotle's treatment of rejected Q-formulae. With these moods, Aristotle hardly ever reduces the rejection of one mood to that of another, dealing instead with a large number of counter-examples to individual cases. We consider formulae with just one Q-premiss (by itself, or paired with either an X- or an L-premiss).

§15.1 Rejected 1-premissed Q moods

Aristotle takes as axiomatic the rejection of

$$\textit{Qa-conv} \qquad \frac{Qab^{a}}{Qba^{a}}$$

For if this is possible for all of that, it is not necessary that that [is possible for all] of this. (A17, 37a2-3)

Qe-conversion is rejected at A3, 25b16-17; and A17, 37a4-8 gives a counter-example to Qe-conversion, using the terms $\langle \text{white}, \text{man} \rangle$. In this counter-example

Aristotle takes it as true that $Q(\text{white})(\text{man})^e$ and false that $Q(\text{man})(\text{white})^e$. The latter is false, he says, because $L(\text{man})(\text{white})^o$ is true. Perhaps this is because he is taking it that swans exist, swans being necessarily not men. But Aristotle knew that Qe-conversion did not have to be rejected axiomatically, since its rejection follows from that of Qa-conversion:

Qe-conversion

$$\left.\begin{array}{c} \dfrac{Qab^a}{Qab^e} \\[3ex] \dfrac{Qab^e}{Qba^e} \\[3ex] \dfrac{Qba^e}{Qba^a} \end{array}\right\} \xrightarrow{T1} \dfrac{Qab^a}{Qba^a}$$

If a is possible for no b, it is not necessary that b is possible for no a. For assume this, and let b possibly belong to no a. Since possibility-affirmatives convert with negatives, both contraries and opposites [*antikeimenai*], and b possibly belongs to no a, it is clear that it is possible for it to belong to all a. But this is false. (A17, 36b36-37a2)

At A17, 37a9-31 he considers and refutes an argument which purports to establish the validity of Qe-conversion, namely the argument

$$\begin{array}{rc} x & \dfrac{\sim Qba^e}{Lba^i} \\ Li-conv & \dfrac{}{Lab^i} \\ y & \dfrac{}{\sim Qab^e} \end{array} \xrightarrow{C} \dfrac{Qab^e}{Qba^e}$$

What is wrong with the argument, says Aristotle, is that x is invalid.[8]

Łukasiewicz labels Aristotle's rejection of Qe-conversion a 'curious statement'.[9] He thinks the convertibility of Qe-forms can be proved as follows:

---

[8] As pointed out in §2.2, he denies that Lab^i is contradictory to Qab^e. In §28.4 we shall see that, on one semantic analysis, it is not even contrary to it, and therefore that y is also invalid.

[9] ŁUKASIEWICZ §59.

$$\dfrac{\dfrac{\dfrac{Qab^e}{Mab^e}\,x}{Mba^e}\,Me-conv \qquad \dfrac{\dfrac{Qab^e}{Mab^i}\,x}{Mba^i}\,Mi-conv}{Qba^e}\,y$$

M-conversion is not controversial. The same cannot be said for steps x and y. These are supposed to be justified by reference to Aristotle's definition of the contingent as two-sided possibility (A13, 32a18-20). But that definition would justify these inferences only if the mode of contingency played a uniform rôle in all contingency-propositions so that Qab^e was a substitution in Qp. It turns out that on the best interpretations of the modal syllogistic, this is not so. It will become clear in §39 that the mode of contingency, like that of necessity, plays different rôles in different forms of proposition, these different rôles being related to a central one by means of structural principles. Łukasiewicz can maintain the position that the mode of contingency plays a uniform rôle in all propositions only by rejecting Complementary Conversion. Complementary Conversion is a principle frequently used by Aristotle. So Łukasiewicz's position is both ill-founded and inconsistent with a basic tenet of Aristotle's text.

§15.2 Rejected 2-premissed Q-X moods

Aristotle rejects a number of two-premissed formulae by a method of modally contrasted counter-instances which is a development of the method of contrasted instances used in the assertoric syllogistic. There, he showed that pairs of premisses yield no assertoric conclusion by showing that they are compatible both with an a- and with an e-relation between major and minor terms. Here, he shows that pairs of premisses yield no conclusion at all by showing that they are compatible both with an La- and with an Le-relation. The method is valid given that if the premisses leave it open that Lab^a they do not imply that Lab^e or Lab^o or ab^e or ab^o or Qab^i or Qab^a or Mab^o (these being incompatible with Lab^a), and given that if they leave it open that Lab^e they do not imply that Lab^a or Lab^i or ab^a or ab^i or Qab^i or Qab^o or Mab^i (these being incompatible with Lab^e). Thus, given the standard laws of subalternation and subordination, the method presupposes the oppositions $La \parallel o$, $Le \parallel i$, $La \mid Mo$, $Le \mid Mi$ and either $La \parallel Qi$ or $Le \parallel Qi$.

The premiss-pairs to which Aristotle intends to apply this technique are just those in one of the three Figures which he has not shown to yield a syllogistic conclusion. (But occasionally he overlooks a premiss-pair, as noted below.)

In §7.2, §10.1 and §12 we attributed to Aristotle a number of 2-premissed QX theses. Considering only affirmative Q-forms, we have the moods in Table 15.1. The remaining premiss-pairs in a syllogistic Figure (with weakened moods bracketed) are given in Table 15.2.

TABLE 15.1

Figure 1	QaXaQa QaXiQi
Figure 3	QaXaQi QaXiQi QiXaMi QiXaMo XaQaQi XaQiMi XiQaQi

TABLE 15.2

Figure 1	QaXe, (QaXo), (QiXa), QiXe, QiXi, (QiXo), XaQa, (XaQi), XeQa, (XeQi), (XiQa), (XiQi), (XoQa), (XoQi)
Figure 2	QaXa, QaXe, (QaXi), (QaXo), (QiXa), (QiXe), (QiXi), (QiXo), XaQa, (XaQi), XeQa, (XeQi), (XiQa), (XiQi), (XoQa), (XoQi),
Figure 3	QaXe, (QaXo), (QiXe), QiXi, (QiXo), XeQa, (XeQi), XiQi, (XoQa), (XoQi).

Aristotle argues in most of these cases that no conclusion follows; and where he provides no such argument, he can be shown to be committed to the result anyway.

§15.2.1 Rejected Q-X moods: First Figure

The following QX premiss-pairs yield neither a possibility- nor a necessity- nor an assertoric conclusion. In displaying the counter-examples, I place above the syllogistic line the 'premisses' which Aristotle takes to be true, and below the line the problematic 'conclusions' which he takes to be false.

QaXe-1

$$\frac{Q(\text{white})(\text{animal})^{a,e} \quad (\text{animal})(\text{snow})^{e}}{M(\text{white})(\text{snow})^{o}}$$

QeXe-1

$$\frac{Q(\text{white})(\text{animal})^{a,e} \quad (\text{animal})(\text{pitch})^{e}}{M(\text{white})(\text{pitch})^{i}}$$

If *b* is taken as not belonging (not: possibly not belonging) to all *c*, there will never be a syllogism, whether the *ab* premiss is negative or affirmative. Common terms for 'belongs necessarily': 'white'-'animal'-'snow'. For 'not possibly': 'white'-'animal'-pitch'. (A15, 35a20-24)

Smith comments pertinently:[10]

The counter model triple used here takes 'White belongs to every animal' and 'White belongs to no animal' as possible premises: but since Aristotle regards some animals (e.g. swans) as necessarily white, others (e.g. ravens) as necessarily not white, these must be possible *at a time* in the way seemingly forbidden in 34b7-18.

But, as I have argued elsewhere,[11] what is forbidden there is not *all* uses of premisses true only at a given time but only their use in a particular type of syllogism. The point of 34b7-18 is that while Barbara X'QM is valid Barbara XQM is not. The present passage is one of many in the *Analytics* where Aristotle takes as true assertorics that can at best be true for a limited time (*e.g.* A9, 30a23-25,28-33, 30b2-6; A10, 30b31-40, 30b7-9; A11, 31b4-8, 31b8-10, 31b27-33, 32a4-5, 31b40-32a4).

QiXa-1
QoXa-1
QiXe-1
QoXe-1
QiXi-1
QoXi-1
QiXo-1
QoXo-1

$$\frac{Q(\text{animal})(\text{white})^{i,o} \quad (\text{white})(\text{man})^{a,e,i,o}}{M(\text{animal})(\text{man})^{o}}$$

$$\frac{Q(\text{animal})(\text{white})^{i,o} \quad (\text{white})(\text{cloak})^{a,e,i,o}}{M(\text{animal})(\text{cloak})^{i}}$$

And if the universal is proposed in relation to the minor extreme, and the particular in relation to the major, be they negative or affirmative, possible or assertoric, there will never be a syllogism. When the premisses are proposed as particular or indefinite, be they taken as possible or assertoric or mixed, there will not be a syllogism. Common terms where the belonging is necessary: 'animal'-'white'-'man'; where it is impossible: 'animal'-'white'-'cloak'. (A15, 35b11-19)

It is false that some man may not be an animal, but it is conceivable according to Aristotle that all or no men be white; and given that white men exist, some whites are contingently animals because some animals are contingently white. Thus the first example works. The second works because we may suppose that some whites are contingently animals - by supposing that some animals (namely men) are contingently white - but we cannot suppose that some cloak may be an animal.

(It was not necessary for Aristotle to go through all these counter-examples, since (i) the sterility of QiXa-1 and QoXa-1 is implied by that of the stronger QaXa-2 and QeXa-2, and (ii) that of QiXe-1 and QoXe-1 is implied by that of the stronger QaXe-1 and QeXe-1. Similarly, the following cases discussed by him, need not have been:

---

[10] SMITH *ad* 35a20-24.
[11] THOM 1994, p.108.

QaXo-1
QeXo-1

$$\frac{Q(\text{white})(\text{animal})^{a,e} \quad (\text{animal})(\text{snow})^{o}}{M(\text{white})(\text{snow})^{o}}$$

$$\frac{Q(\text{white})(\text{animal})^{a,e} \quad (\text{animal})(\text{pitch})^{o}}{M(\text{white})(\text{pitch})^{i}}$$

When the not belonging is taken as a particular, there will be no syllogism. Terms for belonging: 'white'-'animal'-'snow'. For not belonging: 'white'-'animal'-'pitch'. For, the proof is through the indeterminate. (A15, 35b8-11)

The first of these examples works because it is false that some snow may not be white; and consistent with that, we may suppose that it is true that every animal may or may not be white. Our supposition would be true if there were no ravens, swans or other animals for which whiteness is not contingent. The second example works under the same hypothesis. Strictly speaking, it was not necessary for Aristotle to give this counter example, since the sterility of QaXo-1 and QeXo-1 is implied by that of the stronger QaXe-1 and QeXe-1, which has already been shown.)

No XQ-premisses in Figure 1 yield a Q-conclusion.

XaQa-1

$$\frac{(\text{man})(\text{moving})^{a} \quad Q(\text{moving})(\text{horse})^{a}}{M(\text{man})(\text{horse})^{i}}$$

$$\frac{(\text{animal})(\text{moving})^{a} \quad Q(\text{moving})(\text{man})^{a}}{M(\text{animal})(\text{man})^{o}}$$

Since if the premisses are stated about the present,[12] there will be no syllogism. For nothing prevents man belonging to all that is moving, for example if nothing else is moving. But moving is possible for all horse. But man is not possible for any horse. Further, let the first be 'animal', the middle 'moving', and the last 'man'. The premisses being thus, the conclusion is necessary, not possible. For it is necessary that man is an animal. (A15, 34b10-17)

Aristotle is saying on the basis of these counter-examples that premisses XaQa-1 leave it open that there is either an affirmative or a negative apodeictic relation

[12] The contrast here is with 'unqualified' propositions. These will be discussed in §34.

between the minor and the major, and that consequently no apodeictic, assertoric or problematic conclusion follows. Presumably the first example works because it is false that some horse may be a man, and it is true that every horse naturally has the possibility of either moving or not moving; and consistent with all this we may suppose that only men are moving. The second example works because it is false that some man may not be an animal, and it is true that all men naturally have the possibility of either moving or not moving; and consistent with all this we may suppose that only animals are moving.[13] In §19.2 we shall find reason to question the first of these examples.

(The same counter-examples count against XaQi-1, a premiss-pair which Aristotle does not comment on.[14] Indeed, the rejection of all moods with premisses XaQi-1 is implied by the rejection of all moods with the stronger premisses XaQa-1. In particular, Aristotle must be understood as being committed to the rejection of Darii XQQ and XQM. Yet, as we saw in §12, he is also committed to accepting Darii XQM. More of this in §19.2, where we shall find reason to query his counter-example and accept this mood.)

At A15,34b31-35a2 Aristotle uses the terms 'raven' 'reasoning' 'man' to show that premisses X'eQa-1 yield no Q-conclusion and no affirmative L-conclusion. (We will discuss this passage in §17.) In the same passage Aristotle uses the terms 'moving' 'knowing' 'man' to show that premisses X'eQa-1 yield no negative L-conclusion. Again assuming that X'e implies Xe, both trios of terms can be used to prove these results for XeQa-1.

What about XeQaMi-1 and XeQaMo-1? Aristotle is committed to the invalidity of these moods since they are equivalent to LeQaXi-3 and LaQaXi-3:

$$\frac{ab^e \quad Qbc^a}{Mac^i} \overset{C}{\longleftrightarrow} \frac{Lac^e \quad Qbc^a}{ab^i} \qquad \frac{ab^e \quad Qbc^a}{Mac^o} \overset{C}{\longleftrightarrow} \frac{Lac^a \quad Qbc^a}{ab^i}$$

A22, 40a9-11 implies the invalidity of these Third Figure moods. Thus Aristotle is committed to the uselessness of XeQa-1 in general.

(While not needing to, Aristotle shows the invalidity of

---

13 PATTERSON 1995 p.175 takes this to be "a second counterexample" to Barbara XQM and as a blunder. The truth is that this term-trio is the second of a pair which *together* show that premisses XaQa-1 yield no conclusion.

14 If premisses XaQi-1 are sterile, so are QiXa-3 (which are equivalent by Qi-conversion), and so are QoXa-3 (equivalent by Q-conversion). Thus the counter examples to XaQi-1 also show that Bocardo QXQ is invalid, contrary to a claim of Mignucci's which we discussed in §10.1.

XiQa-1
(XiQe-1)
XoQa-1
(XoQe-1)
XiQi-1
(XiQo-1)
XoQi-1
(XoQi-1)

$$\frac{(\text{animal})(\text{white})^{i,o} \quad Q(\text{white})(\text{man})^{a,e,i,o}}{M(\text{animal})(\text{man})^{o}}$$

$$\frac{(\text{animal})(\text{white})^{i,o} \quad Q(\text{white})(\text{cloak})^{a,e,i,o}}{M(\text{animal})(\text{cloak})^{i}}$$

(A15, 35b11-19 quoted above.)

It is false that some man may not be an animal, and it is true according to Aristotle that whiteness is contingent for all men; and consistent with all this we may suppose that both swans and snow exist, so that some whites are animals and some are not. Thus the first example works. Similarly for the second. It was not necessary for Aristotle to go through all these counter-examples, since (i) the sterility of XiQa-1 and XiQe-1 is implied by that of the stronger XaQa-1 and XaQe-1, shown above; and (ii) that of XoQa-1 and XoQe-1 is implied by that of the stronger XeQa-1 and XeQe-1, shown above. Again, the inutility of XiQi-1 and XiQo-1 is implied by that of the discredited XiQa-1 and XiQe-1; and that of XoQi-1 and XoQo-1 by that of XoQa-1 and XoQe-1.)

§15.2.2 Rejected Q-X moods: Second Figure

No pair of QX-premisses in this Figure yields a Q-conclusion.

QaXa-2

$$\frac{Q(\text{healthy})(\text{animal})^{a} \quad (\text{healthy})(\text{man})^{a}}{M(\text{animal})(\text{man})^{o}}$$

$$\frac{Q(\text{healthy})(\text{horse})^{a} \quad (\text{healthy})(\text{man})^{a}}{M(\text{horse})(\text{man})^{i}}$$

If both are stated in the affirmative, there will be no syllogism. Terms for belonging: 'health'-'animal'-'man'. For not belonging: 'health'-'horse'-'man'. (A18, 37b35-38)

QeXi-2
$$\frac{Q(\text{healthy})(\text{animal})^{e} \quad (\text{healthy})(\text{man})^{i}}{M(\text{animal})(\text{man})^{o}}$$

$$\frac{Q(\text{healthy})(\text{horse})^{e} \quad (\text{healthy})(\text{man})^{i}}{M(\text{horse})(\text{man})^{i}}$$

For, when the affirmative is assertoric, whether it is stated as universal or particular, there will be no syllogism (for, this is proved through the same terms as before). (A18, 37b40-38a2)

Aristotle does not say so, but the inutility of QeXi-2 (*i.e.* Festino QX) follows formally from that of QaXa-2.

QaXo-2
(QeXo-2)
QiXo-2
(QoXo-2)
$$\frac{Q(\text{healthy})(\text{animal})^{a,e,i,o} \quad (\text{healthy})(\text{man})^{o}}{M(\text{animal})(\text{man})^{o}}$$

$$\frac{Q(\text{healthy})(\text{horse})^{a,e,i,o} \quad (\text{healthy})(\text{man})^{i,o}}{M(\text{horse})(\text{man})^{i}}$$

If the assertoric is negative, stated as a particular, there will be no syllogism. Nor when both are stated as indefinites, whether affirmative or negative, or as particulars. The proof is the same through the same terms. (A18, 38a8-12)

All these Second Figure examples assume natural inclusions and exclusion among men, animals and horses. They also assume that it is in the nature of all animals, including men and horses, to be either healthy or unhealthy. And consistent with all this, they suppose some men, animals or horses, to be, or not to be, healthy. Some of these premiss-pairs have already been implicitly ruled out by the exclusion of other pairs: QiXo-2 and QoXo-2 by QiXe-1 and QoXe-1.

What then of QaXe-2? Aristotle overlooks this premiss-pair, probably because he discusses the stronger pair QaX'e-2, and acknowledges that they yield an M-conclusion. (See §13.2.) However, he is committed to believing that such premisses are sterile. QaXeMi-2 is equivalent to QaLeXi-1, and QaXeMo-2 to QaLaXi-1:

$$\frac{Qab^{a} \quad ac^{e}}{Mbc^{i}} \overset{C}{\longleftrightarrow} \frac{Qab^{a} \quad Lbc^{e}}{ac^{i}}$$

$$\frac{Qab^a \quad ac^e}{Mbc^o} \overset{C}{\longleftrightarrow} \frac{Qab^a \quad Lbc^a}{ac^i}$$

These First Figure moods are stated to be invalid at A16, 36a27-31 and 36a5.

With regard to XQ-2 moods, none of these yields a Q-conclusion.

XaQa-2 $$\frac{(\text{healthy})(\text{animal})^a \quad Q(\text{healthy})(\text{man})^a}{M(\text{animal})(\text{man})^o}$$

$$\frac{(\text{healthy})(\text{horse})^a \quad Q(\text{healthy})(\text{man})^a}{M(\text{horse})(\text{man})^i}$$

(A18, 37b35-38 quoted above.)

XaQo-2 $$\frac{(\text{healthy})(\text{animal})^a \quad Q(\text{healthy})(\text{man})^o}{M(\text{animal})(\text{man})^o}$$

$$\frac{(\text{healthy})(\text{horse})^a \quad Q(\text{healthy})(\text{man})^o}{M(\text{horse})(\text{man})^i}$$

(A18, 37b40-38a2 quoted above.)

Aristotle does not say so, but the inutility of XaQo-2 (*i.e.* Baroco XQ) follows from that of XaQa-2.

XoQa-2
(XoQe-2)
XiQi-2
(XiQo-2)
XoQi-2
(XoQo-2)

$$\frac{(\text{healthy})(\text{animal})^{i,o} \quad Q(\text{healthy})(\text{man})^{a,e,i,o}}{M(\text{animal})(\text{man})^o}$$

$$\frac{(\text{healthy})(\text{horse})^{i,o} \quad Q(\text{healthy})(\text{man})^{a,e,i,o}}{M(\text{horse})(\text{man})^i}$$

(A18, 38a8-12 quoted above.)

Some of these premiss-pairs have already been implicitly ruled out by the exclusion of other pairs: XiQi-2 and XiQo-2 by XiQi-1 and XiQo-1, XoQi-2 and XoQo-2 by XoQa-2 and XoQe-2.

What about XeQa-2? Aristotle is committed to believing that such premisses are sterile.[15] XeQaMi-2 is equivalent to QaLeXi-3, and XeQaMo-2 to QaLaXi-3:

$$\frac{ab^e \quad Qac^a}{Mbc^o} \overset{C}{\longleftrightarrow} \frac{Qac^a \quad Lbc^a}{ab^i}$$

$$\frac{ab^e \quad Qac^a}{Mbc^i} \overset{C}{\longleftrightarrow} \frac{Qac^a \quad Lbc^e}{ab^i}$$

These Third Figure moods are stated to be invalid at A22, 40a15 and 40a35-38.

At A18, 37b24-29 Aristotle apparently considers premisses XeQa-2 and QaXe-2, saying that they yield an Me-conclusion by conversion to the First Figure. But the First Figure syllogism in question can only be Celarent X'QM, as is clear from A15, 34b19-27; so it is actually X'eQa-2 and QaX'e-2 that he here claims yield M-conclusions.

§15.2.3 Rejected Q-X moods: Third Figure

A21, 39b7-9 asserts without proof that in Figure 3 there are no QXX or XQX syllogisms. This assertion can be proved by a consideration of universal Figure 3 premiss-pairs, since if it is true of them it is true of the weaker premiss-pairs containing particular forms. Of these, QaXe-3 and XeQa-3 reduce to QaXe-1; but QaXaXo-3 and XaQaXo-3 have to be rejected axiomatically along with either QaXaXi-3 or XaQaXi-3 (which are equivalent by i-conv). Counter-examples to QaXaXo-3 (and XaQaXo-3) and QaXaXi-3 can be supplied:

$$\frac{Q(\text{white})(\text{man})^a \quad (\text{white})(\text{man})^a}{(\text{white})(\text{white})^o}$$

$$\frac{Q(\text{white})(\text{man})^a \quad (\text{black})(\text{man})^a}{(\text{white})(\text{black})^i}$$

---

[15] STAHL 1977, p.213, mistakenly claims that Aristotle asserts Cesare XQM, Camestres QXM and Festino XQM.

These examples suppose that men are all white, or all black, consistent with whiteness being contingent for all men (as Aristotle believed).

Seemingly omitted from Aristotle's discussion of XQ-3 moods are Felapton, Ferison and Bocardo (premisses XeQa-3, XeQi-3 and XoQa-3). In fact, Ferison XQM is equivalent by rule *C* to Darii LQX, which Aristotle rejects at A16, 36a37-b2. Felapton and Bocardo XQM are equivalent by rule *C* to Barbari and Barbara LQX, which we find reason to reject in §15.3.1. Thus Aristotle is committed to rejecting Ferison, Felapton and Bocardo XQM.

40a1-3 proves that pairs of particular Q-X premisses yield no conclusion at all, by reference to the terms ⟨animal, man, white⟩, ⟨horse, man, white⟩. This deals with XiQi-3. It can also be shown that the sterility of all these pairs reduces to that of First Figure pairs.

TABLE 15.3

Figure 1	XaQaMi-1, XaQaMo-1, XeQaMi-1, XeQaMo-1, QaXeMi-1, QaXeMo-1, QiXaMi-1, QiXaMo-1
Figure 2	XaQaMi-2, XaQaMo-2, QaXaMi-2, QaXaMo-2, QaXeMi-2, QaXeMo-2, XeQaMi-2, XeQaMo-2
Figure 3	QaXaXi-3, QaXaXo-3, XaQaXo-3.

It is possible to reduce all of Aristotle's Q-X rejected formulae to those in Table 15.3.

§15.3 Rejected Q-L moods

In §10.2 and §12 we attributed to Aristotle the 2-premissed QL theses shown in Table 15.2. This leaves the formulae in Table 15.4.

Aristotle argues in most of these cases that no conclusion follows; and where he provides no such argument - with one exception - he can be shown to be committed to the result anyway. The exception is LoQa-2, which we shall discuss in its place. Aristotle's conclusions from the remaining Q-L premiss-pairs are shown in Table 15.5.

TABLE 15.4

Figure 1	QaLe, (QaLo), QiLa, (QiLe), (QiLi), (QiLo), LiQa, (LiQi), LoQa, (LoQi)
Figure 2	QaLa, (QaLi), (QiLa), QiLe, (QiLi), (QiLo), LaQa, (LaQi), (LiQa), (LiQi), LoQa, (LoQi)
Figure 3	QaLe, (QaLo), (QiLe), QiLi, (QiLo), LiQi, (LoQi).

TABLE 15.5

Figure 1	QaLaQa, QaLiQi, LaQaMa, LaQiMi, LeQaXe, LeQiXo
Figure 2	QaLeXe, QaLoXo, LeQaXe, LeQiXo
Figure 3	QaLaQi, QaLiQi, QiLaMi, QiLaMo, LaQaQi, LaQiMi, LeQaXo, LeQiXo, LiQaQi, LoQaMo

§15.3.1 Rejected Q-L moods: First Figure

Premiss-sequences QaLe and the equivalent QeLe in Figure 1 are shown not to yield any syllogistic consequence (A16, 36a27-31) by reference to the terms ⟨white, animal, snow⟩ and ⟨white, animal, pitch⟩. The specific assumption needed to make these counter-examples work is that all existing animals are contingently white, an assumption which is satisfied so long as there are no swans, ravens or other animals for which whiteness is not contingent.

(QiLe and QoLe are shown useless (36b7-10) by the terms ⟨animal, white, raven⟩ and ⟨animal, white, pitch⟩. The required assumption is that for some animals whiteness is contingent, and thus (by Qi-conversion) for some whites being an animal is contingent. This condition is met, according to Aristotle, so long as men exist.)

QiLa and QoLa are defeated (36b10-12) by the terms ⟨animal, white, swan⟩ and ⟨animal, white, snow⟩. The assumption here is the same as in the previous example.

QaLaX-1 is said without proof to be invalid at A16, 36a5, the equivalent QeLaX-1 at 36a20. This passage reduces the rejection of Celarent QLX to that of Festino QXM:

$$\frac{Qab^e \quad Lbc^a}{ac^e} \xleftrightarrow{C} \frac{Qab^e \quad ac^i}{Mbc^o}$$

But, while the latter's invalidity is implied by the inutility of QeLa-2, Aristotle's demonstration of this last is flawed (as we shall see in §19.1). A counter-example to Celarent QLX can be given:

$$\frac{Q(\text{white})(\text{man})^{a,e} \quad L(\text{man})(\text{man})^a}{(\text{white})(\text{man})^{i,o}}$$

Consistent with Aristotelian doctrines about the necessity of humanity for its possessors, and the contingency of whiteness, we may suppose either that all or that no men are white.

LiQa, LiQe, LoQa and LoQe are discredited (36b3-7) by the terms ⟨animal, white, man⟩ and ⟨animal, white, cloak⟩. Here the required assumption is that for some whites being an animal is necessary, for others impossible. The assumption is satisfied so long as there are swans and snow.

Aristotle does not expressly consider LaQaQ-1. His silence on this matter, in a context where A16, 35b38-36a2 affirms the validity of Barbara LQM, suggests that he thinks LaQa-1 do not imply a Q-conclusion. If so, he was right, as the following counter-example shows. Suppose all men are contingently white, and everything white is necessarily a man. These premisses would be true if all and only men were white. But it is not true that some man is contingently a man.

LaQiX-1, and QeLiX-1 are declared invalid at 36a37-b2. Aristotle says the proof will be the same as before.

Aristotle does not expressly consider LaQiQi-1. But if, as argued above, LaQaQi-1 is invalid, so is LaQiQi-1.

Aristotle does not discuss whether LeQaQi-1 is valid. He maintains at A16, 36a15-17 that Celarent LQM is valid, so his silence on LeQaQi-1 suggests that he thought it was invalid. That it is invalid is shown by the following counter-example. Take the terms 'pitch' 'white' 'man'. Nothing white is pitch (necessarily so), and all men are contingently white. But it is not the case that some man is contingently pitch.

LiQi, LiQo, LoQi, LoQo, QiLi, QiLo, QoLi and QoLo are vanquished (36b12-18) by the terms ⟨animal, white, man⟩ and ⟨animal, white, inanimate⟩. Non-modal assumptions that make these examples work are as follows:

LiQi, LiQo: there are swans, white men and white stones.

LoQi, LoQo: there is snow, there are white men and white stones.

QiLi, QoLi: there are white men and snow. (If there are white men then some animal is contingently white, so some white is contingently an animal. If there are white men then some white is necessarily a man, so some men are necessarily

white.[16] If there is snow then something inanimate is necessarily white.)
QiLo, QoLo: this case presents a problem. If there are white horses then some animal is contingently white, so some white is contingently an animal. If there is pitch then something inanimate is necessarily not white. But, under what conditions is some *man* necessarily not white? It is not clear that there is such a condition. (Aristotle frequently says that all men are contingently white.) However, we can supply an alternative trio of terms for the necessary affirmative relation, namely 'animal' 'white' 'animal'. What is required is then that there be white men and that there be ravens.

§15.3.2 Rejected Q-L moods: Second Figure

In Figure 2, Aristotle argues, premisses QeLa or QaLa yield no conclusion, but his argument is flawed. The argument deals with three cases.

(i) Such premisses are shown not to yield a Q-conclusion (A19, 38a26-36) by the terms ⟨white, man, swan⟩. So far so good. The assumptions here are the familiar ones that all men are contingently white, that all swans are necessarily white, and that all swans are necessarily not men. These terms also show that the premisses imply no affirmative assertoric or apodeictic conclusion.

(ii) 38a36-38 argues that QeLa-2 yield no negative L-conclusion by appealing to the principle that a necessary conclusion follows only when there is a negative necessary premiss. Smith comments:[17]

> Aristotle's appeal to the requirement that a necessary conclusion can be obtained only from two necessary premisses or one necessary and one assertoric premise is illegitimate, since he has not offered any justification for that requirement (or for that matter even asserted it except in application to the first figure).

This is true; nonetheless, Aristotle is right to deny that QaLa-2 or QeLa-2 yield a negative L-conclusion. This can perhaps be shown by a consideration of the terms 'white' 'sleeping' 'swan'. Suppose only men and horses are asleep. Then everything asleep is contingently white and every swan necessarily so. But all swans are possibly asleep. For further discussion of this example see §28.

(iii) 38a38-b5 argues that QeLa-2 yield no negative X-conclusion (*i.e.* that Cesare QLX is invalid), using the terms ⟨moving, animal, awake⟩.

---

16 STRIKER p.48 detects an inconsistency between the assertion that white necessarily belongs to some man and the assertion (regarding QeLe-1) that white contingently (she says "possibly") belongs to no animal. She is mistaken. The first proposition is true if there are white men, the second if there are no swans, ravens or other animals for which whiteness is not contingent; and these conditions can be met simultaneously.

17 SMITH *ad* 38a36-38.

$$\frac{Q(\text{moving})(\text{animal})^a \quad L(\text{moving})(\text{awake})^a}{(\text{animal})(\text{awake})^o}$$

He says that the premisses may be true while the conclusion is false. This example must be challenged.[18] If, as Aristotle supposes, everything awake is an animal and all animals contingently move, then (by Barbara QXQ) everything awake contingently moves. But Aristotle says that everything awake *necessarily* moves, and that the contingent is not necessary. So, the suppositions in this example conflict with Aristotelian doctrine. The odd one out among the suppositions is the minor premiss. There is nothing odd about supposing everything waking to be an animal; and Aristotle elsewhere supposes all animals to move contingently. But this is the only place where he supposes everything waking to be necessarily in motion. More of this in §19.1, where reasons will be given for holding that QaLa-2 *do* imply a negative X-conclusion. Certainly they must imply an M-conclusion, as we saw in §12.2.

Though he does not deal with them explicitly, Aristotle is committed to rejecting both QiLeMi-2 and QiLeMo-2 because they are equivalent to moods elsewhere rejected:

$$QiLeMi-2 \xleftrightarrow{C} QiLeMi-1$$

$$QiLeMo-2 \xleftrightarrow{C} QiLaMi-1$$

Premisses LaQa-2, says 38b13-17, yield no negative L- or X-conclusion, appealing to the principle that negative conclusions require negative premisses. With regard to negative L-conclusions, the argument is the same as that used at 38a36-38. It is inadequate. But Aristotle's statement can perhaps be established by reference to the terms we suggested in §14.1 for Baroco LXL. If only a potential man is white then every man may be necessarily an animal and everything white contingently an animal, though something white is not necessarily not a man. The question will be taken up in §28. With regard to negative X-conclusions, the argument is inconclusive, and in §19.1 we shall find reason to disagree with Aristotle's statement that no such conclusion follows from LaQa-2. Rather, we shall hold, LaQaXe-2 (= Camestres LQX) is valid.

38b17-22 shows that LaQa-2 yield no Q-conclusion, and no affirmative L- or X-conclusion, by reference to the terms ⟨white, swan, man⟩. Here the assumptions are again that whiteness is necessary for swans and contingent for men, men being necessarily not swans.

---

[18] See STAHL 1977, p.217. n.4.

(The same terms are appealed to at 38b27-29 to show that LaQo-2 and QoLa-2 are sterile. The same procedures are appealed to at 38b29-31 to show that LiQa-2, LaQi-2, QiLa-2 and QaLi-2 are sterile. 38b35-37 says that earlier counter examples can be adapted to show that pairs of particular L-Q premisses in Figure 2 yield nothing syllogistically. In all these cases, the rejected premisses are weaker than premisses already rejected.)

He overlooks premisses LoQa-2. These cover the two moods LoQaMi-2 and LoQaMo-2. LoQaMi-2 is equivalent to QaLeMa-3, whose invalidity follows from that of QaLeMi-3. LoQaMo-2 can be shown invalid by the terms ⟨white, animal, man⟩. Assuming that there exists a raven and a man, then some animal (the raven) is necessarily not white, and every man is contingently white, but every man is necessarily an animal.

§15.3.3 Rejected Q-L moods: Third Figure

A22, 40a35-38 argues that premisses QaLe-3 are sterile by reference to the terms ⟨sleeping, sleeping horse, man⟩ and ⟨sleeping, waking horse, man⟩. The assumptions here are that contingently all men sleep, that necessarily no men are sleeping (waking) horses, that necessarily all sleeping horses sleep, and that necessarily no waking horses sleep. We shall see in §18.5 that these last two assumptions are doubtful, to say the least. We shall also see there that other terms can be supplied to achieve Aristotle's stated purpose.

He overlooks QiLi-3 and LiQi-3; but these moods are equivalent to already rejected First Figure moods, using the equivalences in Figure 15.1.

$$QiLiMi-3 \xleftrightarrow{Li-conv,\ T_1} QiLiMi-1$$
$$QiLiMo-3 \xleftrightarrow{Li-conv,\ T_1} QiLiMo-1$$
$$LiQiMi-3 \xleftrightarrow{Qi-conv,\ T_1} LiQiMi-1$$
$$LiQiMo-3 \xleftrightarrow{Qi-conv,\ T_1} LiQiMo-1$$

Fig. 15.1

In Figure 3 Aristotle states without proof that all QLL and LQL moods are invalid (A22, 40a9-11). A proof can be supplied as follows. The strongest premisses of the required kind are (i) QaLa, (ii) LaQa, (iii) QaLe, (iv) LeQa. Terms showing that no necessary conclusion follows from these pairs are:

(i) 'sleeping' 'horse' 'sleeping horse';

(ii) 'white' 'sleeping' 'swan';

(iii) 'sleeping' 'sleeping horse' 'man';
(iv) 'black' 'sleeping' 'swan'.
The required assumptions are:
(i) that sleep is contingent for all horses including sleeping horses, and that sleeping horses are necessarily horses (true if there are sleeping horses);
(ii) that swans are necessarily white and contingently asleep, while sleepers are contingently white (true if there are swans and only non-swans are asleep);
(iii) that all men are contingently asleep and are necessarily not sleeping horses, but all sleeping horses are contingently asleep (true if there are men and sleeping horses);
(iv) that all swans are necessarily not black and are contingently asleep, while all sleepers are contingently black (true if there are swans and only non-swans are asleep).

40a15 states the invalidity of QaLaX-3 and LaQaX-3 without proof. QaLa-3 can be shown not to yield an X-conclusion by the trio 'sleeping' 'man' 'man'. All men are contingently asleep but necessarily men, and consistent with this they may be all asleep or all awake. LaQa-3 may be shown not to yield an X-conclusion by the trios 'white' 'sleeping' 'swan' and 'black' 'sleeping' 'swan'. Swans must be white and cannot be black but may or may not be asleep; and consistent with all this, sleepers may be supposed to be all white or all black.

40b1 excludes the validity of LaQiX-3, QaLiX-3, QiLaX-3 and LiQaX-3 without proof. 40b2-3 excludes QeLiX-3 and QoLaX-3 without proof. 40b8-12 states that QiLe-3 yield no conclusion, and that this can be proved by the same terms. In all these cases, except for LiQaX-3, the premiss-pairs are weaker than or equivalent to pairs already shown to be sterile. See Figure 15.2.

$$LaQiX-3 \longrightarrow LaQaX-1$$
$$QaLiX-3 \longleftrightarrow QeLiX-3 \longrightarrow QaLaX-1$$
$$QiLaX-3 \longleftrightarrow QoLaX-3 \longrightarrow QaLaX-3$$
$$QiLe-3 \longleftrightarrow QiLe-1$$

Fig. 15.2

(The mood to the right of an arrow is provable if that to the left is.) Counter-examples to LiQaX-3 can be constructed from the terms 'white' 'white' 'man' and 'black' 'white' 'man'. Given that there are white men, some men are necessarily white (since some whites are necessarily men) and all men are contingently white, but all whites are whites. Given that there are black men and whites, some men are necessarily black (since some blacks are necessarily men) and all men are contingently white, but no whites are blacks.

Aristotle's Q-L rejections can all be reduced to the formulae in Table 15.6.

TABLE 15.6

Figure 1	QaLaXi-1, QaLaXo-1, QaLeMi-1, QaLeMo-1, LaQaQi-1, LaQiXi-1, LaQiXo-1, LiQaMi-1, LiQaMo-1, LeQaQi-1, LoQaMi-1, LoQaMo-1
Figure 2	QaLaMi-2, QaLaMo-2, LaQaMi-2, LaQaMo-2
Figure 3	QaLaXi-3, QaLaXo-3, LaQaXi-3, LaQaXo-3, QaLeMi-3, QaLeMo-3, LeQaLi-3, LeQaLo-3.

And in addition we need LoQaMo-2.

Given L-subordination, the Q-L and Q-X rejections can all be reduced to the following thirty four:

TABLE 15.7

Figure 1	XaQaMi-1, XaQaMo-1, XeQaMi-1, XeQaMo-1, QaLaXi-1, QaLaXo-1, QiXaMi-1, QiXaMo-1, QaLeMi-1, QaLeMo-1, LaQaQi-1, LaQiXi-1, LaQiXo-1, LiQaMi-1, LiQaMo-1, LeQaQi-1, LoQaMi-1, LoQaMo-1
Figure 2	QaLaMi-2, QaLaMo-2, QaXeMi-2, QaXeMo-2, LaQaMi-2, LaQaMo-2, XeQaMi-2, XeQaMo-2
Figure 3	QaLaXi-3, QaLaXo-3, LaQaXi-3, LaQaXo-3, QaLeMi-3, QaLeMo-3, LeQaLi-3, LeQaLo-3

plus LoQaMo-2.

Given the implications shown in Figure 15.3,

$$
\begin{gathered}
LeQaLi-3 \xleftrightarrow{C} MeQaMi-1 \xrightarrow{M-sub} XeQaMi-1 \\
LeQaLo-3 \xleftrightarrow{C} MaQaMi-1 \xrightarrow{M-sub} XaQaMi-1 \\
XeQaMi-2 \xrightarrow{L-sub} LeQaMi-2 \xleftrightarrow{C} QaLeMi-3 \\
LaQaMi-2 \xleftrightarrow{C} QaLeMo-3 \\
LaQaXi-3 \xleftrightarrow{C} XeQaMo-1 \\
LaQaXo-3 \xleftrightarrow{C} XaQaMo-1 \\
QaLaXi-3 \xleftrightarrow{C} XeQaMo-2 \\
QaLaXo-3 \xleftrightarrow{C} XaQaMo-2 \xrightarrow{L-sub} LaQaMo-2 \\
QaXeMi-2 \xleftrightarrow{C} QaLeXi-1 \xrightarrow{M-sub} QaLeMi-1 \\
QaXeMo-2 \xleftrightarrow{C} QaLaXi-1 \\
QaLeMo-1 \xleftrightarrow{C} QaLaMi-2 \\
QaLaXo-1 \xleftrightarrow{C} QaXaMo-2 \xrightarrow{L-sub} QaLaMo-2
\end{gathered}
$$

Fig. 15.3

the list can be reduced to twenty two:

TABLE 15.8

Figure 1	XaQaMi-1, XaQaMo-1, XeQaMi-1, XeQaMo-1, QaLaXi-1, QiXaMi-1, QiXaMo-1, QaLeMi-1, LaQaQi-1, LaQiXi-1, LaQiXo-1, LiQaMi-1, LiQaMo-1, LeQaQi-1, LoQaMi-1, LoQaMo-1
Figure 2	QaLaMi-2, QaLaMo-2, LaQaMo-2, XeQaMo-2
Figure 3	QaLeMi-3, QaLeMo-3

plus LoQaMo-2.

§16. AMPLIATED CONTINGENCY MOODS

§16.1 Rejected Q' moods: First Figure

Aristotle rejects all the following moods by a single argument:

Q'i,oQ'a,eQ'i-1
Q'i,oQ'i,oQ'i-1

If that relating to the major extreme is stated as a particular, and that relating to the minor extreme as a universal, if both are stated in the affirmative, or both negative, or of dissimilar form, or both indefinite, or both particular, there will never be a syllogism. For nothing prevents *b* extending beyond *a*, and not being predicated of the same. Let the *b* beyond *a* be *c*. For this, it is not possible that *a* belongs either to all or to none or to some or not to some, since possibility-premisses convert and *b* possibly belongs to more than *a*. (A14, 33a34-b3)

In showing that premisses $\langle Q'ab^i, Q'bc^a \rangle$ yield no conclusion, Aristotle uses a procedure which Smith describes as 'a variant on *ekthesis*, a technique which also emphasizes the extensions of terms'.[19] To aid understanding the supposed situation is represented in Figure 16.1.

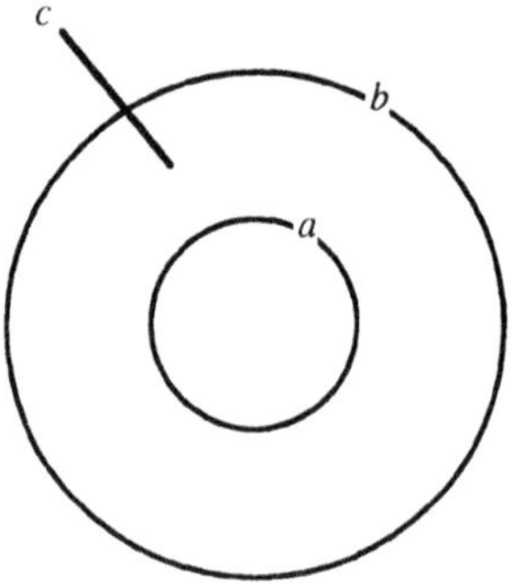

Fig. 16.1

But what do the letters 'a', 'b' and 'c' stand for here? If they stand for what simply *is a, b,* and *c*, then we have a counter-example to iai-1; but the model will not count against Q'iQ'aQ'i-1 unless a further condition is met. This condition is identified by Smith:[20]

If the argument is to work, it is necessary to suppose (as I have) that C is the part of B to which A *necessarily* does not belong.

---

[19] SMITH *ad* 33a38-40.
[20] SMITH *ad* 33a40-b3.

Aristotle's argument then is that, if we let this b be c, it is not the case that a is contingent for c, even if a is contingent for some b.[21]

Once Q'iQ'aQ'i-1 is rejected, the rejection of the other seven moods follows easily. Some are equivalent to the iai mood by Q'-conversion; the others have weaker premisses (by Q'a-subalternation).

Aristotle (33b3-17) also uses two counter-examples to show that no conclusion (not just no Q'-conclusion) follows from premisses Q'iQ'a-1:

$$\frac{Q'(\text{animal})(\text{white})^i \quad Q'(\text{white})(\text{man})^a}{L(X,Q')(\text{animal})(\text{man})^o}$$

$$\frac{Q'(\text{animal})(\text{white})^i \quad Q'(\text{white})(\text{cloak})^a}{L(X,Q')(\text{animal})(\text{cloak})^i}$$

It is assumed here that every possible man is an animal necessarily and not contingently; and that no possible cloak can be an animal. The first example supposes every possible man to be contingently white, and at the same time supposes some possible white to be an animal contingently. Since Q'i-forms convert, the latter supposition will be verified so long as some possible animal (*e.g.* a man) is contingently white. The second example supposes every possible cloak to be contingently white, and some possible white to be an animal contingently. This situation is realised if there are possible cloaks and possible men.

§16.2 Rejected Q' moods: Second Figure

Nothing follows from two Q'-premisses in Figure 2. A17, 37a32-35 points out that Cesare Q'Q'Q' cannot be proved valid by Direct Reduction to Celarent Q'Q'Q':

$$\textit{Celarent } Q'Q'Q' \frac{x\,\dfrac{Q'ab^e}{Q'ba^e} \quad Q'ac^a}{Q'bc^e}$$

because step x is invalid. Likewise 37a35-37 points out that the mood is not provable by Indirect Reduction:

---

21 According to SMITH *ad* 33a40-b3 Aristotle's reasoning is: Suppose that a is contingent for some b. Let c be "just those Bs to which A does not (even possibly) belong". Then it is not the case that a is possible for some c. So (by Q'-conv) no Q'-conclusion follows. The procedure is sound, and will be given a justification in §29.3.

$$x\frac{Q'bc^{a} \quad Q'ab^{e}}{Q'ac^{e}} \longrightarrow \frac{Q'ab^{e} \quad Q'ac^{a}}{Q'bc^{e}}y$$

because x's conclusion is not incompatible with y's minor premiss (and, one may add, because y's conclusion is not incompatible with x's minor premiss).[22]

Of course, the failure of these attempted proofs does not establish that Cesare Q'Q'Q' is invalid. 37a38-b10 does the latter by means of the terms ⟨white, man, horse⟩: since it is contingent that white belongs to no possible man and to every possible horse, but man is not contingent (but impossible) for every possible horse, the premisses can be true and the conclusion false. 37b10-18 points out that the same trio of terms shows that any other combination of pure contingency-premisses in Figure 2 yield no contingency-conclusion. (This follows, given Q-subordination.) And since (as 37a37-40 says) only contingency-conclusions follow from pure contingency-premisses, it is clear (as stated at A17, 36b26-29) that no conclusion follows from pure contingency-premisses in Figure 2. (The principle that only the contingent follows from the contingent has not been proved by Aristotle; but the desired result can be demonstrated using the terms 'white' 'man' 'horse' along with 'white' 'man' 'man'. The first trio show that no affirmative apodeictic or assertoric conclusion follows from Q'iQ'i or any stronger combination of contingency premisses in Figure 2. The second trio establish a like result for negative apodeictic or assertoric conclusions.)

§16.3 Rejected Q' moods: Third Figure

TABLE 16.1

Figure 1	Q'iQ'aQ'i-1, Q'iQ'aXi-1, Q'iQ'aXo-1
Figure 2	Q'aQ'aQ'i-2, Q'aQ'aXi-2, Q'aQ'aXo-2
Figure 3	Q'iQ'iQ'i-3, Q'iQ'iXi-3, Q'iQ'iXo-3.

In Figure 3, A20, 39b2-6 argues, no combination of two particular Q'-premisses yields a syllogistic conclusion; Aristotle appeals to the terms ⟨animal, man, white⟩ and ⟨horse, man, white⟩. He assumes that animal, man and

22 ROSS *ad* 37a35-37 finds the text unsatisfactory (because Aristotle mentions only one of the two flaws in the purported proof). He follows Maier's emendation which has the effect of replacing $Q'bc^{a}$ in x by Lbc^{i}. But it is doubtful whether Maier's suggestion really helps, because it also leaves Aristotle remarking on only one of two flaws in the 'proof'. The unremarked flaw on his reading is that Lbc^{i} is not incompatible with $Q'bc^{e}$.

horse each belongs contingently to some (possible) white. These assumptions are correct given that whiteness belongs contingently to some possible animal, some possible man, and some possible horse.

In sum, Aristotle's rejections in the pure Q' fragment can all be reduced to those in Table 16.1.

§17. POSSIBILITY AND 'UNQUALIFIED' MOODS

Mo-conversion is rejected at A3, 25b13-14, being compared with the conversion of assertoric o-forms. Presumably the comparison is meant to extend to the counter-example $\langle \text{animal}, \text{man} \rangle$ given to the latter at A2, 25a12-13. If so, Aristotle is asserting here that man possibly does not belong to some animal, and denying that animal possibly does not belong to some man.

The discussion of XaQa-1 at A15, 34b10-17 can reasonably be taken as intending the exclusion of an M-conclusion, since it expressly deals with the case where the major is an assertoric with a temporal qualification, and occurs in a context where Aristotle has maintained that if the major does *not* have any temporal qualification then an M-conclusion follows. Similarly, the discussion of XeQa-1 at A15, 34b31-35a2 can reasonably be taken as intending the exclusion of XeQaM-1 syllogisms. Such may be Aristotle's intention, but in §19.1 we shall call its soundness into question.

At A16, 36a17-25 Aristotle rejects Celarent QLX on the ground that it cannot be proved by Indirect Reduction. For, he says, from ac^i and Qab^e nothing follows. (In particular, $\sim Lbc^a$ does not follow.) This amounts to a reduction of the rejection of Celarent QLX to that of Festino QXM:

$$\textit{Celarent QLX}\ \frac{Qab^e \quad Lbc^a}{ac^e} \xrightarrow{\ C\ } \frac{Qab^e \quad ac^i}{Mbc^o}\ \textit{Festino QXM}$$

The latter's rejection is implied by that of QaLaXo-1:

$$QaXiMo-2 \xleftrightarrow{\ C\ } QaLaXe-1 \xrightarrow{\ sub\ } QaLaXo-1$$

A16 36a5, rejects Barbara QLX but gives no reason. We can construct a disproof parallel to the one just considered in relation to Celarent QLX:

$$\textit{Barbara QLX}\ \frac{Qab^a \quad Lbc^a}{ac^a} \xrightarrow{\ C\ } \frac{Qab^a \quad ac^o}{Mbc^o}\ \textit{Baroco QXM}$$

The rejection of Barbara QLX is thus consequential on that of Baroco QXM, a mood whose rejection follows from that of QaLaXa-1:

$$QaXoMo-2 \xleftrightarrow{\;C\;} QaLaXa-1$$

As noted earlier, the technique of modally contrasted instances shows that no M-, as well as no L-, X- or Q-, conclusion follows from the given premisses. For, in showing that an La-relation between the extremes is consistent with the premisses, it shows that no Mo-conclusion follows. And, in showing that an Le-relation between the extremes is consistent with the premisses, it shows that no Mi-conclusion follows. And, if there is no particular there is no universal M-conclusion.

But none of this gives us any need for axiomatic M-rejections beyond Mo-conversion.

Mignucci[23] maintains, in the face of the text, that Baroco LXL is valid. His reasoning is:

$$\cfrac{\textit{Camestres LXL}\,\cfrac{Lab^a \quad ad^e}{Lbd^e} \qquad a-conv\,\cfrac{cd^a}{dc^i}}{Lbc^o}\,\textit{Ferio LXL} \qquad \xrightarrow{\;o-ecthesis\;} \qquad \frac{Lab^a \quad ac^o}{Lbc^o}$$

This argument assumes the validity of Camestres LXL, and thus of Celarent XLL, which Mignucci indeed maintains, again contrary to the text. His argument for this appears to be:

$$\textit{Barbara LXL}\,\frac{Lac^a \quad cd^a}{Lad^a} \xrightarrow[La\,\|\,Me]{\textit{Celarent XQM}\;K} \cfrac{\cfrac{ab^e \quad Qbd^a}{Mad^e} \qquad cd^a}{Mac^o}$$

$$\xrightarrow{\;Qi-ecthesis\text{ AA}\;} \frac{ab^e \quad Qbc^i}{Mac^o} \xrightarrow{\;x\;} \frac{ba^e \quad Lac^a}{Lbc^e}$$

This argument is open to three objections: (i) Celarent XQM is, according to our reasoning above, excluded by Aristotle (unlike Celarent X'QM), (ii) the assumed rule of Qi-ecthesis (for which see §32.2) must be invalid if Qi-forms convert (because it is not symmetrical in *b* and *c*), and (iii) the transformation *x* is invalid, depending on an assumed contradiction $Le \mid Qi$.

Aristotle discusses only one invalid inference involving 'unqualified' assertorics:

---

23 MIGNUCCI *ad* 31a15.

X'eQa-1
$$\frac{X'(\text{raven})(\text{reasoning})^{e} \quad Q(\text{reasoning})(\text{man})^{a}}{L(X,Q)(\text{raven})(\text{man})^{i}}$$

$$\frac{X'(\text{moving})(\text{knowing})^{e} \quad Q(\text{knowing})(\text{man})^{a}}{L(X,Q)(\text{moving})(\text{man})^{o}}$$

And it is clear through terms that the conclusion will not be possible. For, let *a* be 'raven', *b* 'reasoning' [*dianooumenon*][24] and *c* 'man'. *a* then belongs to no *b*, for nothing reasoning is a raven. And *b* is possible for all *c*, for all man think. But *a* is necessary for no *c*. So the conclusion is not possible. But it is not always necessary. For, let *a* be 'moving', *b* 'knowledge' and *c* 'man'. Then *a* belongs to no *b*, and *b* is possible for all *c*. But the conclusion will not be necessary. For it is not necessary that no man moves, nor is it necessary for some. Clearly then the conclusion is 'to none does it necessarily belong'. But one must take better terms. (A15, 34b31-35a2)

Smith comments:[25]

Aristotle is evidently unhappy with his counter example, perhaps (as Alexander suggests) because the major premise is not assertoric but necessary (or perhaps just because the two triplets share only one term instead of his preferred two).

More likely, Aristotle may have noticed that, while *No ravens are reasoning* is plausible as an 'unqualified' proposition, *Nothing with knowledge is moving* is not since it could only be true for a limited time.[26]

(Again, the counter-example serves to discredit a premiss-pair not directly discussed by Aristotle, this time XeQi-1. And again, the rejection of this premiss-pair is formally implied by the rejection of the stronger XeQa-1.)

By way of summarising our survey of the Aristotelian text I present Table 17.1, which lists:

(i) the moods explicitly accepted by Aristotle - these are marked '+',

(ii) the moods explicitly rejected by him - these are marked '-'.

It will be seen that there are many places where the Table is incomplete: Aristotle explicitly states results for only 133 out of 196 cases. Relations of subordination are shown at the top of the table. Q/M subordinations are marked with a dotted line, not all of them being valid. XQX and QXX are omitted as not having any valid Aristotelian moods.

---

[24] SMITH's translation.

[25] SMITH *ad* 34b31-35a2.

[26] See PATTERSON 1995 p.184 for an alternative interpretation.

TABLE 17.1

	XLL	LLL	LXL	QLQ	QXQ	QLM	QXM	QLX	QQQ	LQQ	XQQ	LQM	XQM	LQX
Barbara		+		+				-				+		
	-		+		+				+		-		-	
Celarent		+		+				-				+		+
	-		+		+				+					
Darii		+		+								+		-
	-		+		+				+		-			
Ferio		+		+		+		-						+
	-		+		+				+					
Cesare		+		-		-		-				+		+
	-		+		-		-		-					
Camestres		+				+		+		-		-		-
	+		-						-		-		-	
Festino		+		-		-		-				+		+
	-		+		-		-		-					
Baroco		+								-		-		-
	-		-		-		-		-		-		-	
Darapti		+		+				-				+		-
	+		+		+				+					
Felapton		+		+				-				+		+
	-		+		+				+					
Disamis		+				+		-		+				-
	+		-				+		+		+			
Datisi		+		+				-				+		-
	-		+		+				+					
Bocardo		+				+		-						
	-		-				+		+					
Ferison		+		+				-						+
	-		+		+				+					

CHAPTER FIVE. FLAWS IN THE FABRIC

Everyone agrees that Aristotle's modal syllogistic is flawed. But I do not share the view of some scholars that its flaws all reduce to a single error - that of applying to *de dicto* inferences principles that are sound only *de re*.[1] I think the flaws in Aristotle's fabric are more various. I distinguish substantial from incidental mistakes, and both from inconsistencies. By incidental mistakes I mean mistakes which can be rectified without altering the body of theses in the system. By substantial mistakes I mean irremediable mistakes, ones whose rectification entails altering the body of theses in the system. Inconsistencies arise when different parts of the body of theory come into conflict with one another.

The incidental mistakes dealt with in this Chapter are (1) Aristotle's rejection of Barbara XLL on the basis of rejecting *Hintikka 1* (a valid formula); (2) his rejection of Datisi XLL and Disamis LXL on the basis of an example which overlooks the convertibility of Li-forms; (3) his rejection of QiLo and QoLo moods in Figure 1 on the basis of an example about men that are necessarily not white, which appears to conflict with Aristotle's repeated assertion that all men are contingently white; (4) his rejection of QaLaLo-2 and QeLaLo-2 moods on the basis of an unproven principle that Lo-conclusions require a negative L-premiss; (5) his rejection of QaLe moods in Figure 3 on the basis of an example which requires, contrary to Aristotle's statements elsewhere, that sleeping horses are necessarily asleep; (6) his assertion of Barbara X'QM, seemingly on the basis of modal *reductio* (an unsound procedure).

The substantial mistakes are (7) the rejection of QaLaXe in Figure 2 on the basis of an unproven doctrine that there can be no conclusion from a pair of affirmative premisses in that Figure; and (8) the rejection of Barbara XQM on the basis of a questionable counter-example.

The inconsistencies are between (9) the use of modal ecthesis and the rejection of Baroco XLL and Bocardo LXL; and (10) between the doctrine of oppositions of Q- to L-forms and the theory of Qi-conversion.

Aristotle's mistakes are due to a variety of causes. In some cases he thinks only of *metaphysical* modal propositions, true because they state the nature of what is included in the subject, thus overlooking those modals deducible from the metaphysical cases by conversion (1,2,10). In some cases he thinks only of *dynamic* possibilities (what things are or can *become*), thus overlooking those *static* possibilities (what things can *be*) that are opposable to the static necessities that connect natures (8). In other cases he fails to work out the details of counter-examples (3,9). In others he relies on unproven adaptations of principles borrowed from the assertoric syllogistic (4,7). In yet others he seems to confuse *de re* and *de dicto* modals (5,7). In one case the extreme difficulty of the subject matter prevents

[1] See STAHL 1976 p.207; STAHL 1977 p.211.

him from giving a sufficiently precise statement of the principle on which he relies (6).

§18. INCIDENTAL MISTAKES

We begin with six incidental mistakes.

§18.1 Barbara XLL and the Hintikka Formulae

The argument of 30a23-27, that Barbara XLL must be rejected on pain of accepting *Hintikka 1* (see §14.1) assumes that that formula must be rejected. But Hintikka[2] argues that if Aristotle's method demonstrates the invalidity of Barbara XLL, it can also be used to demonstrate the invalidity of syllogisms claimed by Aristotle as valid. Hintikka points out that an analogous argument can be constructed for the rejection of Barbara XLX:

$$\text{Darapti XLL}\ \frac{\text{Barbara XLX}\ \dfrac{ab^a \quad Lbc^a}{ac^a} \qquad Lbc^a}{Lab^i}$$

The only difference is that instead of arguing from Barbara XLL and Darapti LLL, we argue here from Barbara XLX and Darapti XLL. Both Darapti LLL and Darapti XLL are asserted by Aristotle to be valid. Both lines of reasoning appear to show that a syllogistic form must be rejected on pain of allowing Hintikka's formula to be valid. But, whereas Aristotle thinks Barbara XLL invalid, he presumably would have to accept the validity of Barbara XLX, as this syllogism appears to be nothing but a strengthened version of Barbara XXX:

$$\text{Barbara}\ \frac{ab^a \qquad \dfrac{Lbc^a}{bc^a}\ La-subord}{ac^a}$$

It seems therefore that no interpretation can simultaneously concur with Aristotle's tally of theses, and adopt all his methods of rejection. (This despite Smith's judgment[3] that Aristotle's procedure here is "sophisticated and flawless".) In particular, the interpreter must choose between rejecting Barbara XLX or Darapti XLL, and abandoning the Aristotelian method of rejecting Barbara XLL which depends on rejecting Hintikka's formula. The rejection of Barbara XLX would involve rejecting modal subordination; and as we saw in §6.2.2, there is reason for

---

[2] HINTIKKA, ch.7 sect.4.
[3] SMITH *ad* 30a25-28.

accepting this inference. The rejection of Darapti XLL flies directly in the face of the text. The least disruptive path therefore is the last. It is the one we shall follow for the present. I argue that, despite its odd appearance, Hintikka's formula is valid. If *b* belongs necessarily to all *c*s then there are necessary *b*s, so since *a* belongs to all *b*s *a* belongs to all necessary *b*s, so some necessary *b*s are *a*s, thus (by the convertibility of Li-forms) *a* belongs necessarily to some *b*.

The consequence of this interpretation is that neither Barbara XLL nor Barbara XLX can be rejected on the ground that their validity would imply that of Hintikka's formula. But Barbara XLL can be rejected for other, better, reasons; Aristotle himself supplied a counter-example to it.

Aristotle's mistake was to conclude that because ab^a is compatible with the denial of Lab^i, the conjunction of ab^a with Lbc^a must be compatible with the denial of Lab^i. Most likely, he did not see this because he overlooked the convertibility of Li-forms.

The above argument does not apply *mutatis mutandis* to *Hintikka 2* and *3*, which are taken as obviously false in Aristotle's disproofs of Celarent XLL (A9, 30a32-33) and Camestres LXL (A10, 30b24-31). For, Hintikka-like analogues of Aristotle's derivations in those places would depend on Felapton XLL or Ferio XLL in the place of Darapti XLL; but Felapton XLL and Ferio XLL are invalid. We shall argue in §22.5 that *Hintikka 2* and *3* are invalid, and therefore that Aristotle's proofs of rejection for Celarent XLL and Camestres LXL are sound.

§18.2 Datisi XLL, Disamis LXL and necessarily awake bipeds

We saw in §14.3 that Aristotle's counter-example to Datisi XLL and Disamis LXL takes it that *Some biped is necessarily awake* is false when every animal (every necessary biped included) is awake; but that he should not do so if he accepts the convertibility of Li-forms, since the given premisses imply that something waking is necessarily a biped, and this in turn implies that some biped is necessarily awake. We also saw that this mistake can be remedied by taking the term 'approaching' instead of 'biped'.

Aristotle may have overlooked the convertibility of Li-forms here and elsewhere because he had in mind only metaphysical Li-propositions, which are not convertible. There is a sense in which a proposition like *Some thing that goes grey is necessarily a man* does not convert to *Some man necessarily goes grey*. The former's predicate gives the essential nature of what falls under the subject; the latter's predicate states an accident of what falls under the subject. In concentrating exclusively on metaphysical Li-propositions, Aristotle forgets that there are derivative Li-truths, obtained from the metaphysical ones by conversion.

§18.3 QiLo-1, QoLo-1 and necessarily non-white men

We saw in §15.3.1 that 36b12-18 attempts to show eight premiss-pairs non-concludent by means of a single argument involving the term-trios 'animal' 'white' 'man' and 'animal' 'white' 'inanimate'. The argument is free of trouble except for the premiss-pairs QiLo-1 and QoLo-1, where counter-examples with the stated terms would require that *Some men are necessarily not white* be true, contrary to Aristotle's frequent assertion that whiteness is contingent for all men. There may be an error here. It may be that it cannot be the case that some men are necessarily not white. The explanation for such an error might be that, in his desire to produce a compendious example, Aristotle was careless in checking that all eight cases could be dealt with by his term-trios. We have already seen that, if this is so, then an alternative example can be constructed to deal with QiLo-1 and QoLo-1.[4]

§18.4 QaLaLo-2, QeLaLo-2 and negative conclusions from affirmative premisses

38a36-38 uses the principle that Lo-conclusions follow only from premisses that include a negative L-proposition, to show that QaLa-2 and QeLa-2 yield no negative L-conclusion. Aristotle gives no proof for this principle. Nor do we need to assume precisely this principle in order to prove the desired result. If we assume that (i) negative L-propositions are a species of negative assertorics, (ii) negative assertorics are provable only from premisses that include a negative, and (iii) Qe-propositions are affirmative, then we can show that QaLa-2 and QeLa-2 do not of themselves imply any Lo-conclusion. The first of these assumptions will be shown to hold in §24, the second is known from the theory of the assertoric syllogism,[5] and the third is Aristotelian doctrine. We saw in §15.3.2 that counter-examples can be used to demonstrate the inutility of the stated premisses.

§18.5 QaLe-3, QeLe-3 and necessarily sleeping horses

At A22, 40a35-38 Aristotle uses the terms ⟨sleep, sleeping horse, man⟩, ⟨sleep, waking horse, man⟩ to show that nothing follows from QaLe premisses in Figure 3. Now, these terms do indeed give us examples of universal attribution and

---

[4] Maybe Aristotelian metaphysical theory allows for a situation in which some men are necessarily not white. Suppose that all that is actually white is snow. And suppose that there are men (not white men of course). Then some men belong to a kind (animals) whose nature excludes the nature which belongs to all actual whites. In that sense, some men are necessarily not white. The trouble is, however, that under the stated conditions it is not true (as Aristotle's counter-example requires) that some white is contingently an animal; on the contrary, no white (i.e. no snow) is possibly an animal.

[5] See THOM 1981 §46 Theorem, Corollary 4.

universal non-attribution, thus showing that the stated premisses imply no assertoric conclusion. *A fortiori* they show that the premisses imply no apodeictic conclusion. But they do not show that no contingency-conclusion follows, unless we assume that (i) every sleeping horse is necessarily asleep, and (ii) every waking horse is necessarily not asleep. The difficulty is that it seems impossible to give any but a *de dicto* reading of these conditions. This difficulty has been felt by various commentators.[6] Some of Aristotle's other examples concerning sleeping and waking also strike a false note, *e.g.* at A19, 38a41-b4 where it is said that everything waking is necessarily moving. Compare A11, 31b8-10, 31b40-32a4 where it is said (i) that it is not necessary that some animal is not awake (every animal is possibly awake), and 31b27-31 where it is said not to be necessary that some biped is awake.

It is conceivable that in claiming that sleeping horses are necessarily asleep, that waking horses are necessarily not asleep, and that the waking are necessarily moving, Aristotle made a confusion between *de re* and *de dicto* readings of modals. Or perhaps, in the case of the first claim, there is a confusion with the sense outlined in *On sleep* 455b25 in which sleep is necessary for animals:

> Again, sleep belongs of necessity to each animal. I use the term 'necessity' in its conditional sense, meaning that if an animal is to exist and have its own proper nature, it must have certain endowments; and, if these are to belong to it, certain others likewise must belong to it.

There is a textual dispute over the wording of 40a35-38.[7] But whatever the correct reading of the text, it is possible to supply triples of terms that do the job Aristotle wanted his triples to do, *e.g.* $\langle$white, swan, man$\rangle$ and $\langle$white, raven, man$\rangle$. Whiteness is contingent for men, and men cannot be either swans or ravens; but whiteness is necessary for swans and impossible for ravens.

---

[6] SMITH *ad* 40a37-38 describes the examples as "curious", noting that "on Aristotle's usual understanding of necessity and possibility, no horse is of necessity awake or of necessity asleep: these are, instead, good examples of things that are 'capable of being otherwise' ...."

[7] On the (traditional) Gohlke reading, this trio of terms is: 'sleep'-'horse'-'sleeping man'. The relations between these terms are supposed to be that (i) sleep is contingent for all (no) horses; (ii) horse is impossible for all sleeping man; (iii) sleep is contingent for all (no) sleeping man; (iv) horse is necessary for some (no) sleeping - depending of whether any horses are sleeping; (v) sleeping man is impossible for all horses; (vi) sleeping man is contingent for any sleepers that are men, and is impossible for any sleepers that are not men - so it is contingent for all (no) sleepers if all sleepers are men, and impossible for all sleepers if no sleepers are men, and both contingent for some sleepers and impossible for some sleepers if some and only some sleepers are men. On the Waitz reading the trio is: 'sleep'-'sleeping horse'-'man'. The relations between these terms, according to Aristotelian metaphysics, is that (i) sleep is contingent for all (no) sleeping horses; (ii) sleeping horse is impossible for all man; (iii) sleep is contingent for all (no) man; (iv) sleeping horse is contingent for (all, no) sleeping if all sleepers are horses, it is impossible for all sleepers if no sleepers are horses, and it is contingent for some and impossible for some sleepers in the remaining case; (v) man is impossible for all sleeping horse; (vi) man is necessary for all (no) sleepers that are men, and impossible for the rest..

§18.6 Barbara X'QM and 'unqualified' assertorics

Anyone who accepts rule K', as Aristotle seems to do, makes a mistake. But we do not have to make that mistake in order to validate the X'QM moods which Aristotle uses this rule to prove. They can all be proved in other ways. There is a valid inference which can be used to justify Barbara X'QM. Following Angelelli's interpretation of unqualifieds as ampliated assertorics, it is clear that if every c is a contingent (and thus a possible) b, and every possible b is an a (and thus a possible a) then every c may be an a. Similarly with Darii X'QM. If some c is a contingent (and thus a possible) b, and every possible b is an a, then some c may be an a; if on the other hand some b is a contingent (and thus a possible) c, and every possible (and thus every actual) b is an a, then some a may be a c.

The situation with Celarent X'QM and Ferio X'QM is a little more complex. We leave its consideration until later when we develop a formal semantics. Our aim will be to have all four perfect X'QM moods come out valid on that semantics, and to formulate a sound version of K' by means of which they can be proved. If this can be achieved, we shall be able to excuse Aristotle's failure to give a precise statement of the rule on which he was relying. He will have stated the right rule in too general terms, an understandable mistake in such difficult subject-matter.

§19. SUBSTANTIAL MISTAKES

Some of Aristotle's mistakes are substantial in the sense that their rectification entails altering the body of theses in his system.

§19.1 QaLaXe-2 and negative conclusions from affirmative premisses

38a38-b5 attempts to show that premisses QeLa-2 yield no negative X-conclusion by reference to the terms $\langle$movement, animal, waking$\rangle$:

$$\frac{Q(\text{moves})(\text{animal})^e \quad L(\text{moves})(\text{awake})^a}{(\text{animal})(\text{awake})^o}$$

Aristotle says that the premisses may be true while the conclusion is false. I argue that premisses QeLa-2 are not sterile but yield an assertoric e-conclusion.[8]

---

[8] MIGNUCCI *ad* 38b4 argues that QaLa-2 yield an Me-conclusion:

$$\textit{Datisi LLL}\frac{Lac^a \quad Lbc^i}{Lab^i} \xrightarrow{K} \frac{}{Li \parallel Qa} \quad \frac{Qab^a \quad Lac^a}{Mbc^e}$$

But the Xe-conclusion is stronger.

QeLaXe-2 is invalid iff QaLaXe-2 is invalid. As against the claim of invalidity, it is clear that any semantics would validate QaLaXe-2 according to which (i) the same individual cannot be both necessarily and contingently a, (ii) Qab^a expresses a universal quantification of singular contingency-forms, and (iii) Lab^a expresses a universal quantification of necessity-forms. Very plausibly, these are all Aristotelian conditions. He repeatedly contrasts the necessary with the contingent. That Qab^a and Lab^a express universal quantifications of singular contingency-/necessity-forms is strongly suggested by the validity of Barbara QXQ and Barbara LXL. I conclude that Aristotle made a mistake about QaLaXe-2. The reason he gives for rejecting it is that two affirmatives in the second figure yield no assertoric conclusion; but this reason, though it may have swayed him, is actually very weak since it is based merely on an enumeration of syllogisms with assertoric premisses.[9]

Along with QaLaXe-2 we should accept LaQaXe-2, QeLaXe-2 (Cesare QLX) and LaQeXe-2 (Camestres LQX).[10]

In fact, QaLaXe-2 is indirectly reducible to Darii LXL, if we assume the opposition $Li \parallel Qa$:

$$\text{Darii } LXL \frac{Lac^a \quad i-conv\,\dfrac{bc^i}{cb^i}}{Lab^i} \quad \xrightarrow[Li \parallel Qa]{K} \quad \frac{Qab^a \quad Lac^a}{bc^e}$$

QaLaXe-2 is also indirectly reducible to Darii QXQ, if we assume the opposition $Qi \parallel La$:

$$\text{Darii } QXQ \frac{Qac^a \quad cb^i}{Qab^i} \quad \xrightarrow[Qi \parallel La]{K} \quad \frac{Qac^a \quad Lab^a}{cb^e}$$

The first of these proofs runs parallel to the one discussed in §12 whereby Alexander proved the validity of QaLaMe-2. Though Aristotle accepts Darii LXL and the opposition $Li \parallel Qa$, he argues (A19, 38b13-14) that premisses LaQa-2 yield no negative conclusion, on the ground that negative conclusions require negative premisses. But the premiss of that reasoning is nowhere established as a general principle, or one that is applicable to modal syllogisms. If our proof of QaLaXe-2

[9] For an excellent discussion of this, see PATTERSON 1995 pp.197-198.

[10] ANSELM discusses what might seem to be a counter-example to Cesare MLX: "No animal is necessarily rational, every man is necessarily rational ... no man is animal" (3.231-3.233). But the first premiss here is false. The reason given in support of it (that rationality is not part of what it is to be an animal - see 3.21) does not show what is required, namely that there is no animal for which being rational is necessary.

is accepted - or QaLaXe-2 is valid for some other reason - then this principle, sound though it may be for simple assertoric syllogisms, must be abandoned in its general form.

QaLoXo-2 (Baroco QLX) is also provable:

$$\textit{Barbara QXQ}\frac{Qab^a \quad bc^a}{Qac^a} \xrightarrow[Qa \,\|\, Lo]{K} \frac{Qab^a \quad Lac^o}{bc^o}$$

In the semantics to be developed presently, I shall treat these QL-2 moods as theses, replacing QaLaMo-2 and LaQaMo-2 in Aristotle's list of rejections by QaLaLo-2, QaLaQi-2, QaXaMo-2, LaQaLo-2, LaQaQi-2 and XaQaMo-2; later we shall consider alternative strategies.

These additions to the list of rejections are justified in §28.3 below.[11]

§19.2 Barbara XQM and ephemeral contingencies

We saw in §15.2.1 that Aristotle says no Q-conclusion follows from XaQa-1. He also wants to say that no M-conclusion follows (assuming always that the major premiss is not 'unqualified'). Aristotle's counter-example to Barbara XQM asks us to suppose that only men are moving and that movement is contingent for all horses; even in these circumstances, says Aristotle, it is impossible, not contingent, that horses should be men.

An objection might be raised to this. Let us assume that all men are necessarily men. Then if everything moving is a man it follows by Barbara LXL that everything moving is necessarily a man:

$$\textit{Barbara LXL}\frac{L(\text{man})(\text{man})^a \quad (\text{man})(\text{moving})^a}{L(\text{man})(\text{moving})^a}$$

But then if it is necessary that no horse is a man, it follows by Camestres LLL that it is necessary that no horse is moving:

$$\textit{Camestres LLL}\frac{L(\text{man})(\text{moving})^a \quad L(\text{man})(\text{horse})^e}{L(\text{moving})(\text{horse})^e}$$

Hence it is not contingent that every horse is moving. Barbara LXL and Camestres LLL are asserted by Aristotle. His counter-example to Barbara XQM assumes the truth of $(\text{man})(\text{moving})^a$ and $L(\text{man})(\text{horse})^e$, and the falsity of

11 For QaLaLo-2 see model 28.19; for QaLaQi-2 see model 28.20; for QaXaMo-2, XaQaMo-2 see model 28.21; for LaQaLo-2 see models 28.22, 28.23; for LaQaQi-2 see model 28.24.

$L(\text{moving})(\text{horse})^e$. Therefore, either the terms of Aristotle's counter-example are ill-chosen, or he cannot accept $L(\text{man})(\text{man})^a$ as true.

Now, while the proposition that all men are necessarily men has no direct Aristotelian sanction, it does seem plausibly true whether the necessity is taken *de dicto* or *de re*. Prudence therefore suggests that we investigate the hypothesis that Aristotle's terms are ill-chosen. This hypothesis could be developed along the following lines.

The validity of Barbara LXL shows that what is necessary at one time may not be necessary at another. For example, the ephemeral assertoric fact that $(\text{man})(\text{moving})^a$ - whose existence depends not just on the existence of movers, but also on their inclusion among men - along with the long-term apodeictic fact that $L(\text{animal})(\text{man})^a$ - whose existence depends merely on the existence of men - entails the ephemeral apodeictic fact that $L(\text{animal})(\text{moving})^a$. Now, if there are ephemeral necessities there must also be ephemeral contingencies, assuming that contingency and necessity are related by the kind of equipollences discussed in §2.2. What our discussion shows is that *Every horse is contingently moving* is just such an ephemeral contingency. Its truth depends, not only on the existence of horses, but also on what things are currently moving. So, the fact that horses naturally have the possibility of either moving or not moving does not entail that *under all conditions* movement is contingent for horses. Under the condition that only men are moving it is not true that every horse is contingently moving. Such, at any rate, is what must be said of the kind of contingency occurring in Aristotle's contingency-syllogisms.

Aristotle's mistake in his counter-example to Barbara XQM was to confuse those Q-propositions - which I shall call *static* - with what I shall call *dynamic* Q-propositions. This distinction can be explained with reference to M-propositions. A static M-proposition states what the subject can *be*; a dynamic M-proposition states what the subject is or can *become*. Whether horses can *be* movers depends not just on the existence of horses, but also on what actually is a mover. Whether horses can *become* movers depends just on the existence of horses. Thus, relative always to there being some horses, the static possibilities for horses are ephemeral in a way the dynamic possibilities for horses are not. The dynamic possibilities for horses are fixed, we might say, simply by the nature of the horse. An analogous account can be given of the distinction between static and dynamic Q-propositions.

The semantics to be presented in the next Chapter will take Barbara XQM (like Barbara X'QM) as valid, where possibility and contingency-forms are taken as static. Dynamic possibility-propositions will receive some attention in §45.

Along with Barbara XQM, we shall accept Darii XQM and Bocardo LQX. We saw in §12.1 how Darii XQM can be proved. Bocardo LQX is provable for me but not for Aristotle, following from either Barbara XQM or Baroco XLL, moods he rejects:

$$\textit{Barbara XQM}\frac{ab^a \quad Qbc^a}{Mac^a} \xrightarrow{C} \frac{Lac^o \quad Qbc^a}{ab^o}$$

$$\textit{Baroco XLL}\frac{ab^a \quad Lac^o}{Lbc^o} \xrightarrow[Lo \,\|\, Qa]{K} \frac{Lac^o \quad Qbc^a}{ab^o}$$

However, we will not completely reject Aristotle's reasoning concerning affirmative conclusions from XaQa-1 premisses. That such premisses imply no contingency-conclusion is demonstrated by the terms 'animal' 'moving' 'man', That they imply no affirmative apodeictic or assertoric can be shown using the terms 'animal' 'moving' 'representation'. We may consistently suppose that only animals move, and that all representations contingently move, without supposing that any representation is an animal. In developing a formal semantics we shall drop XaQaMi-1 from Aristotle's list of rejections, adding XaQaQi-1 and XaQaXi-1.

§20. INCONSISTENCIES

Two inconsistencies in Aristotle's modal syllogistic will be outlined in this section, and I shall present an interpretive strategy for dealing with them.

§20.1 Baroco XLL, Bocardo LXL and ecthesis

No interpretation can concur with Aristotle's tally of theses and non-theses, while adopting all his methods of validation, in particular his use of ecthesis as a method of validating syllogisms.

The problem is this. Aristotle rejects Baroco XLL and Bocardo LXL. And he proves Baroco LLL and Bocardo LLL by ecthesis. But, if the ecthetic proofs of Baroco LLL and Bocardo LLL are accepted, then they can be adapted to prove Baroco XLL and Bocardo LXL.[12] We noted in §4.2 that there are two possible readings of the ecthetic proofs of Baroco and Bocardo LLL, either using Lo-ecthesis *A* or Lo-ecthesis *AA*. But on either reading, these deductions can easily be adapted so as to prove the validity of Baroco XLL and Bocardo LXL, moods rejected by Aristotle. Following Rule Lo-ecthesis *A*, these moods can be proved as follows:

12 This is argued by HENLE p.99 n.27, and by VAN RIJEN pp.195, 198.

$$\text{Felapton LXL}\;\dfrac{\text{Camestres XLL}\;\dfrac{ab^a \quad Lad^e}{Lbd^e} \qquad cd^a}{Lbc^o} \qquad \xrightarrow{\text{Lo-ecthesis A}} \qquad \dfrac{ab^a \quad Lac^o}{Lbc^o}$$

$$\text{Felapton LXL}\;\dfrac{Lad^e \qquad \text{Barbara}\;\dfrac{cd^a \quad bc^a}{bd^a}}{Lab^o} \qquad \xrightarrow{\text{Lo-ecthesis A}} \qquad \dfrac{Lac^o \quad bc^a}{Lab^o}$$

Following Lo-ecthesis *AA* they are provable thus.

$$\text{Felapton LLL}\;\dfrac{\text{Camestres XLL}\;\dfrac{ab^a \quad Lad^e}{Lbd^e} \qquad Lcd^a}{Lbc^o} \qquad \xrightarrow{\text{Lo-ecthesis AA}} \qquad \dfrac{ab^a \quad Lac^o}{Lbc^o}$$

$$\text{Felapton LXL}\;\dfrac{Lad^e \qquad \text{Barbara XLX}\;\dfrac{Lcd^a \quad bc^a}{bd^a}}{Lab^o} \qquad \xrightarrow{\text{Lo-ecthesis AA}} \qquad \dfrac{Lac^o \quad bc^a}{Lab^o}$$

If, contrary to the text, we should accept Baroco XLL and Bocardo LXL as theses, what then of Aristotle's reasons for rejecting Baroco XLL and Bocardo LXL? His purported disproof of Baroco XLL is summary and indirect. He says at A10, 31a16-17 that Baroco XLL may be invalidated "by the same terms" as he used for invalidating Baroco LXL. However, when we turn to 31a12-15 for the disproof of Baroco LXL, we find that these terms are not identified. Instead, we are told that the same terms may be used here as were used for "the universal syllogisms". Presumably, by this phrase Aristotle means Camestres LXL

$$\dfrac{Lab^a \quad ac^e}{Lbc^e}$$

which is shown invalid by letting *a* be 'animal', *b* 'man' and *c* 'white'. On the supposition that animal belongs to nothing white (where it is assumed that animal necessarily belongs to all man), man will belong to no white:

> ... but this will not be necessarily so, for it is possible that man be white, even though he could not be so as long as animal belongs to no white.

This argument can indeed be extended to show the invalidity of Baroco LXL, by adding "it is possible that *every* man be white" *etc.* But it cannot similarly be extended to show the invalidity of Baroco XLL

$$\frac{ab^a \quad Lac^o}{Lbc^o}$$

A parallel demonstration would have to take it, if we let *a* be 'animal', *b* 'man' and *c* 'white', that animal *necessarily* belongs to not all white. As Van Rijen notes,[13] the major premiss here is not just true but necessarily true; and this creates a problem for Aristotle, because it means that the example counts against Baroco LLL if it counts against Baroco XLL; but Aristotle regards Baroco LLL as valid. Further, if the minor is taken to be true with these terms, then the conclusion will be true also. For, let the minor be supposed true on the Aristotelian ground that something white, *e.g.* snow, is necessarily not animal. Then the conclusion will also hold necessarily, since the same class of white things which made the minor true will also make the conclusion true: something white, *viz.* snow, is necessarily not a man. It seems, then, that Aristotle did not pursue in detail his statement that the terms ⟨animal, man, white⟩ invalidate Baroco XLL; and it seems that if one does so, then (i) if these terms invalidate Baroco XLL they also invalidate Baroco LLL, and (ii) on Aristotelian assumptions, they do not invalidate Baroco XLL.[14]

Aristotle's rejection of Bocardo LXL at A11, 32a4-5 can also be questioned. He states that a counter-example can be constructed with the terms 'biped' 'moving' 'animal', with 'animal' as middle. Thus we are asked to suppose the truth of $L(\text{biped})(\text{animal})^o$ and $(\text{moving})(\text{animal})^a$ without supposing that of $L(\text{biped})(\text{moving})^o$. How is not explained. On the other hand, if every animal is moving and some animal *e.g.* a goat is necessarily not a biped, then it would seem that something moving, *viz.* that goat, is necessarily not a biped.[15] In this way, Aristotle's intuitions regarding these two moods are open to challenge. Acceptance of Bocardo LXL is also implied by Aristotle's acceptance, noted in §8.2, of Barbara MXM

---

13 VAN RIJEN (footnote 1), 195.

14 MIGNUCCI *ad* 31a17 agrees that Baroco XLL can be shown valid by ecthesis and that Aristotle has not demonstrated its invalidity here. But Mignucci thinks Aristotle's counter-example is designed to show the validity (!) of inferring both bc^o and $\sim Lbc^o$ from the premisses $\langle ab^a, \sim Lab^a, ac^o \rangle$. This is wrong. A counter-example cannot demonstrate validity. And, contrary to Mignucci's statement, Aristotle's example does not make ab^a true but not necessary; for, the statement *Every white is man*, if true is for Aristotle necessary (provable by Darii LXL).

15 MIGNUCCI *ad* 32a5 accepts Aristotle's counter-example, but maintains that it is valid to infer both ab^o and $\sim Lab^o$ from the premisses $\langle Lac^o, bc^a, \sim Lbc^a \rangle$. On my view, the apodeictic as well as the assertoric version of ab^o follows whether bc^a is apodeictic or assertoric, provided that ac^o is apodeictic.

$$\frac{Mab^a \quad bc^a}{Mac^a} \xrightarrow{C} \frac{Lac^o \quad bc^a}{Lab^o}$$

His rejection of Baroco XLL and Bocardo LXL must be put down to carelessness. The sources of Aristotle's mistake are (i) his failure to give explicit formulations of rules for ecthesis and thus fully to integrate ecthesis into the modal syllogistic, and (ii) his casual attitude to the registering of counter-examples.[16]

Van Rijen rightly says that one cannot have a consistent system containing Aristotle's axioms *and* rules of ecthesis.[17] The cases of Baroco XLL and Bocardo LXL oblige the interpreter to revise Aristotle's system, either by accepting rules or theses which he rejects, or by rejecting rules or theses which he accepts. Interpreters may want these revisions to be minimal; but revisions there must be. In the semantics about to be presented, we shall retain Ecthesis as a method of proof, and accept the unAristotelian theses Baroco XLL and Bocardo LXL.[18] Later, alternative strategies will be considered.

§20.2 Q-opposition and Qi-conversion

A second inconsistency in the Aristotelian system arises from the fact that the doctrines of Qi-conversion and of Q-opposition cannot both be brought into harmony with some of Aristotle's most pervasive examples. The examples I have in mind are:

(i) $Q(\text{white})(\text{man})^a$ (A14, 33b3-17; A16, 36b3-7; A17, 37a4-8, 37b1-16),

(ii) $L(\text{animal})(\text{man})^a$ (A9, 30a23-25, 28-33; A10, 30b31-40, 31a10-15; A15, 34b10-17, 35b11-14, 14-19; A18, 37b35-38, 40-38a2; A16, 36b3-7, 36b12-18).

From these two propositions it follows by Darapti QLQ that $Q(\text{white})(\text{animal})^i$, and hence by Qi-conversion that $Q(\text{animal})(\text{white})^i$. Aristotle acknowledges the truth of this proposition at A14, 33b3-17. But, consistent with its being true that animal is contingent for some white, it might be the case that animal is necessary

16 The latter, at any rate, is recognised by van Rijen (footnote 1), 201.

17 VAN RIJEN p.198. See THOM 1991 §3, and PATTERSON 1995 §4.3.5, for further discussion of VAN RIJEN.

18 Acceptance of Baroco XLL coheres with our decision to accept Barbara XQM. For it implies this:

$$\frac{ab^a \quad Lac^o}{Lbc^o} \xrightarrow[Lo \,\|\, Qa]{K} \frac{ab^a \quad Qbc^a}{Mac^a}$$

for all white; for it might be the case that only men are white (in which case, since animal is necessary for all men, and all whites are men, animal is necessary for all whites, by Barbara LXL).

By similar reasoning it can be shown that we cannot accept both Li-conversion and the opposition $Li \parallel Qa$. Compatible with Aristotle's statement that white is contingent for all men, white can be necessary for some man. Suppose some white is a man; this is compatible with the statement that white is contingent for all men. Then given that all men are necessarily men, it follows (in Darii LXL) that some white is necessarily a man, and thence by Li-conversion that some man is necessarily white.

Thus, accepting the truth of Aristotle's examples, and accepting Barbara LXL, Darii LXL and Darapti QLQ (all perfect according to Aristotle), we cannot accept both Qi-conversion and the opposition $Qi \parallel La$, and we cannot accept both Li-conversion and the opposition $Li \parallel Qa$. Either we reject Qi-conversion or we hold that Qi-forms are not opposed to La-forms: Qi-forms, we must say, are either convertible or opposable but not both. And either we reject Li-conversion or we hold that Li-forms are not opposed to Qa-forms: Li-forms, we must say, are either convertible or opposable but not both. Whatever we decide, we go contrary to Aristotle's explicit statements.

A solution is not hard to find. For, the oppositions $Qi \parallel La$ and $Li \parallel Qa$ are never used by Aristotle in proofs of Q-X or Q-L theses. So we can keep his counter-examples along with Li-conversion and Qi-conversion without having to abandon any of his Q-X or Q-L theses.

But this easy solution runs into trouble when we consider the M-layer. For, the opposition $Qi \parallel La$, not used in the Q-layer, is used in Aristotle's proofs of Bocardo QXM and Bocardo QLM; and $Li \parallel Qa$ (or more exactly $Li \parallel Qe$), also a stranger to the Q-layer, is used in the proof of Ferio QLM.

§28 will present an alternative proof for Ferio QLM, which does not make use of this opposition. However, Bocardo QXM and Bocardo QLM cannot be similarly rescued. And there is reason to doubt their validity.

Consider Bocardo QLM. Suppose all whites are swans, and suppose there are non-white men. Because there are non-white men (and all men are contingently white), some animal is contingently white, and thus some white is contingently an animal. Because all whites are swans (and all swans are necessarily swans), all whites are necessarily swans. So we have the premisses of a Bocardo QLM. But the conclusion, that some swans are possibly not animals, is false. Thus, if we use familiar Aristotelian examples, Bocardo QLM is invalid; and if it, then Bocardo QXM.

(The assertion of Bocardo QLM and QXM is also inconsistent with Aristotle's doctrine (38a26-b5) that pairs of affirmative Q-L premisses in Figure 2 yield no conclusion, if that doctrine is meant to rule out M-conclusions. For this would commit him to rejecting

$$\frac{Lab^a \quad Qac^o}{Mbc^o}$$

and hence to rejecting Bocardo QLM $\dfrac{Qac^o \quad Lbc^a}{Mab^o}$. However, as we saw in §19.1, that doctrine is mistaken.)

Our easy solution has become a little harder. It is now clear, taking the M layer into account, that if we wish to keep Aristotle's common counter-examples along with Li-conversion and Qi-conversion without having to abandon any of his Q-X or Q-L theses, we can do so, and we can also agree with him about the theses in the M layer *except* for Bocardo QLM and QXM, which we must judge to be invalid. In the semantics for the modal syllogistic which we are about to present, we shall take Bocardo QLM (and consequently Bocardo QXM) as invalid, adding QiLaMo-3 to our list of rejections. Later we shall investigate the possibility of accepting the oppositions $Qi \parallel La$ and $Li \parallel Qa$ while excluding Qi- and Li-conversion and moods whose proof requires either Li- or Qi-conversion.

(Wieland also thinks the oppositions $Qi \parallel La$ - or the equivalent $Qo \parallel La$ - and $Li \parallel Qa$ - or the equivalent $Qe \parallel Li$ - should be rejected:

Wollte man aber positive Modi vom Typus NM [*i.e.* LQ] mit assertorischer Conclusio beweisen oder aber auf der Grundlage positiver Modi vom Typus NO [*i.e.* LX] Syllogismen vom Typus NM mit assertorischer Conclusio ableiten, so müßte man auf jeden Fall auch die Gültigkeit von Intermodalgesetzen wie "$MBoc \longrightarrow \neg NBaC$" [*i.e.* $\dfrac{Qab^o}{\sim Lab^a}$] und "$MBec \longrightarrow \neg NBiC$" [*i.e.* $\dfrac{Qab^e}{\sim Lab^i}$] voraussetzen. Man ist somit zu der Annahme berechtigt, daß diese beiden Ausdrücke keine im aristotelischen System gültigen logischen Gesetze bezeichnen.[19]

But his reasons for this position are inadequate. In the first place, it is not true that the oppositions in question must be used if the affirmative mood Barbara LQX is to be proved as Celarent LQX is: see note 10 to §10.2.1. But, even if this were true,

[19] WIELAND 1975 p.88.

it would not be sufficient to discredit these oppositions; for the 'proof' of Barbara LQX relies on the unAristotelian Baroco LXL.)

The philosopher-interpreter does not rest content with reporting the mistakes and inconsistencies in the Aristotelian system, but seeks to articulate approximations to that system which lack these mistakes and inconsistencies. I have argued that any such approximation, if it is to exclude logical mistakes, must hold *Hintikka 1* valid, and must reject rules C' and K'. I have also argued that in the interests of logical truth, such an approximation must hold QaLaXe-2 and Barbara XQM valid.

Beyond that, I have argued, we have two choices to make. First, we can either include proof by modal ecthesis and the theses Baroco XLL and Bocardo LXL, or else exclude both these moods along with modal ecthesis. Second, we can either include Qi- and Li-conversion but not the oppositions $Qi \parallel La$ and $Li \parallel Qa$ (or moods like Bocardo QLM which depend on them), or else include those oppositions but exclude Qi- and Li-conversion and moods whose proof requires either of these conversions.

CHAPTER SIX. SYSTEM *AL*

We proceed to develop logical systems intended to provide formal interpretations for the different layers of the modal syllogistic. In this Chapter, we construct an axiomatic system and formal semantics for an apodeictic system which includes ecthetic proofs of Baroco XLL and Bocardo LXL. In our system, Aristotle's restrictions on Substitution and Cut are dropped. And the problematic moods are integrated with the apodeictic. A related system is formulated in which only assertoric forms appear, with or without modalised terms. We then give a classification of alternative approaches. The axiomatic system of McCall, with Johnson's semantics - which rejects Baroco XLL and Bocardo LXL - is located within this classification.

§21. ASSERTORIC FORMS

A semantic interpretation for the system *A* of assertoric syllogistic[1] is constructed by first defining an *A* model as an ordered triple $\langle D, F, V\rangle$ satisfying the following conditions:

21.1 *D* is a non-empty set of individuals.

21.2 *F* is a non-empty set of non-empty sets whose members are members of *D*. (These sets will be the extensions of the terms in formulae.)

V assigns members of *F* to the term-variables of *A*'s wffs, and evaluates forms as true or false in accordance with **21.3-21.6**. Truth in a model is defined accordingly. I suppose that to *a,b,c,...* are assigned the sets $f_1, f_2, f_3, \ldots$ respectively, where $f_1, f_2, f_3 \in F$.

21.3 ab^a is true iff $f_2 \subseteq f_1$, *i.e.* iff the *a*s include the *b*s.

21.4 ab^e is true iff $f_2 \cap f_1 = \varnothing$, *i.e.* iff the *a*s exclude the *b*s.

21.5 ab^i is true iff $f_2 \cap f_1 \neq \varnothing$, *i.e.* iff the *a*s intersect the *b*s.

21.6 ab^o is true iff $f_2 \not\subset f_1$, *i.e.* iff the *a*s do not include the *b*s. See Figures 21.1-21.4.

21.7 A formula $\frac{R}{p}$ is true iff some member of *R* is false or *p* is true.

21.8 A formula $\frac{R}{p}$ is valid iff it is true in all models.

---

[1] THOM 1981 §18.

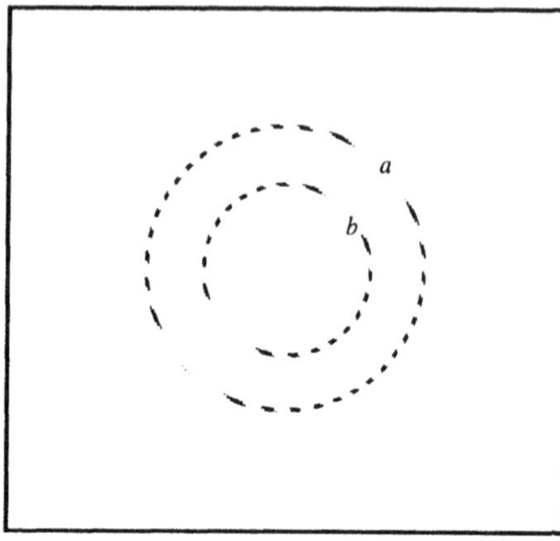

Fig. 21.1

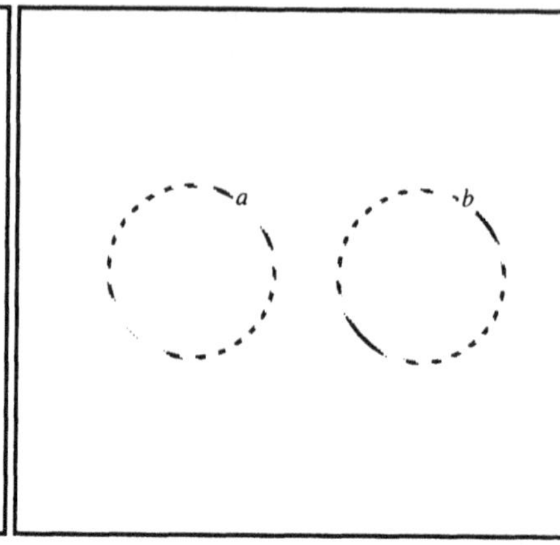

Fig. 21.2

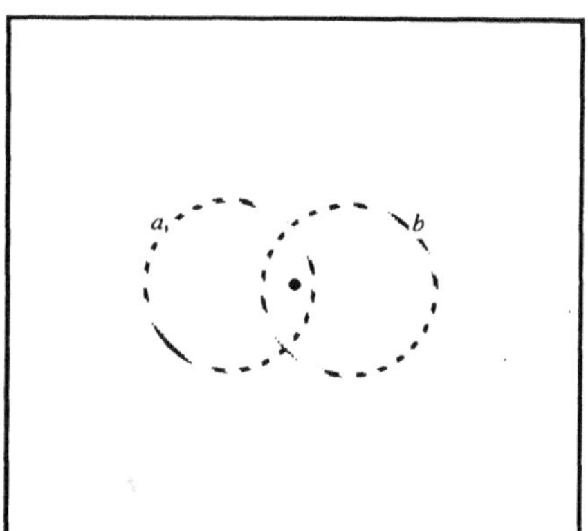

Fig. 21.3

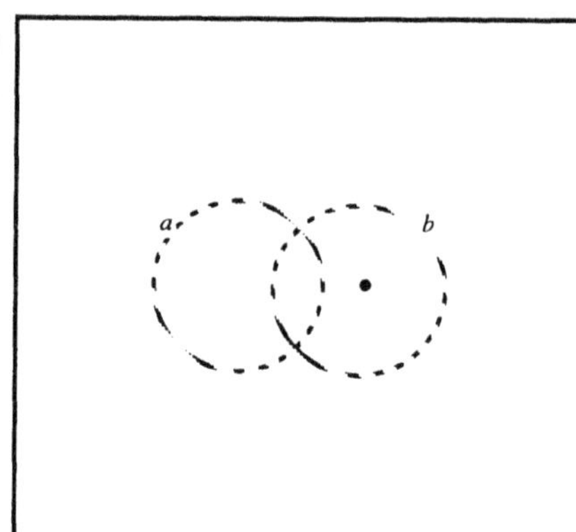

Fig. 21.4

Given these semantics, the axioms of §6 are all valid:

e-conv If $f_2 \cap f_1 = \varnothing$ then $f_1 \cap f_2 = \varnothing$.

a-conv If $f_2 \subseteq f_1$ then $f_1 \cap f_2 \neq \varnothing$. (By **21.2**, f_2 is non-null.)

i-conv If $f_2 \cap f_1 \neq \varnothing$ then $f_1 \cap f_2 \neq \varnothing$.

a-rep If $f_2 \subseteq f_1$ then $f_2 \subseteq f_1$.

Barbara If $f_2 \subseteq f_1$ and $f_3 \subseteq f_2$ then $f_3 \subseteq f_1$.

Celarent If $f_2 \cap f_1 = \varnothing$ and $f_3 \subseteq f_2$ then $f_3 \cap f_1 = \varnothing$.

Darii If $f_2 \subseteq f_1$ and $f_3 \cap f_2 \neq \varnothing$ then $f_3 \cap f_1 \neq \varnothing$.

Ferio If $f_2 \cap f_1 = \varnothing$ and $f_3 \cap f_2 \neq \varnothing$ then $f_3 \not\subset f_1$.

Preservative of truth is

o-ecthesis If $f_2 \not\subset f_1$
then for some f_4, $f_4 \neq \varnothing$ and $f_4 \subseteq f_2$ and $f_4 \cap f_1 = \varnothing$
$\because$ if $f_2 \not\subset f_1$
then $f_2 \cap \overline{f_1} \neq \varnothing$ and $f_2 \cap \overline{f_1} \subseteq f_2$ and $f_2 \cap \overline{f_1} \cap f_1 = \varnothing$.

§22. NECESSITY FORMS

Given the axiomatic structure of Aristotle's L layer we can state constraints the satisfaction of which would be sufficient for a formally adequate semantics for the system. These constraints are that La-repetition, Lo-repetition, L-subordination, Le-conv, Li-conv, Barbara LXL, Celarent LXL, Darii LXL and Ferio LXL be valid; that Barbara XLL, Celarent XLL, Darii XLL, Felapton XLL, Baroco LXL, Bocardo XLL and Lo-conv be invalid; and that *Lo-ecthesis A* preserve validity. (There is no need to consider First Figure LLL moods separately if we have their LXL analogues plus L-subordination; neither need we take partial La-conversion as axiomatic if we have rule *K*.)

In constructing a formal semantics for the necessity-forms that figure in Aristotle's modal syllogistic, one must consider the distinction between *de dicto* and *de re* modals. A *de dicto* modal is a singular proposition stating of a proposition that it is necessary; a *de re* modal is a singular, particular or universal proposition, stating of an individual or of some or all members of a subject-class, that it or they necessarily belong to a predicate-class.

The *de dicto* reading of universal apodeictics, though it suits the pure L moods and the principles of L-conversion,[2] does not work as an interpretation of the L moods as a whole, since it is incompatible with the validity of Barbara LXL and Celarent LXL. What the validity of those moods amounts to is that the subject-position in La-forms, and both subject- and predicate-positions in Le-forms, are referentially transparent: any subterm can be substituted there *salva veritate*. By contrast, these positions in *de dicto* La- and Le-forms are referentially opaque: if it is necessary that all (no) *b*s are *a*, and it is the case that all *c*s are *b*, it does not follow that it is necessary that all *c*s are *a*. What is needed, therefore, for each of the four L-forms, is a *de re* reading which satisfies the above constraints.

Not that the story stops with *de re* analysis. All that *de re* analysis does is to reduce the meaning of universal and particular apodeictics to the meaning of assertoric propositions containing modalised terms ("necessarily-*a*", "possibly-*a*"). The question remains, What is the meaning of such terms? In the present section, a *de re* analysis of Aristotelian modals is sketched following the basic idea of Johnson.[3] Johnson assigns more than one class to each term '*a*', namely a class of *a*s, a class of 'essential' *a*s, and a class of 'essential' *non-a*s. Thus he cashes out "necessarily-*a*" as "essentially *a*". The effect is that Aristotle's apodeictic propositions become non-modal propositions stating class-relations among ordinary classes and classes of essential *a*s *etc*. The semantics to be developed here is similar to Johnson's, but differs from his in (i) not requiring that 'essential' classes be non-null and (ii) not separately postulating a class of 'essential' *non-a*s. Further, no particular essentialist theory is here presupposed in relation to what we call 'necessary' *a*s.

§22.1 Formal Semantics

We proceed to give formal semantics for a system *AL*, whose base is to be specified in the next section. We build on the semantics for assertorics given in §21. We

---

[2] OCKHAM II.24.27-48 argues that "the Philosopher in the *Prior Analytics* speaks of the conversion of necessity-propositions only when they are taken in the composite sense [*i.e. de dicto*] or equivalently."

[3] JOHNSON 1989.

define an AL model as an ordered quadruple $\langle D, V, F, F^* \rangle$ such that V assigns ordered pairs whose members are respectively members of F and F^* to the term-variables of AL's wffs, and evaluates forms as true or false in accordance with **22.1.1-22.1.8**.

22.1.1 F^* is a non-empty set of possibly empty sets whose members are members of D. The members of F^* will be referred to as star sets. (There is no assumption that subsets, or unions, of star sets are themselves star sets.)

22.1.2 An assignment of values to the variables in a wff associates with each term variable v in the wff an ordered pair consisting of a member of F followed by a member of F^*. These sets I will refer to as the vs and the v^*s respectively. The v^*s can be thought of as a privileged set of individuals standing in some ontological relation to the vs. On the present interpretation, the v^*s are in every case included in the vs. Thus:

22.1.3 If the pair $\langle \alpha_1, \alpha_2 \rangle$ is assigned to a variable then $\alpha_2 \subseteq \alpha_1$.[4] See Figure 22.1.

Since the non-*a*s are a set if the *a*s are, they must have a star set too (though it may be null). More important than this star set for our purposes is what I shall the *sun set* of the *a*s. In terms of Johnson's motivating idea, these are the 'essential' *non-a*s, or the impossible *a*s. There are different ways in which this set could be specified, and we shall look at various definitions in §25.2. For the present I propose to define a set's sun set as the union of all star sets whose originals exclude the original of any star set that includes the given set:

22.1.4 $$f^{¤} = \bigcup_{g^*} \exists h^* \left[f \subseteq h^* \wedge g \subseteq \bar{h} \right]$$[5]

The membership of a given set's sun set is thus dependent on what star sets are non-null. Without making any assumption about whether the sun set is itself a star set, or whether it is non-empty, I shall represent it as a single set outside a star set including the *a*s, as in Figure 22.2.[6]

---

[4] This is weaker than the requirement in THOM 1991 p.139 that $\alpha_1 = \alpha_2 \vee \alpha_2 = \varnothing$.

[5] This is the approach of THOM 1991.

[6] STAHL 1976 p.209 uses similar diagrams.

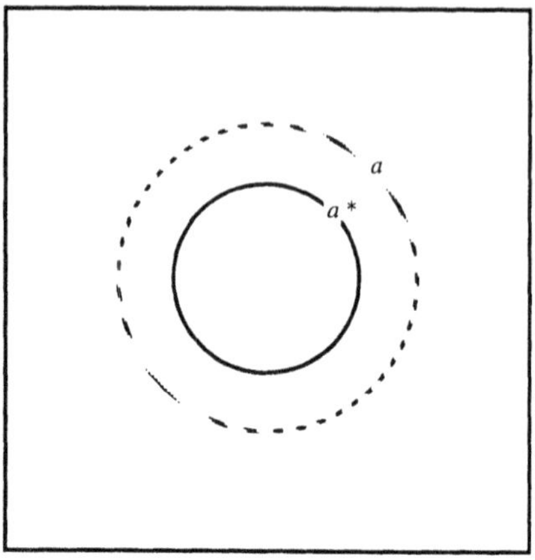

Fig. 22.1

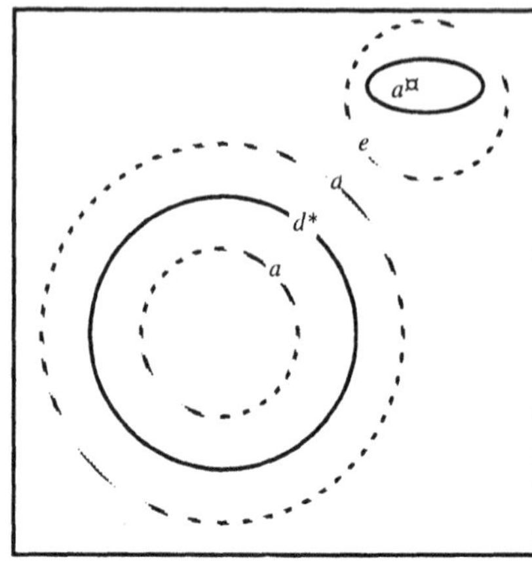

Fig. 22.2

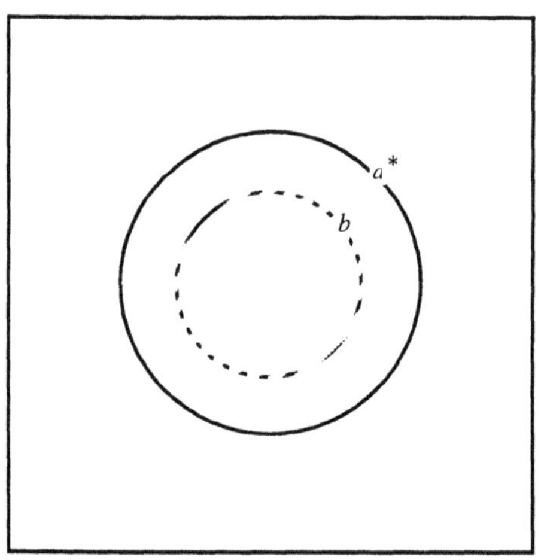

Fig. 22.3

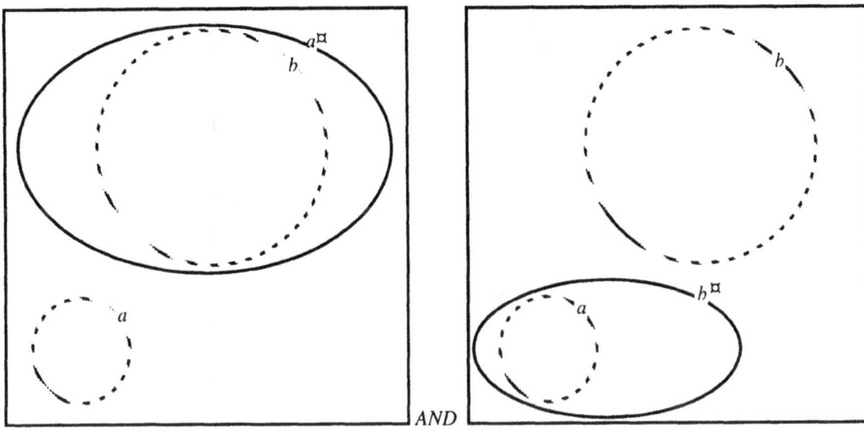

Fig. 22.4

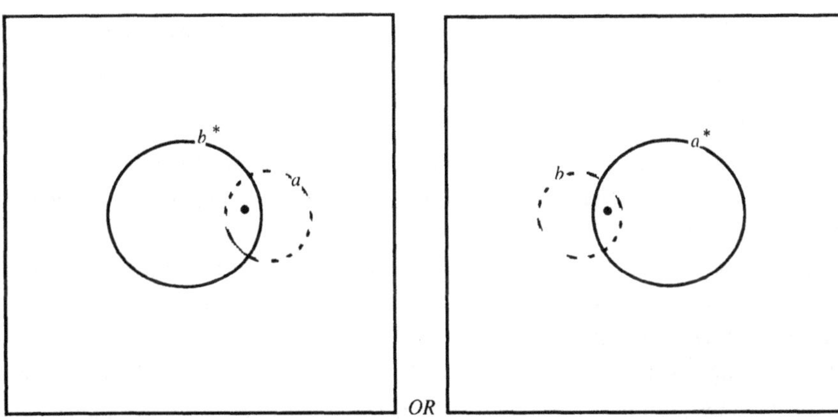

Fig. 22.5

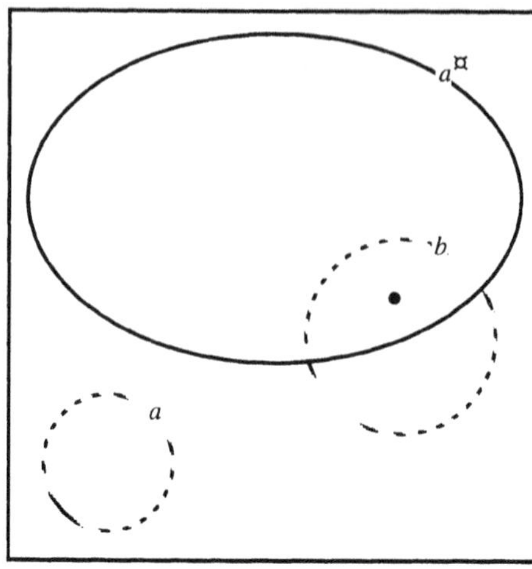

Fig. 22.6

I suppose that to D belong individuals $d, d_1, d_2, \ldots$, to a is assigned the pair of sets $\langle f_1, f_1^* \rangle$, to b the pair of sets $\langle f_2, f_2^* \rangle$, and to c the pair of sets $\langle f_3, f_3^* \rangle$, where $f_1, f_2, f_3 \in F$, $d \in D$, $f_1^*, f_2^*, f_3^* \in F^*$.

22.1.5 Lab^a is true iff $f_2 \subseteq f_1^*$, *i.e.* iff the *a*s' star set includes the *b*s. (In terms of Johnson's motivating idea, the *b*s are here represented as being among the necessary *a*s.) See Figure 22.3.

22.1.6 Lab^e is true iff $f_2 \subseteq f_1^{¤} \wedge f_1 \subseteq f_2^{¤}$, *i.e.* the *b*s are among the *a*s' sun set and the *a*s are among the *b*s' sun set. (The *b*s are impossible *a*s and the *a*s impossible *b*s.) See Figure 22.4.

22.1.7 Lab^i is true iff $f_2 \cap f_1^* \neq \varnothing \vee f_1 \cap f_2^* \neq \varnothing$ *i.e.* iff the *a*s' star set intersects the *b*s, or the *b*s' star set intersects the *a*s. (Either some *a* is a necessary *b* or some *b* is a necessary *a*.) See Figure 22.5.

22.1.8 Lab^o is true iff $f_2 \cap f_1^{¤} \neq \varnothing$, *i.e.* some of the *b*s are in the *a*s' sun set. (Some *b* is an impossible *a*.) See Figure 22.6.

22.1.9 *AL* validity is defined as truth in all *AL* models.[7]

---

[7] The analysis of BRENNER p.3 is similar in important ways to ours. He analyses apodeictic forms as follows:

Lab^a $\forall x(bx \supset \Box ax)$

§22.2 *The Sun set Theorem*

As a consequence of the definition **22.1.4** of sun sets, we have the **Sun set Theorem**

22.2.1 $(f \subseteq g) \supset (g^{¤} \subseteq f^{¤})$

If one set is included in another, the sun set of the latter is included in the sun set of the former.

Proof Suppose $f \subseteq g$ and $x \in g^{¤}$
then $x \in \bigcup_{h^*} \exists j^* [g \subseteq j^* \wedge h \subseteq \bar{j}]$
so $x \in \bigcup_{h^*} \exists j^* [f \subseteq j^* \wedge h \subseteq \bar{j}]$
so $x \in f^{¤}$.

The Sun set Theorem should be contrasted with the following non-theorems:

Lab^e $\forall x(bx \supset \forall y(ay \supset \Box y \neq x))$

Lab^i $\exists x(bx \wedge \Box ax) \vee \exists x(ax \wedge \Box bx)$

Lab^o $\exists x(bx \wedge \forall y(ay \supset \Box y \neq x)) \vee \exists x(\Box(bx) \wedge {\sim}ax \wedge \forall y(\Box ay \supset \Box y \neq x))$

This analysis assumes some quantified modal logic with identity; Brenner does not make it clear just which modal system is assumed. It should be noted that in even the weakest such system the formula $x = y \supset \Box x = y$ is valid, and in quantified S5 with identity $x \neq y \supset \Box x \neq y$ is valid. Thus, if the analysis of Le-forms is not to collapse into $\forall x(bx \supset \forall y(ay \supset y \neq x))$, *i.e.* $\forall x[bx \supset {\sim}ax]$, then the underlying quantified modal logic must be weaker than quantified S5 with identity. Relative to some such system, Brenner's interpretation gives fully Aristotelian results for apodeictic forms. In particular it rejects Baroco XLL and Bocardo LXL.

$(f_1{}^{¤} \subseteq f_2{}^{¤}) \supset (f_2 \subseteq f_1)$

22.1		*	¤
f_1	d_1, d_2	d_1	d_4
f_2	d_2, d_3	d_2, d_3	d_1, d_4
f_3	d_1	d_1	d_2, d_3, d_4
f_4	d_1, d_2, d_3	d_1, d_2, d_3	d_4
f_5	d_4	d_4	d_1, d_2, d_3

$f_1{}^{¤} \subseteq f_2{}^{¤}$ but $f_2 \not\subset f_1$.

In model 22.1 we have chosen a domain D (consisting of d_1, d_2, d_3, d_4) and a set F^* of star sets ($\{d_1\}, \{d_2, d_3\}, \{d_1, d_2, d_3\}, \{d_4\}$), and sets $f_1 - f_5$ of elements from D in such a way that definition **22.1.4** is observed but $(f_1{}^{¤} \subseteq f_2{}^{¤}) \supset (f_2 \subseteq f_1)$ is false. If one wanted to turn this into a counter-example one would have to pick actual terms to which $f_1 - f_5$ and the various star sets were assigned. Terms fitting the bill are 'white' 'man' 'swan' 'animal' and 'pitch'. The model then supposes that there are four individuals:

d_1 is a swan which is necessarily a swan, necessarily white and necessarily an animal;

d_2, d_3 are men both of whom are necessarily men and necessarily animals, the first white, the second not white;

d_4 is some pitch which is necessarily pitch.

The model also supposes that there are four star sets:

f_2 are the men;

f_3 are the swans;

f_4 are the animals;

f_5 is the pitch.

Given these suppositions, the sun set of the white things (the union of all star sets excluding some star set including the white things) consists of the pitch. (The narrowest star set to include both white things is the animals.) The sun set of the men consists of the pitch plus the swan. Thus the sun set of the men includes that of the white things. But the white things do not include the men, since one of the men is not white.

$\overline{f_1} \subseteq f_1^{¤}$

22.2		*	¤
f_1	d_1, d_2	–	–
$\overline{f_1}$	d_3, d_4	–	–
f_2	d_1	d_1	d_2, d_3, d_4
f_3	d_2	d_2	d_1, d_3, d_4
f_4	d_3	d_3	d_1, d_2, d_4
f_5	d_4	d_4	d_1, d_2, d_3

$\overline{f_1} \not\subset f_1^{¤}$.

In model 22.2 we have a domain consisting of four individuals d_1, d_2, d_3, d_4, the star sets being their singletons. We have chosen sets $f_1 - f_5$ from among these individuals in such a way that $\overline{f_1} \not\subset f_1^{¤}$ is false. Again we could turn this structure into a counter-example by picking terms to which $f_1 - f_5$ and the various star sets were assigned. The terms could be 'beautiful' 'non-beautiful' 'man' 'music' 'fear' 'day'. The model then supposes that there are four individuals: a man, some music, a fear and a day. Each of these is supposed to be necessarily of its own kind (the man necessarily a man, the music necessarily music, *etc.*). Each, it is supposed, cannot be of any of the other kinds (the day cannot be a fear *etc.*). The man and the music are taken to be beautiful, the fear and the day not so. The sun set of the beautiful is null, because there is no star set including all beautiful things. But the non-beautiful is a non-null set. Thus, the non-beautiful is not included the sun set of the beautiful.

In model 22.3 we suppose a domain of just one individual which is the sole member of both f_1 and f_2, and is the sole member of f_1's star set, while f_2's star set is null. Given this f_1 is included in f_2 but f_1's star set is not included in f_2's. Terms of this sort are 'man' 'walking'. Then if a solitary walking man exists, necessarily a man but not necessarily walking, the men are among those walking but the necessary men are not among the necessarily walking.

$(f_1 \subseteq f_2) \supset (f_1^* \subseteq f_2^*)$

22.3		*
f_1	d_1	d_1
f_2	d_1	–

$f_1 \subseteq f_2$ but $f_1^* \not\subset f_2^*$.

§22.3 Validity of the axioms

That our semantics satisfies the constraints on the interpretation of L-forms is clear. The asserted axioms are all valid. This is trivial for La-rep and Lo-rep. The others work out as follows.

La-subord If $f_2 \subseteq f_1^{*}$ then $f_2 \subseteq f_1$
$\because f_1^{*} \subseteq f_1$.

Le-subord If $f_1 \subseteq f_2^{¤}$ and $f_2 \subseteq f_1^{¤}$ then $f_2 \cap f_1 = \varnothing$
$\because f_1^{¤} \cap f_1 = \varnothing$.

Li-subord If $f_2 \cap f_1^{*} \neq \varnothing$ then $f_2 \cap f_1 \neq \varnothing$
$\because f_1^{*} \subseteq f_1$ and
if $f_1 \cap f_2^{*} \neq \varnothing$ then $f_2 \cap f_1 \neq \varnothing$
$\because f_2^{*} \subseteq f_2$.

Lo-subord If $f_2 \cap f_1^{¤} \neq \varnothing$ then $f_2 \not\subset f_1$
$\because f_1^{¤} \cap f_1 = \varnothing$.

Le-conv If $f_2 \subseteq f_1^{¤}$ and $f_1 \subseteq f_2^{¤}$ then $f_1 \subseteq f_2^{¤}$ and $f_2 \subseteq f_1^{¤}$.

Li-conv If $f_2 \cap f_1^{*} \neq \varnothing$ or $f_1 \cap f_2^{*} \neq \varnothing$ then $f_1 \cap f_2^{*} \neq \varnothing$ or $f_2 \cap f_1^{*} \neq \varnothing$.[8]

Barbara LXL If $f_3 \subseteq f_2$ and $f_2 \subseteq f_1^{*}$ then $f_3 \subseteq f_1^{*}$.

[8] PATTERSON 1990 is mistaken when he says (n.20) that Li-conversion "is not valid on the reading (*de re*) one needs if Darii NAN [our LXL] is to be valid". Our *de re* semantics validates both Li-conversion and Darii LXL.

Celarent LXL $\quad$ If $f_3 \subseteq f_2$ and $\begin{Bmatrix} f_1 \subseteq f_2^{¤} \\ f_2 \subseteq f_1^{¤} \end{Bmatrix}$ then $\begin{Bmatrix} f_1 \subseteq f_3^{¤} \\ f_3 \subseteq f_1^{¤} \end{Bmatrix}$

$\because$ if $f_3 \subseteq f_2$ then $f_2^{¤} \subseteq f_3^{¤}$ - by **22.2.1**.

Darii LXL $\quad$ If $f_3 \cap f_2 \neq \varnothing$ and $f_2 \subseteq f_1^{*}$ then $f_3 \cap f_1^{*} \neq \varnothing$.

Ferio LXL $\quad$ If $f_3 \cap f_2 \neq \varnothing, f_1 \subseteq f_2^{¤}$ and $f_2 \subseteq f_1^{¤}$ then $f_3 \cap f_1^{¤} \neq \varnothing$.

Lo-ecthesis A preserves validity.

Proof $\quad$ If $f_2 \cap f_1^{¤} \neq \varnothing$ then for some non - null $f_4, f_4 \subseteq f_2$ and $f_4 \subseteq f_1^{¤}$
$\because f_2 \cap f_1^{¤} \subseteq f_2$ and $f_2 \cap f_1^{¤} \subseteq f_1^{¤}$.

§22.4 *Invalidity of the axiomatic rejections*

The axiomatic rejections are all invalid.

Lo-conv

22.4		*	¤
f_1	d_1	d_1	d_2
f_2	d_1, d_2	d_1, d_2	–
f_3	d_2	d_2	d_1

$f_2 \cap f_1^{¤} \neq \varnothing$ but $f_1 \cap f_2^{¤} = \varnothing$.

Aristotle's terms 'man' 'animal' (A2, 25a25-26) fit model 22.4. The model supposes that a man and a non-human animal, say a koala, exist. In the model, the men are the set f_1, the animals the set f_2, and the koalas the set f_3; there is just one man d_1 and just one koala d_2. For the man, being a man and being an animal are necessary. For the koala, being an animal is necessary, and being a man is impossible. This model illustrates the fact that it is sometimes necessary, when constructing a model, to assign values to term variables that are not in the formula being modelled. A model must specify all the star sets that there are; and some of these may not correspond to any term variable in the formula.

Barbara XLL
Darii XLL

22.5		*
f_1	d_1	–
f_2	d_1	d_1
f_3	d_1	–

$$f_2 \subseteq f_1 \text{ and } \begin{cases} (i) f_3 \subseteq f_2^* \\ (ii) f_3 \cap f_2^* \neq \varnothing \end{cases}$$

$$\text{but } \begin{cases} (i) f_3 \not\subset f_1^* \\ (ii) f_3 \cap f_1^* = \varnothing \text{ and } f_1 \cap f_3^* = \varnothing. \end{cases}$$

(This is Brenner's model.[9]) Aristotle's terms 'moving' 'animal' 'white' (A9, 30b2-6), used by him to demonstrate the invalidity of Darii and Ferio XLL,[10] fit the model. The model supposes that there is a moving white animal for which being an animal is necessary while being in motion and being white are not necessary.

Celarent XLL
Felapton XLL
Bocardo XLL

22.6		*	¤
f_1	d_1, d_2	–	–
f_2	d_3	d_3	d_1, d_2
f_3	d_3	d_3	d_1, d_2
f_4	d_1	d_1	d_2, d_3
f_5	d_2	d_2	d_1, d_3

$$f_3 \subseteq f_2^* \text{ and } \begin{cases} (i) f_2 \cap f_1 = \varnothing \\ (ii) f_1 \cap f_3 = \varnothing \\ (iii) f_3 \not\subset f_1 \end{cases} \text{ but } \begin{cases} (i) f_3 \not\subset f_1^{¤} \\ (ii, iii) f_2 \cap f_1^{¤} = \varnothing. \end{cases}$$

Aristotle uses the terms 'moving' 'animal' 'man' (A9, 30a23-25,28-33) to show the invalidity of Celarent XLL. These terms can be adapted to fit model 22.6. The model supposes the existence of a stationary man, who is taken to be necessarily a man and necessarily an animal but contingently moving. It also supposes the existence of two moving non-animals, say a rock and a star (f_4 and f_5). Each of these is taken to be necessarily of its own kind, incapable of being a man, and

[9] BRENNER p.8.

[10] The terms 'moving' 'animal' 'man', which he uses to demonstrate the invalidity of Barbara XLL and Celarent XLL, do not fit this model because $d_1 \in f_3^*$ so the model does not invalidate Darii XLL.

contingently moving. The man is taken to be incapable of being either a rock or a star.

Baroco LXL

22.7		*	¤
f_1	d_1	d_1	–
f_2	d_1	d_1	–
f_3	d_2	–	–

$f_2 \subseteq f_1^{*}$ and $f_3 \not\subset f_1$ but $f_3 \cap f_1^{¤} = \varnothing$.

Aristotle's terms 'animal' 'man' 'white' (A10, 31a10-15) can be adapted to fit model 22.7, given appropriate metaphysical assumptions. The model supposes that a man exists, who is necessarily both a man and an animal but not white, and a white thing which is not necessarily anything and so which is not necessarily or impossibly either a man or an animal. The required metaphysical assumption is that there are beings (specifically white non-animals) *which are not necessarily anything*. That such things can be countenanced in an Aristotelian metaphysic is suggested by two types of case. (1) Maybe aggregates (including artefacts) are such, not themselves necessarily anything, though composed of parts each of which is necessarily something: in this case, the required white thing could be a white-man-on-a-white-horse. (2) Or maybe beings such as seeds, which given favourable conditions will naturally develop into members of an Aristotelian species, should be counted as not themselves necessarily anything, yet: the required white thing could be such.

Model 22.7 can be extended in the following way to make a second model falsifying Baroco LXL:

22.8		*	¤
f_1	d_1	d_1	–
f_2	d_1	d_1	–
f_3	d_2	–	–
f_4	d_1, d_2	d_1, d_2	–

Aristotle's terms can be adapted to this model too. The model supposes that a man exists who is necessarily both a man and an animal but is not white. It also supposes that a white thing exists which is not necessarily white. Finally, the model supposes that the man and the white thing both belong necessarily to a kind f_4. On this model, our white non-animal is necessarily something. It necessarily belongs to the genus f_4 (say bodies) of which animals are a species. The peculiar feature of the model is that our white non-animal does not belong necessarily to any species of that genus even though other beings do belong necessarily to such a species. So the required metaphysical assumption is that there are beings (specifically white non-animals) which belong necessarily to a genus *without belonging necessarily to any species of that genus.* While this is a formal possibility, it is not one that Aristotle would accept. He explicitly denies this possibility in the *Topics* (Δ2, 122a26-27).

§22.5 Corollaries

Our semantics also makes Baroco XLL, Bocardo LXL and Hintikka 1 valid.

Baroco XLL

If $f_2 \subseteq f_1$ and $f_3 \cap f_1^{¤} \neq \varnothing$ then $f_3 \cap f_2^{¤} \neq \varnothing$
$\because$ if $f_2 \subseteq f_1$ then $f_1^{¤} \subseteq f_2^{¤}$ - by **22.2.1**.

Bocardo LXL

If $f_3 \cap f_1^{¤} \neq \varnothing$ and $f_3 \subseteq f_2$ then $f_2 \cap f_1^{¤} \neq \varnothing$.

Hintikka 1 $\dfrac{ab^a \quad Lbc^a \quad Lbc^a}{Lab^i}$

If $f_2 \subseteq f_1$ and $f_3 \subseteq f_2^{*}$ then $f_1 \cap f_2^{*} \neq \varnothing$
$\because f_2^{*} \subseteq f_2$.

Not valid however are

Hintikka 2 $\dfrac{ab^e \quad Lbc^a \quad Lbc^a}{Lab^o}$

22.9		*	¤
f_1	d_1	–	–
f_2	d_2	d_2	–
f_3	d_2	–	–

$f_2 \cap f_1 = \varnothing$ and $f_3 \subseteq f_2^{*}$ but $f_2 \cap f_1^{¤} = \varnothing$.

Hintikka 3 $\dfrac{Lab^a \quad ac^e \quad Lab^a}{Lca^o}$

22.10		*	¤
f_1	d_1	d_1	–
f_2	d_1	d_1	–
f_3	d_2	–	–

$f_2 \subseteq f_1^{*}$ and $f_3 \cap f_1 = \varnothing$ but $f_1 \cap f_3^{¤} = \varnothing$.

Also not valid are

McCall $\dfrac{}{Laa^i}$ [11]

22.11		*
f_1	d_1	–

$f_1 \cap f_1^* = \varnothing.$

Thomason $\dfrac{ab^i}{Lab^i}$ [12]

22.12		*
f_1	d_1	–
f_2	d_1	–

$f_2 \cap f_1 \neq \varnothing$ but $f_2 \cap f_1^* = \varnothing$ and $f_1 \cap f_2^* = \varnothing$.

Truth-conditions for particular apodeictics reduce to those for universals. Relative to any AL model $\langle D, V, F, F^* \rangle$: Lab^i is true iff for some d,e in F: Lad^a, bd^a are true or Lbe^a, ae^a are true. Lab^o is true iff for some d,e,f in F: Lda^a, Lef^a, de^e, bf^a are true, *i.e.* iff for some f in F: Laf^e, bf^a are true. Thus the ecthetic rule for Li-propositions will be

Li -ecthesis A

$$\left.\begin{array}{c} \dfrac{R \quad Lad^a \quad bd^a}{p} \\[3em] \dfrac{R \quad Lbe^a \quad ae^a}{p} \end{array}\right\} \longrightarrow \dfrac{R \quad Lab^i}{p} \text{ where } d,e \text{ are not in } R, p.$$

[11] See McCALL 1963 §20. McCall's approach will be discussed in §25.1.

[12] See THOMASON p.112. Thomason's ideas will be discussed in §25.1.

§23. ADDING *C*, *K*, *U* AND *T*

If we add rules *C* and *K* to the apodeictic syllogistic, some simplifications can be made. At the assertoric level we can drop the axiom i-conv, which follows from e-conv by *C*:

$$e-conv\,\frac{ba^e}{ab^e}\ \xrightarrow{C}\ \frac{ab^i}{ba^i}$$

Similarly we can drop a-conv which follows by *K* from e-conv:

$$e-conv\,\frac{ba^e}{ab^e}\ \xrightarrow[e\,\|\,a]{K}\ \frac{ab^a}{ba^i}$$

We can also drop the axioms Darii and Ferio because they are derivable from Celarent and e-conv:

$$e-conv\,\dfrac{\text{Celarent}\,\dfrac{e-conv\,\dfrac{ac^e}{ca^e}\quad ab^a}{cb^e}}{bc^e}\qquad \xrightarrow{C}\ \frac{ab^a\quad bc^i}{ac^i}$$

$$\text{Celarent}\,\dfrac{e-conv\,\dfrac{ab^e}{ba^e}\quad ac^a}{bc^e}\qquad \xrightarrow{C}\ \frac{ab^e\quad bc^i}{ac^o}$$

a-rep follows from Ferio and a-conv by o-ecthesis and *C*:

$$\text{Ferio}\,\dfrac{ad^e\quad a-conv\,\dfrac{bd^a}{db^i}}{ab^o}\qquad \xrightarrow{o-ecthesis}\ \frac{ab^o}{ab^o}\ \xrightarrow{C}\ \frac{ab^a}{ab^a}$$

Finally, Barbara is provable:

$$\text{Felapton}\,\dfrac{e-conv\,\dfrac{\text{Celarent}\,\dfrac{ab^a\quad e-conv\,\dfrac{ad^e}{da^e}}{db^e}}{bd^e}\qquad cd^a}{bc^o}\qquad \xrightarrow{o-ecthesis}\ \frac{ab^a\quad ac^o}{bc^o}\ \xrightarrow{C}\ \frac{ab^a\quad bc^a}{ac^a}$$

Felapton, used here, is provable from Ferio and a-conv:

$$\textit{Ferio}\ \frac{ac^e \quad \dfrac{bc^a}{cb^i}\ a-conv}{ab^o}$$

This leaves us with two axioms (e-conversion, Celarent) and six rules (U^*, *M, C, K,* T_1, *o-ecthesis*).

The introduction of the rules of Indirect Reduction requires that we specify contradictories and contraries for L-forms. Now, we know that Aristotle believed the contradictory pairs to be $La \mid Mo$, $Le \mid Mi$, $Li \mid Me, Lo \mid Ma$. It follows from this (given L-subordination, L-subalternation) that Aristotle must treat the pairs $La \parallel Le$, $La \parallel e$, $La \parallel Me$, $La \parallel Lo$, $La \parallel o$, $Le \parallel a$, $Le \parallel Ma$, $Le \parallel Li$, $Le \parallel i$, $Li \parallel e$, $Lo \parallel a$ as contraries. So, let us introduce the M-operator into our language, stipulating that contradictories and contraries are as above. In this language, we want Me- and Mi-forms to be convertible and Mo-conversion to be rejected.

Further streamlining of our system is achieved by generalising our rule of Substitution U^* to U (which permits proper as well as improper substitutions), by allowing formulae to have zero premisses, and correspondingly by generalising our rule of Cut T_1 to T:[13]

$$\left.\begin{array}{c}\dfrac{R}{q}\\[2em] \dfrac{q \quad S}{r}\end{array}\right\} \longrightarrow \frac{R \quad S}{r}$$

where R may be null.

Theses now include

Ma-conv $\quad Le-conv\,\dfrac{Lba^e}{Lab^e} \xrightarrow[Le \parallel Ma]{K} \dfrac{Mab^a}{Mba^i}$ $\qquad$ *Me-conv* $\quad Li-conv\,\dfrac{Lba^i}{Lab^i} \xrightarrow{C} \dfrac{Mab^e}{Mba^e}$

Mi-conv $\quad Le-conv\,\dfrac{Lba^e}{Lab^e} \xrightarrow{C} \dfrac{Mab^i}{Mba^i}$

[13] These are all features of system *A* in THOM 1981 Chapter Six.

Ma-rep $$\text{Lo-rep}\,\frac{Lab^o}{Lab^o} \xrightarrow{C} \frac{Mab^a}{Mab^a}$$

Me-rep $$\text{Me-conv}\,\dfrac{\text{Me-conv}\,\dfrac{Mab^e}{Mba^e}}{Mab^e}$$

Mi-rep $$\text{Mi-conv}\,\dfrac{\text{Mi-conv}\,\dfrac{Mab^i}{Mba^i}}{Mab^i}$$

Mo-rep $$\text{La-rep}\,\frac{Lab^a}{Lab^a} \xrightarrow{C} \frac{Mab^o}{Mab^o}$$

X/M-subord $$\text{Lo-subord}\,\frac{Lab^o}{ab^o} \xrightarrow{C} \frac{ab^a}{Mab^a}$$

X/M-subord $$\text{Li-subord}\,\frac{Lab^i}{ab^i} \xrightarrow{C} \frac{ab^e}{Mab^e}$$

X/M-subord $$\text{Le-subord}\,\frac{Lab^e}{ab^e} \xrightarrow{C} \frac{ab^i}{Mab^i}$$

X/M-subord $$\text{La-subord}\,\frac{Lab^a}{ab^a} \xrightarrow{C} \frac{ab^o}{Mab^o}$$

L-subalt $$\text{Me-rep}\,\frac{Mab^e}{Mab^e} \xrightarrow[Me\| La]{K} \frac{Lab^a}{Lab^i}$$

L-subalt $$\text{Ma-rep}\,\frac{Mab^a}{Mab^a} \xrightarrow[Ma\| Le]{K} \frac{Lab^e}{Lab^o}$$

M-subalt $$\text{Le-rep}\,\frac{Lab^e}{Lab^e} \xrightarrow[Le\, \|\, Ma]{K} \frac{Mab^a}{Mab^i}$$

M-subalt $$\text{La-rep}\,\frac{Lab^a}{Lab^a} \xrightarrow[La\, \|\, Me]{K} \frac{Mab^e}{Mab^o}$$

Barbara MXM $$\text{Bocardo LXL}\,\frac{Lac^o \quad bc^a}{Lab^o} \xrightarrow{C} \frac{Mab^a \quad bc^a}{Mac^a}$$

If we use rule Lo-ecthesis *A*, Lo-rep is provable:

$$\text{Felapton LXL}\,\frac{Lad^e \quad bd^a}{Lab^o} \xrightarrow{\text{Lo-ecthesis A}} \frac{Lab^o}{Lab^o}$$

La-subord is provable:

$$\frac{ab^o}{ab^o} \xrightarrow[o\, \|\, La]{K} \frac{Lab^a}{ab^a}$$

and the remaining L-subord inferences are provable analogously.

Semantically, we need to note that for each set there is a corresponding *dagger set* (the shaded area in Figure 23.1) namely the complement of the original set's sun set.

23.1 $f^{\dagger} =_{df.} \overline{f^{¤}}$

For all dagger sets, we have the Theorem

23.1.1 $f^{\dagger} \neq \varnothing$

Proof If a given set's dagger set were empty, its sun set would be the universal set. And, since the sun set excludes the given set, the given set would be empty, contrary to our stipulation (**21.2**) that it be nonempty.

Given dagger sets, we can explain the truth-conditions for M-forms as in **23.2-23.5**..

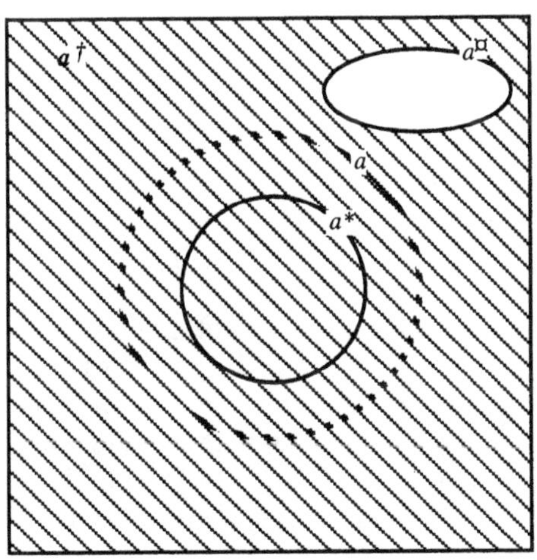

Fig. 23.1

23.2 Mab^a is true iff $f_2 \cap f_1^{¤} = \varnothing$, *i.e.* $f_2 \subseteq f_1^{\dagger}$, every b is a possible a.

23.3 Mab^e is true iff $f_2 \cap f_1^* = \varnothing \wedge f_1 \cap f_2^* = \varnothing$, *i.e.* $f_2 \subseteq \overline{f_1^*} \wedge f_1 \subseteq \overline{f_2^*}$, no b is a necessary a AND no a is a necessary b.

23.4 Mab^i is true iff $f_2 \not\subset f_1^{¤} \vee f_1 \not\subset f_2^{¤}$, *i.e.* $f_2 \cap f_1^{\dagger} \neq \varnothing \vee f_1 \cap f_2^{\dagger} \neq \varnothing$, EITHER some b is a possible a OR some a is a possible b.

23.5 Mab^o is true iff $f_2 \not\subset f_1^*$, *i.e.* $f_2 \cap \overline{f_1^*} \neq \varnothing$, some b is not a necessary a.

Given this, it turns out that the contradictory oppositions are as required. The validity of Me-conversion therefore follows from that of Li-conversion, and that of Mi-conversion from that of Le-conversion. Further, Mo-conversion is invalid.

23.1		*
f_1	d_1	–
f_2	d_1	d_1

$f_2 \not\subset f_1^*$ but $f_1 \subseteq f_2^*$.

Barbara MXM is valid:

Proof If $f_2 \subseteq f_1^{\dagger}$ and $f_3 \subseteq f_2$ then $f_3 \subseteq f_1^{\dagger}$

Celarent MXM, however, is not:

23.2		*
f_1	d_1	–
f_2	d_1	–
f_3	d_1	d_1

$f_2 \cap f_1^* = \varnothing$ and $f_1 \cap f_2^* = \varnothing$ and $f_3 \subseteq f_2$ but $f_1 \cap f_3^* \neq \varnothing$.

System *AL* takes for axioms e-conv, Celarent, La-rep, Li-conv, Le-conv, Barbara LXL, Darii LXL, Celarent LXL, Ferio LXL, plus rules *U, M, C, K, T,* o-ecthesis, Lo-ecthesis *A*. Alternative axiomatisations are possible, as shown in Figure 23.2. The L- and M-moods can be axiomatised in a variety of ways. Any axiomatisation of *AL* based on rules *U, M, C, K, T* needs to include, in addition to a basis for assertoric syllogistic, one axiom from each of the following groups:
(1) La-rep or Mo-rep;
(2) Lo-rep or Ma-rep;
(3) Li-conv or Me-conv;
(4) Le-conv or Mi-conv;
(5-8) La-, Le-, Li- and Lo-subordination, or their M-equivalents;
plus (9-14) one modal syllogism from each of the groups in Figure 23.2.

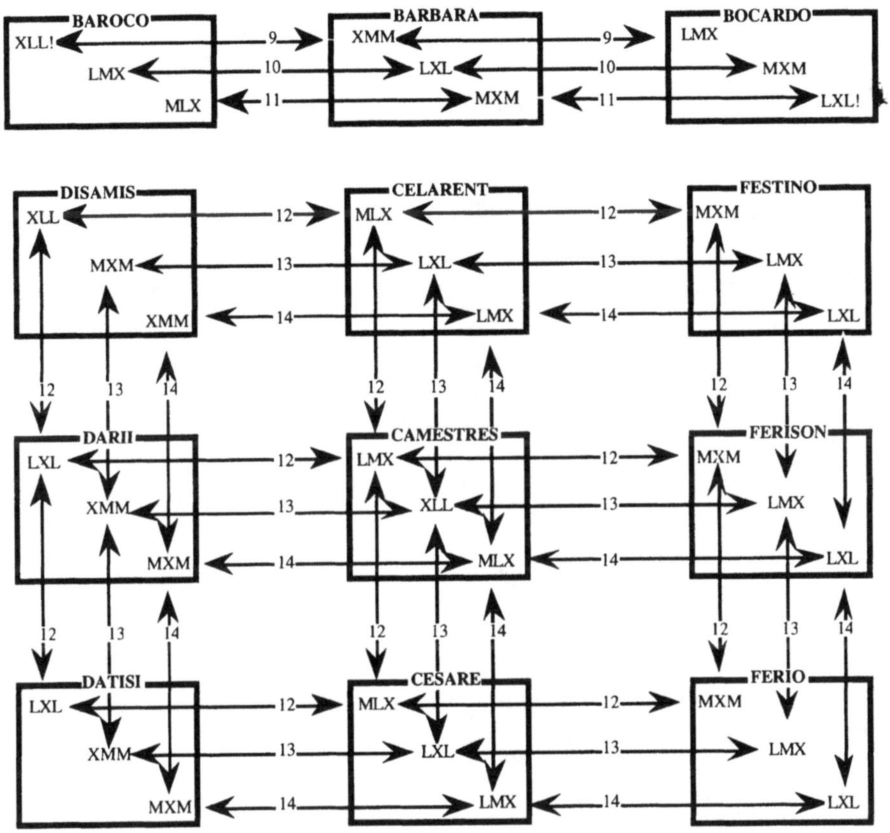

Fig. 23.2

By including rules of modal ecthesis - Li-ecthesis A (§22.5) and Lo-ecthesis A - in our base, we can reduce the number of axioms.

Given Ferio LXL (group 14) we have Felapton LXL by assertoric syllogistic; and given this, we have Lo-rep (group 2) by Lo-ecthesis A.

Lo-subord (8) is provable by Lo-ecthesis A, given Le-subord (6) and Felapton:

$$\text{Felapton}\ \frac{\text{Le-subord}\ \dfrac{Lad^e}{ad^e} \qquad bd^a}{ab^o} \xrightarrow{\text{Lo-ecthesis A}} \frac{Lab^o}{ab^o}$$

Li-subord (7) is provable by Li-ecthesis A, given La-subord (5) and Darapti:

$$\left.\begin{array}{c} \textit{Darapti}\,\dfrac{\textit{La-subord}\,\dfrac{Lad^a}{ad^a}\quad bd^a}{ab^i} \\[2ex] \textit{Darapti}\,\dfrac{ad^a \quad \textit{La-subord}\,\dfrac{Lbd^a}{bd^a}}{ab^i} \end{array}\right\} \xrightarrow{\textit{Li-ecthesis A}} \frac{Lab^i}{ab^i}$$

Given Barbara and Felapton LXL (courtesy of group 14) we can prove Bocardo LXL (group 11) by Lo-ecthesis *A*:

$$\textit{Felapton LXL}\,\frac{Lad^e \quad \textit{Barbara}\,\dfrac{bc^a \quad cd^a}{bd^a}}{Lab^o} \xrightarrow{\textit{Lo-ecthesis A}} \frac{Lac^o \quad bc^a}{Lab^o}$$

Given Camestres XLL (group 13) and Felapton LXL we can prove Baroco XLL (group 9) by Lo-ecthesis *A*:

$$\textit{Felapton LXL}\,\frac{\textit{Camestres XLL}\,\dfrac{ab^a \quad Lad^e}{Lbd^e} \quad cd^a}{Lbc^o} \xrightarrow{\textit{Lo-ecthesis A}} \frac{ab^a \quad Lac^o}{Lbc^o}$$

Thus given our two rules of L-ecthesis we can reduce our axiom-set from 14 to 9 - one each from groups (1), (3)-(6), (10), (12)-(14).

We could instead use rules of Mi- and Mo-ecthesis. We earlier suggested the following ecthetic rule for M:

$$\frac{Mac^a \quad Mbc^a}{p} \longrightarrow \frac{Mab^i}{p}$$

This assumes that Mab^i is true iff for some *c*: Mac^a, Mbc^a are both true; and this is correct.

Proof If either $f_2 \cap f_1^\dagger \neq \varnothing$ or $f_1 \cap f_2^\dagger \neq \varnothing$ then for some $f_4, f_4 \neq \varnothing$ and $f_4 \subseteq f_1^\dagger$ and $f_4 \subseteq f_2^\dagger$

$$\because \begin{cases} \text{if } f_2 \cap f_1^\dagger \neq \varnothing \text{ then } \left(f_2 \cap f_1^\dagger\right) \subseteq f_1^\dagger \text{ and } \left(f_2 \cap f_1^\dagger\right) \subseteq f_2^\dagger \\ \because f_2 \subseteq f_2^\dagger \\ \text{if } f_1 \cap f_2^\dagger \neq \varnothing \text{ then } \left(f_1 \cap f_2^\dagger\right) \subseteq f_2^\dagger \text{ and } \left(f_1 \cap f_2^\dagger\right) \subseteq f_1^\dagger \\ \because f_1 \subseteq f_1^\dagger. \end{cases}$$

However, this rule cannot be used (with our semantics) to prove Mi-conversion, as we earlier suggested:

$$\textit{Darapti MMM}\ \frac{Mad^a \quad Mbd^a}{Mba^i} \longrightarrow \frac{Mab^i}{Mba^i}$$

For Darapti MMM is invalid on our semantics:

23.3		*	¤	†
f_1	d_1	d_1	d_2	d_1, d_3
f_2	d_2	d_2	d_1	d_2, d_3
f_3	d_3	–	–	d_1, d_2, d_3

$f_3 \subseteq f_1{}^\dagger$ and $f_3 \subseteq f_2{}^\dagger$ but $f_1 \cap f_2{}^\dagger = \varnothing$ and $f_2 \cap f_1{}^\dagger = \varnothing$.

In fact, Darapti turns out to be invalid on all but one of the semantical interpretations to be investigated in Chapters Six and Seven, in the absence of additional semantic assumptions. (The exception is Buridan's semantics, discussed in §§26.3 and 32.3. Buridan accepts Darii MMM and partial Ma-conversion, from which Darapti MMM follows.)

§24. REDUCTION TO ASSERTORIC SYLLOGISTIC

The formulae of AL can be reduced to formulae of assertoric syllogistic, and on that basis their validity decided. To show this we construct an assertoric syllogistic that contains modalised terms. Corresponding to any variable v we introduce two terms v_L and v_M. The resulting language will contain, in addition to ordinary assertoric forms, assertoric forms with modalised terms, such as $a_M b_L{}^i$.

We define a system A_L whose axioms are e-conv, Celarent, and an axiom

$$\textit{Campsall}\quad \frac{ab^a}{a_M b_M{}^a}$$

which corresponds to the Sun set Theorem. This is stated in the form of a rule by Richard of Campsall (d. *c.* 1350-1360):

Everything contained under *b* is contained under *a*, so everything that can be contained under *b* can be contained under *a*.[15]

In addition we have the rules *U, C, K, T*, o-ecthesis, and the rules[16]

R_L $$\frac{R \quad a{a_L}^a}{p} \longrightarrow \frac{R}{p}$$

provided *R* contains some member $v{a_L}^a, v{a_L}^i, a_L v^a$ or $a_L v^i$

R_M $$\frac{R \quad a_M a^a}{p} \longrightarrow \frac{R}{p}$$

provided *R* contains some member va^a, va^i, av^a or av^i

The semantic interpretation of assertorics with modalised terms will be similar to that for *AL*'s assertorics, with the following differences. Terms may be assigned null sets; and affirmative forms (only) have existential import. (Without these changes we would be committed to assuming that for every term "*a*" there are necessary *as*.) Thus its semantics will comprise **21.1**, **21.7** and **21.8**, plus

24.1 *F* is a non-empty set of possibly empty sets whose members are members of *D*.

24.2 ab^a is true iff $f_2 \neq \varnothing \wedge f_2 \subseteq f_1$

24.3 ab^e is true iff $f_2 \cap f_1 = \varnothing$

24.4 ab^i is true iff $f_2 \cap f_1 \neq \varnothing$

24.5 ab^o is true iff $f_2 = \varnothing \vee f_2 \not\subset f_1$.

In addition we have the rules

24.6 If f is assigned to v then f^* is assigned to v_L

24.7 If f is assigned to v then $\overline{f^{¤}}$ is assigned to v_M.

A_L validity is defined analogously to *AL* validity.

A_L is sound with respect to this semantics.[17]

A correspondence can be set up between the forms in *AL* and those in A_L as follows. For every form *f* of *AL*, there is a form g - or a pair of forms g_1, g_2 - of

[15] CAMPSALL 14.04.

[16] The basic idea of introducing modalised terms and special principles relating them to unmodalised terms is in STAHL 1976 p.179, and before that in GEACH pp.202-203.

[17] A proof can be adapted from that in THOM 1981 §46.

A_L such that either (i) under any valuation f is true iff g is true, or (ii) under any valuation f is true iff both g_1 and g_2 are true, or (iii) under any valuation f is true iff either g_1 or g_2 is true. This is clear:

Lab^a is true in AL iff $a_L b^a$ is true in A_L

Lab^e is true in AL iff both $a_M b^e$ and $b_M a^e$ are true in A_L

Lab^i is true in AL iff either $a_L b^i$ or $b_L a^i$ is true in A_L

Lab^o is true in AL iff $a_M b^o$ is true in A_L

Analogous rules apply to M forms in AL.

It follows that for every wff x of AL under any valuation, x's truth-value is a truth-function of the truth-values of a set of A_L forms. Let x be $\frac{R}{p}$. x is then true iff either p is true or some member of R is false. But the truth-value of p and of each member of R is, as we have seen, a truth-function of the truth-values of a set of A_L forms. Since validity is truth in all models under all valuations, it follows that for every wff x of AL under any valuation, x's validity is a truth-function of the validity of a set of A_L forms. These truth-functions are:

$\frac{Lab^a \quad R}{Lcd^a}$ is AL valid iff $\frac{a_L b^a \quad R}{c_L d^a}$ is A_L valid

$\frac{Lab^a \quad R}{Lcd^e}$ is AL valid iff both $\frac{a_L b^a \quad R}{c_M d^e}$ and $\frac{a_L b^a \quad R}{d_M c^e}$ are A_L valid

$\frac{Lab^a \quad R}{Lcd^i}$ is AL valid iff either $\frac{a_L b^a \quad R}{c_L d^i}$ or $\frac{a_L b^a \quad R}{d_L c^i}$ is A_L valid

$\frac{Lab^a \quad R}{Lcd^o}$ is AL valid iff $\frac{a_L b^a \quad R}{c_M d^o}$ is A_L valid

$\frac{Lab^e \quad R}{Lcd^a}$ is AL valid iff either $\frac{a_M b^e \quad R}{c_L d^a}$ or $\frac{b_M a^e \quad R}{c_L d^a}$ is A_L valid

$\frac{Lab^e \quad R}{Lcd^e}$ is AL valid iff both $\left\{\text{either } \frac{a_M b^e \quad R}{c_M d^e} \text{ or } \frac{b_M a^e \quad R}{c_M d^e} \text{ is } A_L \text{valid}\right\}$ and $\left\{\text{either } \frac{a_M b^e \quad R}{d_M c^e} \text{ or } \frac{b_M a^e \quad R}{d_M c^e} \text{ is } A_L \text{valid}\right\}$

$\dfrac{Lab^e \quad R}{Lcd^i}$ is AL valid iff either $\dfrac{a_M b^e \quad R}{c_L d^i}$ or $\dfrac{a_M b^e \quad R}{d_L c^i}$ or $\dfrac{b_M a^e \quad R}{d_L c^i}$ or $\dfrac{b_M a^e \quad R}{c_L d^i}$ is A_L valid

$\dfrac{Lab^e \quad R}{Lcd^o}$ is AL valid iff either $\dfrac{a_M b^e \quad R}{c_M d^o}$ or $\dfrac{b_M a^e \quad R}{c_M d^o}$ is A_L valid

$\dfrac{Lab^i \quad R}{Lcd^a}$ is AL valid iff both $\dfrac{a_L b^i \quad R}{c_L d^a}$ and $\dfrac{b_L a^i \quad R}{c_L d^a}$ are A_L valid

$\dfrac{Lab^i \quad R}{Lcd^e}$ is AL valid iff $\dfrac{a_L b^i \quad R}{c_M d^e}$ and $\dfrac{a_L b^i \quad R}{d_M c^e}$ and $\dfrac{b_L a^i \quad R}{c_M d^e}$ and $\dfrac{b_L a^i \quad R}{d_M c^e}$ are A_L valid

$\dfrac{Lab^i \quad R}{Lcd^i}$ is AL valid iff either $\left\{\text{both } \dfrac{a_L b^i \quad R}{d_L c^i} \text{ and } \dfrac{b_L a^i \quad R}{d_L c^i} \text{ are } A_L \text{ valid}\right\}$ or $\left\{\text{both } \dfrac{a_L b^i \quad R}{c_L d^i} \text{ and } \dfrac{b_L a^i \quad R}{c_L d^i} \text{ are } A_L \text{ valid}\right\}$

$\dfrac{Lab^i \quad R}{Lcd^o}$ is AL valid iff both $\dfrac{a_L b^i \quad R}{c_M d^o}$ and $\dfrac{b_L a^i \quad R}{c_M d^o}$ are A_L valid

$\dfrac{Lab^o \quad R}{Lcd^a}$ is AL valid iff $\dfrac{a_M b^o \quad R}{c_L d^a}$ is A_L valid

$\dfrac{Lab^o \quad R}{Lcd^e}$ is AL valid iff both $\dfrac{a_M b^o \quad R}{c_M d^e}$ and $\dfrac{a_M b^o \quad R}{d_M c^e}$ are A_L valid

$\dfrac{Lab^o \quad R}{Lcd^i}$ is AL valid iff either $\dfrac{a_M b^o \quad R}{c_L d^i}$ or $\dfrac{a_M b^o \quad R}{d_L c^i}$ is A_L valid

$\dfrac{Lab^o \quad R}{Lcd^o}$ is AL valid iff $\dfrac{a_M b^o \quad R}{c_M d^o}$ is A_L valid

Similarly for M forms.

A decision procedure exists for A_L validity. To decide whether an A_L formula is valid take its *antilogism sequence*, *i.e.* the premiss-sequence followed by the conclusion's contradictory. See if this can be ordered so that it *interweaves*, *i.e.* each member shares one of its terms with its predecessor and one with its successor (the first member being defined to be the successor of the last). If it cannot, *extend* it in accordance with (some of) the rules

Extension rule L If a variable v in one member is matched by v_L in another, then add $vv_L{}^a$ to the sequence (provided there is some member $wv_L{}^a, wv_L{}^i, v_Lw^a$ or v_Lw^i).

Extension rule M If a variable v in one member is matched by v_M in another, then add v_Mv^a to the sequence (provided there is some member wv^a, wv^i, vw^a or vw^i).

See if the resulting sequence is an *Aristotelian chain, i.e.* (i) in each of the interweaving's sharings at least one member is distributed (distributed terms being the predicates of negative and the subjects of universal forms), (ii) there is exactly one negative member.[18] Thus we are able to decide the validity or invalidity of *AL* formulae. Here are two examples:

(1) Darii LLL

$$\frac{Lab^a \quad Lbc^i}{Lac^i}$$

is AL valid iff either $\left\{\text{both } 1\frac{a_Lb^a \quad b_Lc^i}{a_Lc^i} \text{ and } 2\frac{a_Lb^a \quad c_Lb^i}{a_Lc^i} \text{ are } A_L \text{ valid}\right\}$ or $\left\{\text{both } 3\frac{a_Lb^a \quad b_Lc^i}{c_La^i} \text{ and } 4\frac{a_Lb^a \quad c_Lb^i}{c_La^i} \text{ are } A_L \text{ valid}\right\}$. In fact, *1* and *2* are both A_L valid since both are provable in A_L:

$$\text{Darii}\frac{\text{Barbara}\dfrac{a_Lb^a \quad bb_L{}^a}{a_Lb_L{}^a} \qquad b_Lc^i}{a_Lc^i} \quad \xrightarrow{R_L} \quad \frac{a_Lb^a \quad b_Lc^i}{a_Lc^i}\,1$$

$$\text{Datisi}\frac{a_Lb^a \qquad \text{Darii}\dfrac{c_Lb^i \quad cc_L{}^a}{cb^i}}{a_Lc^i} \quad \xrightarrow{R_L} \quad \frac{a_Lb^a \quad c_Lb^i}{a_Lc^i}\,2$$

(2) Celarent XLL

$$\frac{ab^e \quad Lbc^a}{Lac^e}$$

[18] See SMILEY; THOM 1981 §48.

is *AL* valid iff both $1\frac{ab^e \quad b_L c^a}{a_M c^e}$ and $2\frac{ab^e \quad b_L c^a}{c_M a^e}$ are A_L valid. In fact, *1* is A_L invalid since, while its antilogism sequence $\langle ab^e, b_L c^a, a_M c^i \rangle$ is extendable to $\langle ab^e, bb_L{}^a, b_L c^a, a_M c^i, a_M a^a \rangle$, the latter is not an Aristotelian chain (a_M being twice undistributed).

§25. ALTERNATIVE *DE RE* APPROACHES: BASE

Within the type of *de re* approach to the formal interpretation of apodeictic syllogistic we are pursuing in this Chapter, considerable variation is possible. I distinguish two broad categories of variation, corresponding to two layers in a formal semantics for modal syllogistic. The first of these consists of existential assumptions and assumptions concerning the semantical analysis of necessity and impossibility (the base); the second consists of the semantical analysis of the individual modal forms (the superstructure). In this section we deal with the base, in the next with the superstructure. In §27 we deal with *de dicto* approaches.

§25.1 Existence assumptions

One group of base assumptions concerns the possibility of null star sets. We might assume that all star sets are non-null:

E1 $f^* \neq \varnothing$[19]

Or we might assume that any star set not identical with its original set is null:

E2 $f^* = f \vee f^* = \varnothing$[20]

Or we might assume that if two sets have a non-null overlap then one of them has a non-null overlap with the other's star set:

E3 $(f \cap g \neq \varnothing) \supset (f \cap g^* \neq \varnothing)$[21]

E3 implies E1:

---

19 JOHNSON 1989 p.274.

20 THOM 1991, 1.5.

21 THOMASON Definition 2.

Proof Since $(f \cap g \neq \varnothing) \supset (f \cap g^{*} \neq \varnothing)$, $(f \cap f \neq \varnothing) \supset (f \cap f^{*} \neq \varnothing)$,
i.e., $f \neq \varnothing \supset f^{*} \neq \varnothing$;
but $f \neq \varnothing$,
$\therefore f^{*} \neq \varnothing$.

With E3 we may perhaps group Johnson's
E3.1 $\left(g \subseteq h^{*} \wedge f \cap g \neq \varnothing\right) \supset \left(f^{*} \cap h^{*} \neq \varnothing\right)$ and
E3.2 $\left(g \subseteq h^{¤} \wedge f \cap g \neq \varnothing\right) \supset \left(f^{*} \cap h^{¤} \neq \varnothing\right)$.[22]
Next we mention Thomason's base assumption
E4 $\left(f \subseteq g^{¤}\right) \supset \left(g \subseteq f^{¤}\right)$[23]
Derivable from this (given that $g \subseteq g$) is Johnson's
E4.1 $\left(f \subseteq g^{¤} \wedge h \subseteq g\right) \supset \left(h \subseteq f^{¤}\right)$.[24]

§25.2 *Sun set assumptions*

A second group of base assumptions concern the composition of a given set's sun set. We consider six ways of defining the sun set. Five of these make the sun set a function of already specified star sets. These five definitions all make the notion of impossibility *mediated* via the notion of necessity; all but the first of them, as we shall see, also make the notion of impossibility *heterogeneous*, in that the sun set is defined as a *union* of star sets.

The minimal approach is to specify merely that the sun set is a subset of the given set's complement:
S0 $f^{¤} \subseteq \bar{f}$.

A stronger approach is to define the sun set as the star set of the given set's complement:
S1 $f^{¤} = \bar{f}^{*}$.
We will also refer to this as **25.1.** (The impossibly a is that which is necessarily not a.) This was Johnson's original approach.[25] S1 can be broken down into

[22] JOHNSON 1989 (d,e) p.274.
[23] THOMASON Definition 3.
[24] JOHNSON 1989 (c) p.274.
[25] JOHNSON 1989 p.274.

S1.1 $\overline{f}^{*} \subseteq f^{\text{¤}}$

and

S1.2 $f^{\text{¤}} \subseteq \overline{f}^{*}$.

S1.2 implies S0.

Next, the sun set could be defined as the union of all star sets the star sets of the complements of whose originals include the given set.

25.2 $f^{\text{¤}} = \bigcup_{h^{*}} f \subseteq \overline{h}^{*}$.

(The impossibly *a* is that which is necessarily such that the *a*s are necessarily not thus.) **25.2** implies

S2 $\left(f \subseteq g^{*}\right) \supset \left(g^{\text{¤}} \subseteq f^{\text{¤}}\right)$.[26]

Proof Suppose $f \subseteq g^{*}$

then $f \subseteq g$

$\therefore \left(\bigcup_{h^{*}} g \subseteq \overline{h}^{*}\right) \subseteq \left(\bigcup_{h^{*}} f \subseteq \overline{h}^{*}\right)$

$\therefore g^{\text{¤}} \subseteq f^{\text{¤}}$ – by **25.2**.

25.2 also implies the stronger

S3 $\left(f \subseteq g\right) \supset \left(g^{\text{¤}} \subseteq f^{\text{¤}}\right)$ (the Sun set Theorem).

The above proof also shows this. Derivable from S2 is

S2.1 $\left(f \subseteq g^{*} \wedge h \subseteq g^{\text{¤}}\right) \supset \left(h \subseteq f^{\text{¤}}\right)$.[27]

It is worth noting that **25.1** does not imply S3:

---

[26] THOMASON Definition 3. Thomason points out that the stronger

$$\left(f \subseteq g^{*}\right) \supset \left(\overline{g} \subseteq f^{\text{¤}}\right)$$

has the undesirable feature of validating Celarent XLL, given E4. Also worth noting is that

$$\left(f \subseteq \overline{g}\right) \supset \left(g \subseteq f^{\text{¤}}\right)$$

(another strengthening of E4) would validate the transformation of Xe- into Le-forms.

[27] JOHNSON 1989 (c) p.274.

25.1		*	¤
f	d_1	–	–
g	d_1, d_2	–	d_3
$\bar{f}$	d_2, d_3	–	–
$\bar{g}$	d_3	d_3	–

$(f \subseteq g) \wedge (g^{¤} \not\subset f^{¤})$

Neither does S2 imply S3 since model 25.1 vacuously satisfies S2. **25.2** has the peculiarity of not implying S1.1, as matrix 25.2 shows.

25.2		*	¤
f	d_1	–	–
$\bar{f}$	d_2	d_2	–

$\bar{f}^* \not\subset f^{¤}$.

25.2 also does not imply S1.2 or E4, as is shown by matrix 25.3.

25.3		*	¤
f	d_1	–	d_2, d_3
g	d_2	d_2	d_1
h	d_3	d_3	d_1
$\bar{f}$	d_2, d_3	–	–
$\bar{g}$	d_1, d_3	d_1	–
$\bar{h}$	d_1, d_2	d_1	–

$(i) f^{¤} \not\subset \bar{f}^*$

$(ii) \bar{f} \subseteq f^{¤}$ but $f \not\subset \bar{f}^{¤}$.

However, **25.2** does imply S0:

Proof If $f \subseteq \bar{h}^*$ then $\overline{\bar{h}^*} \subseteq \bar{f}$ so $h^* \subseteq \bar{f}$ – ∵ $h^* \subseteq h \subseteq \overline{\bar{h}^*}$

$$\therefore \left(\bigcup_{h^*} f \subseteq \bar{h}^* \right) \subseteq \bar{f}.$$

Derivable from S3 is

S3.1 $\left(f \subseteq g \wedge h \subseteq g^{¤}\right) \supset \left(h \subseteq f^{¤}\right)$.[28]

S3, as well as being implied by **25.2**, is implied by a definition of the sun set as the union of all star sets whose original sets exclude the given set.

25.3 $f^{¤} = \bigcup_{h^*} f \subseteq \bar{h}$

(The impossibly *a* is what is necessarily such that the *a*s are not thus.)

Proof Suppose $f \subseteq g$

then $\left(\bigcup_{h^*} g \subseteq \bar{h}\right) \subseteq \left(\bigcup_{h^*} f \subseteq \bar{h}\right)$

$\therefore g^{¤} \subseteq f^{¤}$ – by **25.3**.

S3 implies S2:

Proof If $(f \subseteq g) \supset \left(g^{¤} \subseteq f^{¤}\right)$

then $\left(f \subseteq g^*\right) \supset \left(g^{¤} \subseteq f^{¤}\right)$ - $\because \left(f \subseteq g^*\right) \supset (f \subseteq g)$.

25.3 implies S1.1:

Proof $\bar{f} \subseteq \bar{f}$

$\therefore \bar{f}^* \subseteq \bigcup_{h^*} h \subseteq \bar{f}$.

It does not, however, imply S1.2 or E4, as the following model shows.

25.4		*	¤
f	d_1	–	d_2
$\bar{f}$	d_2, d_3	–	–
g	d_2	d_2	–
$\bar{g}$	d_1, d_3	–	–

$(i) f^{¤} \not\subset \bar{f}^*$

$(ii) g \subseteq f^{¤}$ but $f \not\subset g^{¤}$.

28 JOHNSON 1989 (c) p.274.

But it does imply S0:

Proof If $f \subseteq \bar{h}$ then $h \subseteq \bar{f}$ so $h^* \subseteq \bar{f}$

$$\therefore \left(\bigcup_{h^*} f \subseteq \bar{h} \right) \subseteq \bar{f}.$$

Stronger than S3 is the base assumption

S4 $\left(f \subseteq g^{\dagger}\right) \supset \left(g^{¤} \subseteq f^{¤}\right)$:

Derivable from this is

S4.1 $\left(f \subseteq g^{\dagger} \wedge h \subseteq g^{¤}\right) \supset \left(h \subseteq f^{¤}\right)$.[29]

S4 is implied by a definition of the sun set as the union of all star sets outside the given set.

25.4 $f^{¤} = \bigcup_{h^*} f \subseteq \overline{h^*}$

(The impossibly *a* is what is necessarily such that only non-*a*s are necessarily thus.)

Proof Suppose $f \subseteq g^{\dagger}$
then $g^{¤} \subseteq \bar{f}$
so - by **25.4** - every star set outside g is a star set outside f,
so the union of all star sets outside g is included in the union of all star sets outside f
$\therefore g^{¤} \subseteq f^{¤}$ – by **25.4**.

S4 implies S3:

Proof If $\left(f \subseteq g^{\dagger}\right) \supset \left(g^{¤} \subseteq f^{¤}\right)$ then $(f \subseteq g) \supset \left(g^{¤} \subseteq f^{¤}\right)$
$\because (f \subseteq g) \supset \left(f \subseteq g^{\dagger}\right)$.

25.4 implies S1.1:

---

[29] JOHNSON 1989 (c) p.274.

Proof $\bar{f}^* \subseteq \bar{f}$

$$\therefore \bar{f}^* \subseteq \left(\bigcup_{h^*} \left(h^* \subseteq \bar{f} \right) \right).$$

It does not imply S1.2 or E4, as matrix 25.5 shows.

25.5		*	¤
f	d_1	–	d_2
$\bar{f}$	d_2, d_3	–	–
g	d_2	d_2	–

$(i) f^{¤} \not\subset \bar{f}^*$

$(ii) g \subseteq f^{¤}$ but $f \not\subset g^{¤}$.

Finally, we might adopt the base assumption

S5 $\left(f \subseteq h^* \wedge g \subseteq \bar{h} \right) \supset \left(g^* \subseteq f^{¤} \right).$

S5 is implied by a definition of the sun set as the union of all star sets whose originals exclude the original of any superstar set of the given set.

25.5 $f^{¤} = \bigcup_{g^*} \exists h^* \left[f \subseteq h^* \wedge g \subseteq \bar{h} \right]$

(The impossibly *a* is what is necessarily such that the *a*s are necessarily otherwise.)

Proof Suppose $\exists h^* \left[f \subseteq h^* \wedge g \subseteq \bar{h} \right]$

$$\therefore g^* \subseteq \bigcup_{j^*} \exists h^* \left[f \subseteq h^* \wedge j \subseteq \bar{h} \right]$$

$\therefore g^* \subseteq f^{¤}$ – by **25.5**.

25.5 implies S3 (as proved in §22.2). But S5 does not imply S3:

25.6		*	¤
f_1	d_1	–	–
f_2	d_1, d_2	–	d_3
f_3	d_3	–	–

$f_3 \subseteq f_2^{¤}$ but $f_2 \not\subset f_3^{¤}$.

(S5 is trivially true since there is no h^* for which $f_i \subseteq h^*$ $(1 \le i \le 3)$.

And neither S5 nor **25.5** implies S1.1 or S1.2 or E4:

25.7		*	¤
f_1	d_1	–	d_4
f_2	d_1, d_2	d_1, d_2	d_4
$\overline{f_1}$	d_2, d_3, d_4	d_2, d_3	–
f_3	d_4	d_4	d_1, d_2, d_3
f_4	d_1, d_2, d_3	–	–

$(i)\overline{f_1}^* \not\subset f_1^{¤}$

$(ii) f_1^{¤} \not\subset \overline{f_1}^*$

$(iii) f_4 \subseteq f_3^{¤}$ but $f_3 \not\subset f_4^{¤}$.

Nor do they imply S4:

25.8		*	¤	†
f_1	d_1	–	–	d_1, d_2, d_3
f_2	d_2	d_2	d_3	d_1, d_2
f_3	d_3	d_3	d_2	d_1, d_3

$f_1 \subseteq f_2^{\dagger}$ but $f_2^{¤} \not\subset f_1^{¤}$.

But **25.5** does imply S0:

Proof If $f \subseteq h^* \wedge g \subseteq \bar{h}$ then $g^* \subseteq \bar{f}$

$- \because g^* \subseteq g$ and $\bar{h} \subseteq \overline{h^*}$

$$\therefore \left(\bigcup_{g^*} \exists h^* \left[f \subseteq h^* \wedge g \subseteq \bar{h} \right] \right) \subseteq \bar{f}.$$

The relations among our various S-assumptions and definitions are shown in Figure 25.1.

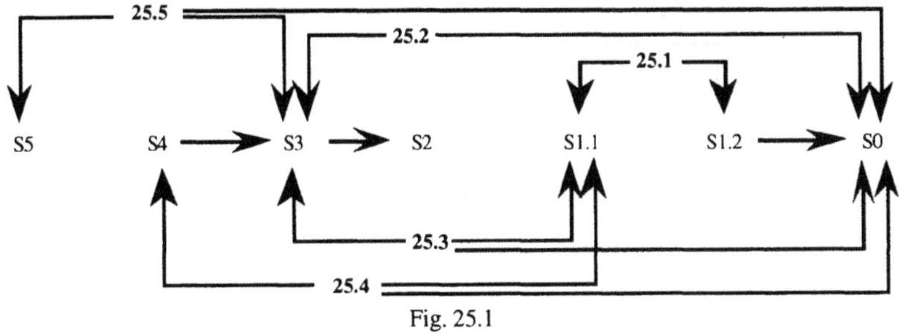

Fig. 25.1

§26. ALTERNATIVE *DE RE* APPROACHES: SUPERSTRUCTURE

We shall consider four different superstructures in combination with various base assumptions. The first two of these superstructures use the semantics for assertoric propositions expounded in **21.1-8**. The remaining two use a semantics consisting of **21.1**, **7** and **8** plus **24.1-5**. This semantics, which allows for empty terms and endows only affirmatives with existential import, was common in medieval times.[30]

§26.1 Superstructure A

We begin with the superstructure used in our system *AL*. The *A* superstructure neither ampliates nor restricts the subjects of apodeictics; but it gives a conjunctive analysis of Le-forms and a disjunctive analysis of Li-forms. The particular version of *A* adopted by us earlier accepts the strong sun set definition **25.5**.

What happens if we drop definition **25.5** in favour of S0? The resulting semantics - *A0* - validates all the axioms of system *AL* with the exception of Celarent LXL. (The proofs of their validity in §22 require nothing stronger than S0; but the validity of Celarent LXL requires the Sun set Theorem.) In fact, in the absence of S3 Celarent LXL is invalid:

26.1		*	¤
f_1	d_1	–	d_2, d_3
f_2	d_2, d_3	–	d_1
f_3	d_2	–	–

$f_2 \subseteq f_1^{¤}$ and $f_1 \subseteq f_2^{¤}$ and $f_3 \subseteq f_2$ but $f_1 \not\subseteq f_3^{¤}$.

Under the same circumstances, Camestres XLL is invalid. The ecthetic proofs of Baroco LLL and Baroco XLL rely on Camestres; and these two Barocos are also invalid:

26.2		*	¤
f_1	d_1	d_1	d_2
f_2	d_1	–	–
f_3	d_2	–	–

$f_2 \subseteq f_1^{*}$ and $f_3 \cap f_1^{¤} \neq \emptyset$ but $f_3 \cap f_2^{¤} = \emptyset$.

---

30 See FREDDOSO p.18.

If we assume S2 then Baroco LLL becomes valid:

Proof If $f_2 \subseteq f_1^{*}$ and $f_3 \cap f_1^{\text{¤}} \neq \varnothing$ then $f_3 \cap f_2^{\text{¤}} \neq \varnothing$
$\because f_1^{\text{¤}} \subseteq f_2^{\text{¤}}$ - by S2.

If we assume S3 - the Sun set Theorem - then Celarent LXL and Baroco XLL are valid, and so are all the axioms of *AL*. See Figure 26.1.

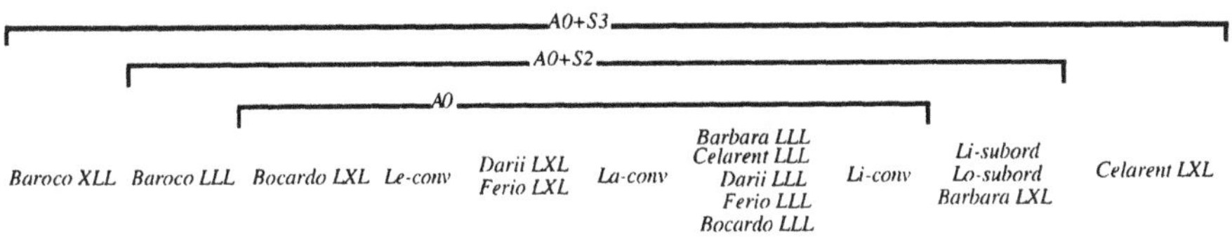

Fig. 26.1

§26.2 *Superstructure* AA

Thomason describes a family of systems based on a semantic superstructure devised by Johnson, according to which Li- and Lo-forms are (to use the medieval term) *restricted* to necessary subjects:[31]

26.2.1 Lab^{a} is true iff $f_2 \subseteq f_1^{*}$

26.2.2 Lab^{e} is true iff $f_2 \subseteq f_1^{\text{¤}}$

26.2.3 Lab^{i} is true iff $f_2^{*} \cap f_1^{*} \neq \varnothing$

26.2.4 Lab^{o} is true iff $f_2^{*} \cap f_1^{\text{¤}} \neq \varnothing$

Consider the effect of combining this superstructure with various base assumptions about the sun set, and various existential assumptions. The combination of S0 with superstructure *AA* - call it *AA0* - validates

L-subord If $f_2 \subseteq f_1^{*}$ then $f_2 \subseteq f_1$
$\because f_1^{*} \subseteq f_1$.

[31] JOHNSON 1989 p.274. THOMASON p.116. GEACH p.202 proposes a superstructure identical with this, except that Lo-forms are read as in superstructure *A*.

If $f_2 \subseteq f_1^{¤}$ then $f_2 \subseteq \overline{f_1}$
$\because f_1^{¤} \subseteq \overline{f_1}$.

If $f_2^* \cap f_1^* \neq \varnothing$ then $f_2 \cap f_1 \neq \varnothing$
$\because f_2^* \subseteq f_2, f_1^* \subseteq f_1$.

If $f_2^* \cap f_1^{¤} \neq \varnothing$ then $f_2 \not\subset f_1$
$\because f_2^* \subseteq f_2, f_1^{¤} \subseteq \overline{f_1}$.

Li-conv If $f_2^* \cap f_1^* \neq \varnothing$ then $f_1^* \cap f_2^* \neq \varnothing$.

Barbara LXL If $f_2 \subseteq f_1^*$ and $f_3 \subseteq f_2$ then $f_3 \subseteq f_1^*$.

Celarent LXL If $f_2 \subseteq f_1^{¤}$ and $f_3 \subseteq f_2$ then $f_3 \subseteq f_1^{¤}$.

Darii LLL If $f_2 \subseteq f_1^*$ and $f_3^* \cap f_2^* \neq \varnothing$ then $f_3^* \cap f_1^* \neq \varnothing$
$\because f_2^* \subseteq f_2$.

Ferio LLL If $f_2 \subseteq f_1^{¤}$ and $f_3^* \cap f_2^* \neq \varnothing$ then $f_3^* \cap f_1^{¤} \neq \varnothing$
$\because f_2^* \subseteq f_2$.

Bocardo LLL If $f_3^* \cap f_1^{¤} \neq \varnothing$ and $f_3 \subseteq f_2^*$ then $f_2^* \cap f_1^{¤} \neq \varnothing$
$\because f_3^* \subseteq f_3$.

The validity of Barbara and Celarent LLL follows from that of Barbara and Celarent LXL and L-subord.

The following are invalid:

Le-conv

26.3		*	¤
f_1	d_1	–	d_2
f_2	d_2	d_2	–

$f_2 \subseteq f_1^{\text{¤}}$ but $f_1 \not\subset f_2^{\text{¤}}$.

La-conv (partial)

26.4		*
f_1	d_1	d_1
f_2	d_1	–

$f_2 \subseteq f_1^*$ but $f_1^* \cap f_2^* = \varnothing$.

Darii LXL

26.5		*
f_1	d_1	d_1
f_2	d_1	–
f_3	d_1, d_2	–

$f_2 \subseteq f_1^*$ and $f_3 \cap f_2 \neq \varnothing$ but $f_3^* \cap f_1^* = \varnothing$.

Ferio LXL

26.6		*	¤
f_1	d_1	–	d_2
f_2	d_2	d_2	–
f_3	d_1, d_2	–	–

$f_2 \subseteq f_1^{\text{¤}}$ and $f_3 \cap f_2 \neq \varnothing$ but $f_3^* \cap f_1^{\text{¤}} = \varnothing$.

Thus, of *AL*'s axioms, *AA0* validates L-subordination, Li-conversion, Barbara LXL and Celarent LXL, Darii LLL, Ferio LLL and Bocardo LLL.

The strengthening of *AA0* can lead to the validation of various other formulae. The addition of E1 to *AA0* validates La-conversion:

Proof If $f_2 \subseteq f_1^*$ then $f_1^* \cap f_2^* \neq \varnothing$
$\because f_2^* \neq \varnothing$ - by E1 - and $f_2^* \subseteq f_2$.

The addition of E3 to *AA0* semantics results in the validation of Darii LXL and Ferio LXL:

Proofs If $f_2 \subseteq f_1^*$ and $f_3 \cap f_2 \neq \varnothing$ then $f_3^* \cap f_1^* \neq \varnothing$
$\because f_3^* \cap f_2 \neq \varnothing$ - by E3.

If $f_2 \subseteq f_1^¤$ and $f_3 \cap f_2 \neq \varnothing$ then $f_3^* \cap f_1^¤ \neq \varnothing$
$\because f_3^* \cap f_2 \neq \varnothing$ - by E3.

Obviously, the addition of E4 will validate Le-conversion.
AA0 by itself makes Baroco LLL invalid.

26.7		*	¤
f_1	d_1, d_2	d_1	d_3
f_2	d_1	–	–
f_3	d_3	d_3	–
$\overline{f_1}$	d_3	d_3	d_1
$\overline{f_2}$	d_2, d_3	–	–
$\overline{f_3}$	d_1, d_2	–	d_3

$f_2 \subseteq f_1^*$ and $f_3^* \cap f_1^¤ \neq \varnothing$ but $f_3^* \cap f_2^¤ = \varnothing$.

But the addition of S2 validates Baroco LLL:

Proof If $f_2 \subseteq f_1^*$ and $f_3^* \cap f_1^¤ \neq \varnothing$ then $f_3^* \cap f_2^¤ \neq \varnothing$
$\because f_1^¤ \subseteq f_2^¤$ - by S2.

The addition of S3 to *AA0* validates Baroco XLL:

Proof If $f_2 \subseteq f_1$ and $f_3^* \cap f_1^¤ \neq \varnothing$ then $f_3^* \cap f_2^¤ \neq \varnothing$
$\because f_1^¤ \subseteq f_2^¤$ - by S3.

But Bocardo LXL is not validated by *AA0* plus S3. In fact, Bocardo LXL remains invalid even relative to the superstructures we get by adopting S4 or S5. These results are shown by matrices 26.8, 26.9 and 26.10.

26.8		*	¤
f_1	d_2	d_2	d_1
$\overline{f_1}$	d_1	d_1	d_2
f_2	d_1, d_2	d_2	–
$\overline{f_2}$	–	–	d_1, d_2
f_3	d_1	d_1	d_2
$\overline{f_3}$	d_2	d_2	d_1

$f_3^* \cap f_1^{¤} \neq \varnothing$ and $f_3 \subseteq f_2$ but $f_2^* \cap f_1^{¤} = \varnothing$.

26.9		*	¤
f_1	d_1, d_2	–	d_3, d_4
f_2	d_1, d_2, d_3, d_4	d_2	–
f_3	d_1, d_3, d_4	d_1, d_3	d_2
f_4	d_3, d_4	d_3, d_4	d_2

$f_3^* \cap f_1^{¤} \neq \varnothing$ and $f_3 \subseteq f_2$ but $f_2^* \cap f_1^{¤} = \varnothing$.

26.10		*	¤
f_1	d_1, d_2	d_1, d_2	d_3, d_4
f_2	d_1, d_2, d_3, d_4	d_2	–
f_3	d_1, d_3, d_4	d_1, d_3	–
f_4	d_3, d_4	d_3, d_4	d_1, d_2

$f_3^* \cap f_1^{¤} \neq \varnothing$ and $f_3 \subseteq f_2$ but $f_2^* \cap f_1^{¤} = \varnothing$.

The extent to which *AL*'s axioms are validated by these various *AA* semantics is shown in Figure 26.2.

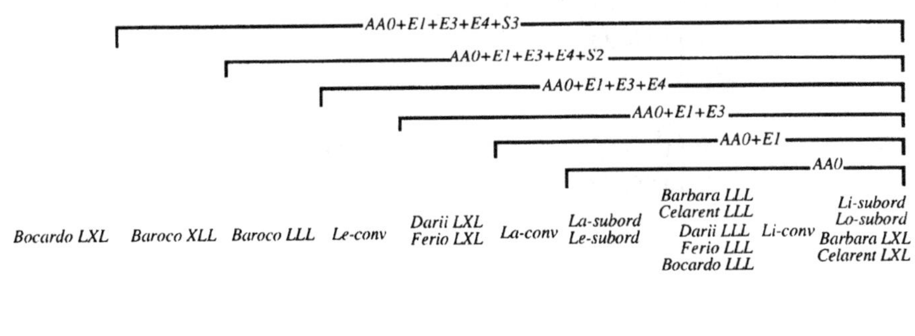

Fig. 26.2

§26.3 *Superstructure* C

We turn to the semantic superstructure used by many of the medieval logicians, including Jean Buridan (before 1300-1358), to account for divided (*i.e. de re*) modals. This rests on **21.1, 7, 8** and **24.1-5** for assertorics, and on a doctrine of ampliation for the subjects of all divided modals. Buridan analyses such modals as follows:[32]

26.3.1 Lab^a is true iff $f_2{}^\dagger \neq \varnothing \wedge f_2{}^\dagger \subseteq f_1{}^*$

26.3.2 Lab^e is true iff $f_2{}^\dagger \subseteq f_1{}^{¤}$

26.3.3 Lab^i is true iff $f_2{}^\dagger \cap f_1{}^* \neq \varnothing$

26.3.4 Lab^o is true iff $f_2{}^\dagger \cap f_1{}^{¤} \neq \varnothing$

Call this superstructure *C*. *C* plus S0 we call *C0*. This elegant semantics validates the following formulae.

Le-subord If $f_2{}^\dagger \subseteq f_1{}^{¤}$ then $f_2 \subseteq \overline{f_1}$
$\because f_2 \subseteq f_2{}^\dagger, f_1{}^{¤} \subseteq \overline{f_1}$.

Le-conv If $f_2{}^\dagger \subseteq f_1{}^{¤}$ then $f_1{}^\dagger \subseteq f_2{}^{¤}$
$\because f_1{}^{¤} = \overline{f_1{}^\dagger}, \overline{f_2{}^\dagger} = f_2{}^{¤}$.

Barbara LLL If $f_3{}^\dagger \subseteq f_2{}^*$ and $f_2{}^\dagger \subseteq f_1{}^*$ then $f_3{}^\dagger \subseteq f_1{}^*$
$\because f_2{}^* \subseteq f_2{}^\dagger$.

Celarent LLL If $f_3{}^\dagger \subseteq f_2{}^*$ and $f_2{}^\dagger \subseteq f_1{}^{¤}$ then $f_3{}^\dagger \subseteq f_1{}^{¤}$
$\because f_2{}^* \subseteq f_2{}^\dagger$.

Darii LLL If $f_3{}^\dagger \cap f_2{}^* \neq \varnothing$ and $f_2{}^\dagger \subseteq f_1{}^*$ then $f_3{}^\dagger \cap f_1{}^* \neq \varnothing$
$\because f_2{}^* \subseteq f_2{}^\dagger$.

---

[32] BURIDAN 2,6,8: "In any divided sentence *de necessario* the subject is ampliated to supposit for those things which can be."

Ferio LLL[33] If $f_3^\dagger \cap f_2^* \neq \varnothing$ and $f_2^\dagger \subseteq f_1^\texttt{¤}$ then $f_3^\dagger \cap f_1^\texttt{¤} \neq \varnothing$
$\because f_2^* \subseteq f_2^\dagger$.

Baroco LLL[34] If $f_2^\dagger \subseteq f_1^*$ and $f_3^\dagger \not\subset f_1^\dagger$ then $f_3^\dagger \not\subset f_2^\dagger$
$\because f_1^* \subseteq f_1^\dagger$.

Bocardo LLL[35] If $f_3^\dagger \cap f_1^\texttt{¤} \neq \varnothing$ and $f_3^\dagger \subseteq f_2^*$ then $f_2^\dagger \cap f_1^\texttt{¤} \neq \varnothing$
$\because f_2^* \subseteq f_2^\dagger$.

Darii LXL If $f_2^\dagger \subseteq f_1^*$ and $f_3 \cap f_2 \neq \varnothing$ then $f_3^\dagger \cap f_1^* \neq \varnothing$
$\because f_3 \subseteq f_3^\dagger, f_2 \subseteq f_2^\dagger$.

Ferio LXL[36] If $f_2^\dagger \subseteq f_1^\texttt{¤}$ and $f_3 \cap f_2 \neq \varnothing$ then $f_3^\dagger \cap f_1^\texttt{¤} \neq \varnothing$
$\because f_3 \subseteq f_3^\dagger, f_2 \subseteq f_2^\dagger$.

C0 does not validate Li-conv:

26.11		*	¤	†
f_1	d_1	d_1	–	d_1, d_2
f_2	d_2	–	–	d_1, d_2

$f_2^\dagger \cap f_1^* \neq \varnothing$ but $f_1^\dagger \cap f_2^* = \varnothing$.

(Buridan recognises as much.[37]) But Li-conv is valid relative to *C0+E3+E4*:

33 BURIDAN 4.2.3: "In the first figure, there is always a valid syllogism from a pair of premisses *de necessario* ... to a conclusion with the same mode ...."

34 BURIDAN 4.2.5: "In the second figure ... there is always a valid syllogism from a pair of premisses *de necessario* ... to a conclusion which is *de necessario* ...."

35 BURIDAN 4.2.8: "In the third figure ... a conclusion *de necessario* always follows from two premisses which are *de necessario* ...."

36 BURIDAN 4.2.53: "In the first figure ... from a major *de necessario* and an assertoric minor there is always a valid syllogism to a particular conclusion *de necessario* ...."

37 BURIDAN 2.6.27: "(a) From no sentence *de necessario* does there follow another sentence *de necessario* by conversion of the terms, except (b) a universal negative follows from a universal negative."

Proof If $f_1^* \cap f_2^\dagger \neq \varnothing$ then $f_2^* \cap f_1^\dagger \neq \varnothing$

$\because (f_1 \cap f_2^\dagger \neq \varnothing) \supset (f_2 \cap f_1^\dagger \neq \varnothing)$ - by E4

and $(f_1^* \cap f_2^\dagger \neq \varnothing) \supset (f_1 \cap f_2^\dagger \neq \varnothing)$ - $\because f_1^* \subseteq f_1$

and $(f_1 \cap f_2^\dagger \neq \varnothing) \supset (f_1^* \cap f_2^\dagger \neq \varnothing)$ - by E3.

C0 semantics does not validate La-subord or Li-subord;[38] and this remains the case even if we assume S5:

26.12		*	¤	†
f_1	d_1	d_1	–	d_1
f_2	–	–	–	d_1

$$\left.\begin{array}{r}(i) f_2^\dagger \neq \varnothing \wedge f_2^\dagger \subseteq f_1^* \\ (ii) f_2^\dagger \cap f_1^* \neq \varnothing\end{array}\right\} \text{but} \left\{\begin{array}{l}(i) f_2 = \varnothing \vee f_2 \not\subseteq f_1 \\ (ii) f_2 \cap f_1 = \varnothing.\end{array}\right.$$

They become valid if we add

E5 $f_1^\dagger \cap f_2^* \neq \varnothing \supset f_1 \cap f_2 \neq \varnothing$.

Lo-subord too is invalid provided the base is as weak as S0+S2:

26.13		*	¤	†
f_1	d_1	–	d_2	d_1
f_2	d_1	d_1	–	d_1, d_2
f_3	d_2	–	–	d_1, d_2

$f_2^\dagger \not\subseteq f_1^\dagger$ but $f_2 \subseteq f_1$.

The addition of S3 validates Lo-subord:

Proof If $f_2^\dagger \not\subseteq \overline{f_1^{¤}}$ then $f_2 \not\subseteq f_1$

$\because$ if $f_2 \subseteq f_1$ then $f_2^\dagger \subseteq f_1^\dagger$ - by S3 - and $f_1^\dagger \subseteq \overline{f_1^{¤}}$ (***23.1***).

This semantics also fails to validate all of Aristotle's LXL axioms: Barbara LXL and Celarent LXL are invalid.[39] Like Lo-subord, Barbara LXL and Celarent LXL become valid if we assume S3:

---

[38] BURIDAN 2.6.14: "From no sentence *de necessario* does there follow an assertoric sentence, or conversely, unless an assertoric universal negative follows from a universal negative *de necessario*."

[39] BURIDAN 4.2.53: "In the first figure ... from a major *de necessario* and an assertoric minor there is always a valid syllogism ... but ... not to a universal conclusion *de necessario*."

Proofs If $f_2{}^\dagger \neq \varnothing \wedge f_2{}^\dagger \subseteq f_1{}^*$ and $f_3 \neq \varnothing \wedge f_3 \subseteq f_2$ then $f_3{}^\dagger \neq \varnothing \wedge f_3{}^\dagger \subseteq f_1{}^*$
$\because f_3 \subseteq f_2, f_2{}^{¤} \subseteq f_3{}^{¤}$ - by S3 - so $f_3{}^\dagger \subseteq f_2{}^\dagger$.

If $f_2{}^\dagger \subseteq f_1{}^{¤}$ and $f_3 \neq \varnothing \wedge f_3 \subseteq f_2$ then $f_3{}^\dagger \subseteq f_1{}^{¤}$
$\because f_2{}^{¤} \subseteq f_3{}^{¤}$ - by S3 - so $f_3{}^\dagger \subseteq f_2{}^\dagger$.

Baroco XLL and Bocardo LXL are invalid if we assume only S0. Buridan gives the following counter-example for Baroco (and Camestres) XLL:

... let us posit that every planetary body shining above our hemisphere is the Sun, and every Moon is necessarily not the Sun. It does not follow that therefore the Moon is necessarily not a planetary body shining above our hemisphere.[40]

And he gives the following counter-example to Bocardo LXL:

... posit that nothing is shining now except a star, and nevertheless that tomorrow there will be a shining fire, which is necessarily not a celestial body. Then this does not follow: Something shining is necessarily not a celestial body, and anything shining is a star; therefore, a star necessarily is not a celestial body.[41]

These counter-examples can be formally represented as follows:

26.14		*	¤	†
f_1	d_2	d_2	d_1	d_2
f_2	d_2	–	–	d_1, d_2
f_3	d_1	d_1	d_2	d_1

f_1: Sun
f_2: planetary body shining on our hemisphere
f_3: Moon
$f_2 \subseteq f_1$ and $f_3{}^\dagger \cap f_1{}^{¤} \neq \varnothing$ but $f_3{}^\dagger \cap f_2{}^{¤} = \varnothing$.

40 BURIDAN 4.2.57.

41 BURIDAN 4.2.65.

26.15		*	¤	†
f_1	d_1	d_1	d_2	d_1
f_2	d_1	d_1	d_2	d_1
f_3	d_1	–	–	d_1, d_2
f_4	d_2	d_2	d_1	d_2

f_1: celestial body
f_2: star
f_3: shining object
f_4: fire
$f_3^{\dagger} \cap f_1^{¤} \neq \varnothing$ and $f_3 \subseteq f_2$ but $f_2^{\dagger} \cap f_1^{¤} = \varnothing$.

But both these moods are valid if we assume S3:

Proofs If $f_3^{\dagger} \cap f_1^{¤} \neq \varnothing$ and $f_3 \subseteq f_2$ then $f_2^{\dagger} \cap f_1^{¤} \neq \varnothing$
$\because f_3^{\dagger} \subseteq f_2^{\dagger}$ - by S3.

If $f_2 \subseteq f_1$ and $f_3^{\dagger} \cap f_1^{¤} \neq \varnothing$ then $f_3^{\dagger} \cap f_2^{¤} \neq \varnothing$
$\because f_1^{¤} \subseteq f_2^{¤}$ - by S3.

The assumption of S3 affects Buridan's counter-examples as follows. In the first example, the supposition that every planetary body shining above our hemisphere is the Sun implies that whatever is necessarily not the Sun is necessarily not [one of the things] shining above our hemisphere; so if the Moon is necessarily not the Sun it *does* follow that the Moon is necessarily not shining above our hemisphere. In the second example, the supposition that everything that is shining is a star implies that whatever is necessarily not a star is necessarily not [one of the things] shining; so tomorrow's shining fire which cannot be a star cannot [now] be shining.

The addition of S1.1+S3 makes Li-subord valid:

Proof If $f_2^{\dagger} \cap f_1^{*} \neq \varnothing$ then $f_2 \cap f_1 \neq \varnothing$
$\because$ if $f_2 \cap f_1 = \varnothing$ then $f_2 \subseteq \overline{f_1}$
so $f_2^{\dagger} \subseteq \overline{f_1}^{\dagger}$ - by S3
so $f_2^{\dagger} \subseteq \overline{f_1^{*}}$ - $\because \overline{f_1}^{\dagger} \subseteq \overline{f_1^{*}}, \because \overline{\overline{f_1}}^{*} \subseteq \overline{f_1}^{¤}$ - by S1.1
i.e., $f_2^{\dagger} \cap f_1^{*} = \varnothing$.

Figure 26.3 shows the extent to which each of these semantics validates the axioms of *AL*.

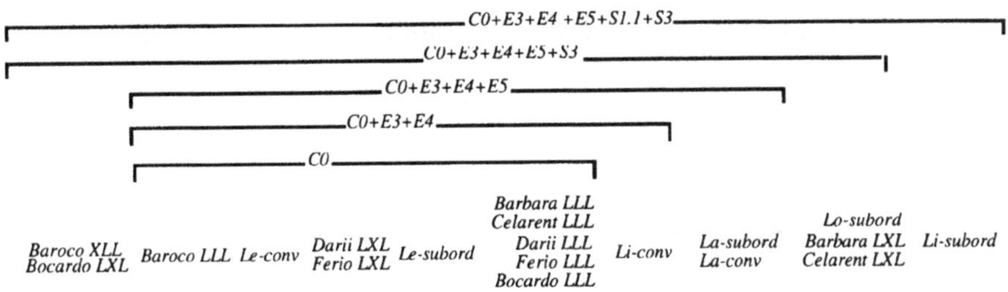

Fig. 26.3

§26.4 Superstructure CC

William Ockham (c.1285-1349) uses an interestingly different modal semantics. Like Buridan, he uses **24.1-5** for assertorics. But whereas Buridan understands apodeictics normally as ampliated to the possible, Ockham takes their normal sense to require neither ampliation nor restriction of their subject-terms:[41]

26.4.1 Lab^a is true iff $f_2 \subseteq f_1^*$

26.4.2 Lab^e is true iff $f_2 \subseteq f_1^{¤}$

26.4.3 Lab^i is true iff $f_2 \cap f_1^* \neq \varnothing$

26.4.4 Lab^o is true iff $f_2 \cap f_1^{¤} \neq \varnothing$

Call this superstructure *CC*. Combining *CC* with definition S0 we get a semantics, *CC0*, which validates L-subordination,[42] Barbara LXL, Celarent LXL, Darii LXL and Ferio LXL.[43] It does not validate partial La-conversion, Li-conversion or Le-conversion.[44] (Proofs obvious.) Also valid is Bocardo LXL:

Proof If $f_3 \cap f_1^{¤} \neq \varnothing$ and $f_3 \subseteq f_2$ then $f_2 \cap f_1^{¤} \neq \varnothing$.

[41] OCKHAM IIIa.31.15-16 *etc.* Buridan allows that the normal implication of an ampliated subject can be cancelled by inserting the phrase 'what is' before the subject; and Ockham allows that the normal implication of a subject that is neither ampliated nor restricted can be cancelled by an explicit ampliation of the subject.

[42] OCKHAM IIIc.11.3-6. These inferences are said to be valid whether the modal propositions are taken '*in sensu compositionis*' or '*in sensu divisionis*'.

[43] OCKHAM IIIa.31.8-43. These syllogisms are said to be valid if (but only if) the modal propositions are taken '*in sensu divisionis*'.

[44] OCKHAM II.24.49-79.

Ockham himself gives an ecthetic proof of Bocardo LXL, which we shall analyse in Chapter Nine.[46]

The addition of E1 to *CC0* validates La-conversion:

Proof If $f_2 \subseteq f_1^{*}$ then $f_1 \cap f_2^{*} \neq \varnothing$
$\because f_2^{*} \neq \varnothing$ - by E1 - and $f_2^{*} \subseteq f_2$.

The addition of E3 to *CC0* semantics results in the validation of Li-conversion:

Proof If $f_2 \cap f_1^{*} \neq \varnothing$ then $f_1 \cap f_2^{*} \neq \varnothing$
$\because$ if $f_2 \cap f_1^{*} \neq \varnothing$ then $f_1 \cap f_2 \neq \varnothing$
and so $f_1 \cap f_2^{*} \neq \varnothing$ - by E3.

The addition of E4 will validate Le-conversion.

The addition of S3 to *CC0* validates Baroco XLL:

Proof If $f_2 \subseteq f_1$ and $f_3 \cap f_1^{¤} \neq \varnothing$ then $f_3 \cap f_2^{¤} \neq \varnothing$
$\because f_1^{¤} \subseteq f_2^{¤}$ - by S3.

The extent to which *AL*'s axioms are validated by these various *CC* semantics is shown in Figure 26.4.

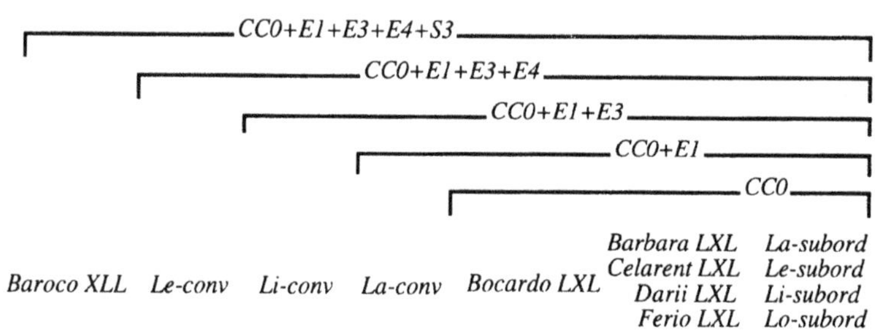

Fig. 26.4

46 §41. See OCKHAM IIIa.33.54-62.

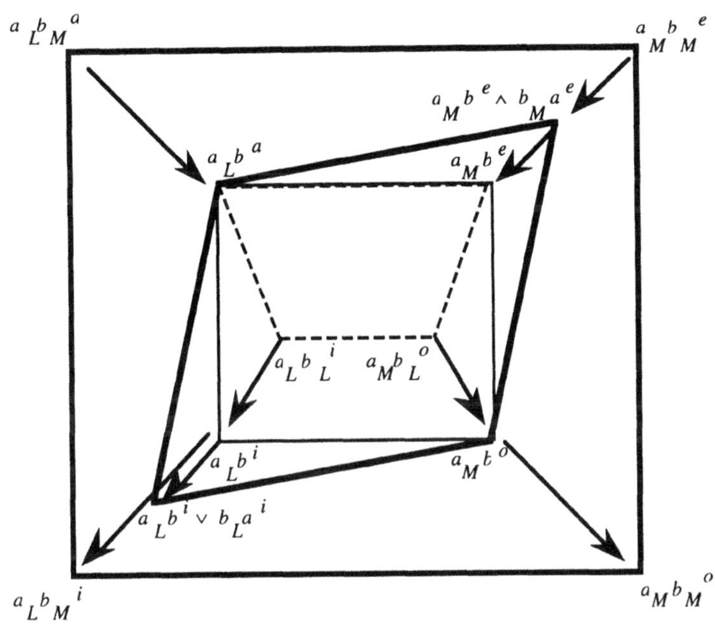

Fig. 26.5

Several conclusions can be drawn from these data. (i) Either *A0+S3* or *A2* validates all the *AL* axioms (as of course does *A5*); no *A* semantics makes Baroco XLL or Bocardo LXL invalid. (ii) No *AA* semantics validates all the *AL* axioms; however, *AA0+E1+E3+E4+S2* validates all but Baroco XLL and Bocardo LXL. (iii) All *AL* axioms are validated by either *C0+E3+E4+S1.1+S3* or *C3+E3+E4*; no *C* semantics validates all *AL* axioms except Baroco XLL and Bocardo LXL. (iv) No *CC* semantics validates all the *AL* axioms without also validating Bocardo LXL; however, *CC0+E1+E3+E4* validates all the *AL* axioms. (v) If one wishes to validate all *AL*'s axioms, one will have to choose between *A* and *C* and *CC* semantics; if one wishes to validate all except Baroco XLL and Bocardo LXL, only an *AA* semantics will do. Our own choice of *A5* will be justified in §43.2-3.

The relations between the four superstructures are shown in Figure 26.5, where *AA* is represented by the central trapezium, *CC* by the inner square, *A* by the diamond, and *C* by the outer square.

§27. ALTERNATIVE APPROACHES: *DE DICTO* THEORIES

Some commentators give a wholly or partly *de dicto* reading of the modal propositions in *Prior Analytics*. We briefly review some of these accounts in this section. I use $\Box$ as a proposition-forming necessity-operator on propositions, $\Diamond$ and ∇ as the corresponding possibility- and contingency-operators.

§27.1 *Pure* de dicto *theories*

The simplest *de dicto* approach is to render apodeictic forms as follows:

Lab^a $\quad \Box\forall x(bx \supset ax)$

Lab^e $\quad \Box\forall x(bx \supset \sim ax)$

Lab^i $\quad \Box\exists x(bx \wedge ax)$

Lab^o $\quad \Box\exists x(bx \wedge \sim ax)$

Nortmann[47] observes that

As much as we know about Theophrastus's work on modal logic, this form of interpretation might be labeled the theophrastine way.

The analysis validates modal subordination, Li-conversion, and the LLL moods; but not La-conversion, the LXL moods or Le-conversion. Its claim to be Theophrastine rests on the fact that it obeys the '*peiorem* rule' - Theophrastus's requirement that the modality of the conclusion be no stronger than that of any of the premisses.[48]

Partial La- and Le-conversion could be rendered valid by changing the reading of universals to incorporate existential import:

Lab^a $\quad \Box[\exists xbx \wedge \forall x(bx \supset ax)]$

Lab^e $\quad \Box[\exists xbx \wedge \forall x(bx \supset \sim ax)]$

The validity of Le-conversion could be secured by changing the analysis of Le-forms to $\Box(\exists x(bx) \wedge \exists x(ax) \wedge \forall x(bx \supset \sim ax))$. But that would make Celarent LLL invalid. And, with all these changes, the LXL moods remain invalid.

Another *de dicto* approach is taken by Łukasiewicz. His formalisation of the apodeictic syllogistic is based on assertoric syllogistic plus axiomatic assertions

$$(\Box p) \supset p \qquad \Box p \equiv \Box \sim\sim p$$

47 NORTMANN 1994, p.117. For a critique of Theophrastus, see PATTERSON 1995 §4.1.

48 See BOCHEŃSKI, Chapter 5 §3.

and axiomatic rejections

$$p \supset \Box p \qquad \sim \Box p$$

with $\Diamond p$ defined as $\sim \Box \sim p$.[49] In this system both Barbara LXL and Barbara XLL are provable. Łukasiewicz's proof of the latter can be paraphrased as follows.[50] From the theses

$$ab^a \supset \left(bc^a \supset ac^a\right)$$

and

$$\left(bc^a \supset ac^a\right) \supset \left(Lbc^a \supset Lac^a\right)$$

we deduce $ab^a \supset \left(Lbc^a \supset Lac^a\right)$, Barbara XLL. The proof of Barbara LXL begins instead from

$$bc^a \supset \left(ab^a \supset ac^a\right)$$

It is clear from this pattern of proof that in this system *all* assertoric syllogisms will have provable XLL and LXL analogues. Thus for Łukasiewicz there are 28 LXL or XLL syllogisms; Aristotle accepts only 13 (see Table 17.1). The prospects for a pure *de dicto* reading are not rosy.

§27.2 *Mixed* de dicto/de re *theories*

Ockham, Buridan and other medievals did not offer any global interpretation of Aristotle's text, instead offering a systematic investigation of which moods are valid on each of these readings. Knuuttila observes:[51]

> When Aristotle's modal syllogistic was viewed from the point of view of this systematic approach, it appeared as a not very well organized fragment of a larger theory which remained beyond Aristotle's purview.

As a development of this line, Becker proposed that, while Aristotle generally used a *de re* understanding of modal syllogisms, some passages in the text which require

49 ŁUKASIEWICZ p.189.

50 ŁUKASIEWICZ p.137.

51 KNUUTTILA p.163.

a *de dicto* sense are later interpolations.[52] Wholesale revision of the text is required by this reading.[53]

Angelelli is another inheritor of the medieval approach. He[54] distinguishes "external" and "internal" readings of modal forms corresponding to our *de dicto* and *de re* readings. Additionally, he distinguishes two types of "internal" form, depending on whether the subject-term is ampliated; and this distinction is applied also to assertoric forms. Thus, the external reading of La-forms is $\Box\forall x(bx \supset ax)$, the internal reading $\forall x(bx \supset \Box ax)$. Clearly, not all moods that Aristotle recognises as valid are valid on the external reading (for example the LXL moods.) Nor are all of them valid on the internal reading (for example Le-conversion.) What Angelelli proposes is that a mood should be judged valid iff there are readings (internal or external, ampliated or unampliated) of the premisses and conclusion according to which the mood is valid. The same form of premiss does not have to be read the same way from one syllogism to another, or even from one premiss to another.[55]

This, rather than an interpretative strategy, seems more like a polite way of saying that Aristotle confused different senses and repeatedly committed the fallacy of equivocation. It also gives imperfect results. As Angelelli himself notes,[56] Disamis LXL is valid by his criterion but is rejected by Aristotle.

There are also moods accepted by Aristotle but not meeting Angelelli's liberal requirements; Nortmann names Ferio LQX.[57] Angelelli proposes that we understand Aristotle as restricting such recalcitrant inferences to cases in which the terms satisfy extra (metaphysically motivated) conditions, *e.g.* $\forall x[\Box ax \vee \Box \sim ax]$.[58] We shall examine assumptions of this type in Chapter Ten. In the meantime, we observe that an interpretation which has to attribute unstated axioms to Aristotle must have a lesser claim on our acceptance than interpretations that do not have to do so. But there is worse in the case of Ferio LQX. This mood's premisses on Angelelli's "internal" reading are $\forall x[bx \supset \Box \sim ax]$, $\exists x[cx \wedge \nabla bx]$; and these premisses are inconsistent with Angelelli's requirement that $\forall x[\Box bx \vee \Box \sim bx]$.

---

52 BECKER, p.90.

53 For critiques of the Becker approach, see McCALL 1963 §8, PATTERSON 1989 §1.

54 ANGELELLI, pp.176-213.

55 ANGELELLI p.207.

56 ANGELELLI p.210.

57 NORTMANN 1994, 120-121. Nortmann mistakenly refers here to Ferio LQM, a mood which, though arguably Aristotelian, is not explicitly proved in the text.

58 ANGELELLI §12.

Nortmann himself uses a combination of external and internal modalities in each form of L-modal.

Lab^a $\forall x\Box(bx \supset \Box ax)$[59]

Lab^e $\forall x\Box(bx \supset \Box \sim ax)$

Lab^i $\exists x\Box(bx \wedge \Box ax)$[60]

Lab^o $\exists x\Box(bx \wedge \Box \sim ax)$

The behaviour of the various modal forms then depends on the logic of the *de dicto* modalities that occur within them. Nortmann takes the underlying modal logic to be quantified S5, where $\Box p \supset \Box\Box p$, $\sim\Box p \supset \Box\sim\Box p$ and $\forall x\Box fx \supset \Box\forall xfx$ are all valid. On this analysis, all axioms of *AL* other than the LXL moods are valid. (Li-conversion is valid since in S5 $\Box(p \wedge \Box q) \supset \Box(q \wedge \Box p)$ is valid.)

To validate the LXL moods, Nortmann proposes that assertoric forms be read $\forall x\Box(bx \supset ax)$ *etc.*[61] This move does indeed rescue the First Figure LXL moods, and also validates Baroco XLL and Bocardo LXL, since in S5

$$[\Box(q \supset p) \wedge \Box(r \wedge \Box \sim p)] \supset \Box(r \wedge \Box \sim q)$$

$$[\Box(r \wedge \Box \sim p) \wedge \Box(r \supset q)] \supset \Box(q \wedge \Box \sim p)$$

are valid. It also validates the unAristotelian moods Barbara XLL and Celarent XLL, since $[\Box(q \supset p) \wedge \Box(r \supset \Box q)] \supset \Box(r \supset \Box p)$ is S5-valid.

Nortmann justifies his reading of assertorics by appealing to Aristotle's injunction (A15, 34b7-9) against using temporally restricted assertorics. But, as argued in §15.2.1, there is no good reason for taking Aristotle's remark there as a general rule covering all assertorics used in syllogisms.

---

59 This analysis occurs in SCHMIDT, p.82.

60 Neither NORTMANN 1990 nor NORTMANN 1994 explicitly includes an analysis of Li- or Lo-forms. The analysis I have suggested implies the convertibility of Li-forms, as shown above.

61 NORTMANN 1990 p.73. For a critique of Nortmann's approach, see PATTERSON 1995 pp.178ff.

CHAPTER SEVEN. SYSTEMS AQ AND AQ'

We pass to the formal interpretation of contingency-syllogistic. In §20.2 we posed the choice between convertible and opposable interpretations of contingency-forms. Convertible interpretations include Qi- and Li-conversion but not the oppositions $Qi \parallel La$ and $Li \parallel Qa$ or the moods Bocardo QLM and Bocardo QXM. Opposable interpretations include those oppositions and moods, but exclude Qi- and Li-conversion and moods whose proof requires either of these conversions. In this Chapter formal interpretations of both types will be developed.

We shall develop both types of interpretations for ampliated as well as for unampliated forms. But it will also be shown that Aristotle's text does not require the Alexandrian interpretation according to which the pure contingency syllogistic deals with ampliated and the mixed contingency syllogistic with unampliated contingency-forms. For the purposes of logical theory, however, ampliated and unampliated forms should still be distinguished, for they have quite different logics. And even though the differences occur mainly outside the areas Aristotle investigated, there are points within his language where the two forms behave differently, *e.g.* in their oppositions to apodeictic forms.

Finally, it will be shown that the adoption of sufficiently strong base assumptions about necessity and impossibility weakens the ampliated/unampliated distinction to the point where ampliated and unampliated universals become semantically equivalent.

§28. UNAMPLIATED CONTINGENCY-FORMS

The contingent is defined at A13, 32a18-20 as that which is not necessary, but which being supposed does not result in anything impossible, *i.e.* as two-sided possibility. Given this, two obvious ways of interpreting Q-forms cohere with the interpretation of L- and X-forms thus far followed. On the first approach, we take Qab^a to mean that every b is a contingent a and Qab^i to mean that either some b is a contingent a or some a is a contingent b. The result is then that Qi-forms are convertible but not opposable, in particular Qi-forms are compatible with La-forms, and Li-forms with Qa-forms. The second approach is to interpret Qab^a as before, but to preserve the classical oppositions and sacrifice Qi-conversion, by taking Qab^i to mean simply that some b is contingently a. For reasons given in §20.2, I prefer the first approach. That is the approach taken in this section; the other will be explored later in the Chapter.

Consider a system AQ, which adds Q-forms to the language of AL. We also add for each Q-form a contradictory:

$$\tilde{Q}ab^a \quad \tilde{Q}ab^e \quad \tilde{Q}ab^i \quad \tilde{Q}ab^o$$

To AL's contraries we add $Qa \parallel La$, $Qa \parallel Le$, $Qa \parallel Lo$, $Qi \parallel Le$.

AQ's axioms are those of *AL* plus Qa-conv (partial), Qi-conv, Q-conv, Barbara QXQ, Darii QXQ, QaLaXe-2 (= Cesare QLX).

We need not take the corresponding QLQ formulae as axiomatic, since they follow from the QXQ axioms by L-subordination; nor do we need to make QaLoXo-2 (= Baroco QLX) axiomatic, since it is provable by methods indicated in §19.1. But QaLaXe-2 must be taken as axiomatic, since the derivations given in §19.1 require one of the oppositions $La \parallel Qi$ and $Li \parallel Qa$, neither of which is endorsed by *AQ*.

QaXiMo-1 (= Ferio QXM) is derivable from QaLaXe-2:

Proof
$$\frac{\dfrac{Qab^e}{Qab^a}\,Q-conv \quad Lac^a}{bc^e}\,QaLaXe-2 \quad \xrightarrow{C} \quad \frac{Qab^e \quad bc^i}{Mac^o}$$

whence Ferio QLM follows. Aristotle's proof of Ferio QLM (given in §12.1), which relies on the opposition $Qe \parallel Li$, is not available to us since the semantics about to be developed does not endorse that opposition.

Our semantics should make the axioms valid. Further, it should make invalid the formulae: Qa-conv, XaQaQi-1 (Barbari XQQ), XaQaXi-1 (Barbari XQX), XaQaMo-1, XeQaMi-1, XeQaMo-1 (Celaront XQM), QaLaXi-1 (Barbari QLX), QiXaMi-1, QiXaMo-1, QaLeMi-1, LaQaQi-1 (Barbari QLQ), LaQiXi-1 (Darii QLX), LaQiXo-1, LiQaMi-1, LiQaMo-1, LeQaQi-1 (= Celaront QLQ), LoQaMi-1, LoQaMo-1, QaLaMi-2, QaLaLo-2 (= Cesaro QLL), QaLaQi-2 (= Cesaro QQL), QaXaMo-2 (= Cesaro QXM), XaQaMo-2 (= Camestrop XQM), LaQaLo-2 (= Camestrop QLL), LaQaQi-2 (= Camestrop QLQ), XeQaMo-2 (= Cesaro XQM), LoQaMo-2, QiLaMo-3 (= Bocardo QLM), QaLeMi-3, QaLeMo-3.

We omit QaLaXo-2, LaQaXo-2 from our earlier list, for reasons given in §19.1, and we add Bocardo QLM for reasons given in §20.2. We do not need to make axiomatic any of the M-rejections signalled in §17, as these are all deducible:

Proof
$$\text{Mo - conv} \xrightarrow{C} \text{La - conv}$$
$$\text{Festino QXM} \xrightarrow{C} \text{QaLaXo - 1}$$
$$\text{Camestres QXM} \xrightarrow{C} \text{QaLiXi - 1.}$$

§28.1 Formal semantics

I proceed to give a semantic theory which does all this, building on the semantics for *AL*. *AQ* models and *AQ* validity are defined analogously to *AL* models and *AL* validity. The distinctive features of *AQ* semantics are as follows.

It is clear that for each set there is a corresponding *double-dagger set* (the shaded area in Figure 28.1), namely the complement of the union of the original set's star set and sun set.

28.1.1 $f^{\ddagger} =_{df.} \overline{f^{*} \cup f^{¤}}$

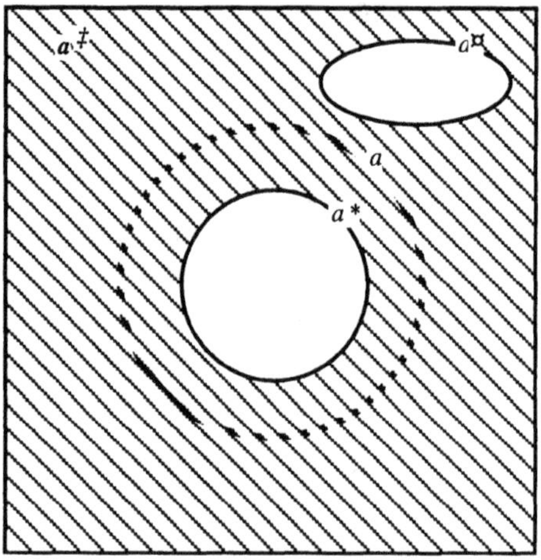

Fig. 28.1

We then stipulate

28.1.2 Qab^{a} is true iff $f_2 \subseteq f_1^{\ddagger}$, *i.e.* iff the *b*s are included among the *a*s' double dagger set.

28.1.3 Qab^{i} is true iff $f_2 \cap f_1^{\ddagger} \neq \varnothing \vee f_1 \cap f_2^{\ddagger} \neq \varnothing$, *i.e.* iff EITHER some *b* is in the *a*s' double-dagger set OR some *a* is in the *b*s' double-dagger set.

As for contingency e- and o-forms, in view of their complementary convertibility with a- and i-forms, the simplest interpretation is to identify their sense with that of their affirmative counterparts, so that contingency-universals (a- and e-forms) are equivalent, as are contingency-particulars (i- and o-forms).

§28.2 *Validity of the axioms*

The following formulae are valid.

Q-conv By definition.

Qa-conv (partial) If $f_2 \subseteq f_1^{\ddagger}$ then $f_2 \cap f_1^{\ddagger} \neq \varnothing$
$\because f_2 \neq \varnothing$.

Qi-conv If either $f_2 \cap f_1^{\ddagger} \neq \varnothing$ or $f_1 \cap f_2^{\ddagger} \neq \varnothing$ then
either $\mathrm{f}_1 \cap \mathrm{f}_2^{\ddagger} \neq \varnothing$ or $f_2 \cap f_1^{\ddagger} \neq \varnothing$.

Barbara QXQ If $f_2 \subseteq f_1^{\ddagger}$ and $f_3 \subseteq f_2$ then $f_3 \subseteq f_1^{\ddagger}$.

Darii QXQ If $f_2 \subseteq \mathrm{f}_1^{\ddagger}$ and $f_2 \cap f_3 \neq \varnothing$ then $f_3 \cap f_1^{\ddagger} \neq \varnothing$.

QaLaXe-2 If $f_2 \subseteq f_1^{\ddagger}$ and $f_3 \subseteq f_1^{*}$ then $f_3 \cap f_2 = \varnothing$
$\because f_1^{\ddagger} \cap f_1^{*} = \varnothing$.

§28.3 *Invalidity of the axiomatic rejections*

The following are invalid formulae.

Qa-conv

28.1		*	¤	‡
f_1	d_1	–	–	d_1
f_2	d_1	d_1	–	–

$f_2 \subseteq f_1^{\ddagger}$ but $f_1 \not\subset f_2^{\ddagger}$.

Aristotle's terms 'white' 'man' (A17, 37a4-8) fit this model. The model supposes that a white man exists. Being a man is taken to be necessary for him, being white contingent. Given all this, we have all men contingently white but not all whites contingently men.

§28.3.1 First Figure

XaQaQi-1
XaQaXi-1

28.2		*	¤	‡
f_1	d_1	d_1	d_2	–
f_2	d_1	–	–	d_1, d_2
f_3	d_2	d_2	d_1	–

$$f_2 \subseteq f_1^{*}, f_3 \subseteq f_2^{\ddagger} \text{ but } \begin{cases} (i) f_3 \cap f_1^{\ddagger} = \varnothing, f_1 \cap f_3^{\ddagger} = \varnothing \\ (ii) f_3 \cap f_1 = \varnothing. \end{cases}$$

Aristotle's terms 'man' 'moving' 'horse' (A15, 34b10-17) fit this model. The model supposes that two things exist - a moving man and a stationary horse. It takes it that the man cannot be a horse and the horse cannot be a man, while either one of them is contingently moving.

XaQaMo-1

28.3		*	¤	‡
f_1	d_1	d_1	–	–
f_2	d_1	–	–	d_1
f_3	d_1	d_1	–	–

$$f_2 \subseteq f_1 \text{ and } f_3 \subseteq f_2^{\ddagger} \text{ but } f_3 \subseteq f_1^{*}.$$

Aristotle's terms 'animal' 'moving' 'man' (A15, 34b10-17) fit our model as follows. The model supposes the only being to be a moving man. It takes this man to be a man necessarily and an animal necessarily, while not necessarily moving.

XeQaMi-1

28.4		*	¤	‡
f_1	d_1	d_1	d_2	–
f_2	d_2	–	–	d_1, d_2
f_3	d_2	d_2	d_1	–

$$f_2 \cap f_1 = \varnothing, f_3 \subseteq f_2^{\ddagger} \text{ but } f_3 \subseteq f_1^{¤}, f_1 \subseteq f_3^{¤}.$$

Aristotle's terms 'animal' 'white' 'cloak' (A15, 35b11-19), used by him to show that no affirmative conclusion follows from XoQa-1, can be adapted to the present model. The model supposes that there are two beings - a non-white animal, and a white cloak. It takes the animal to be contingently white, but necessarily an animal. (It could be a black man.) For the white cloak, being white is taken to be contingent, being a cloak necessary, and being an animal impossible.

XeQaMo-1

28.5		*	¤	‡
f_1	d_1	d_1	d_2, d_3	–
f_2	d_2, d_3	d_2	–	d_1, d_3
f_3	d_1	d_1	d_2, d_3	–
f_4	d_3	d_3	d_1, d_2	–

$f_2 \cap f_1 = \varnothing, f_3 \subseteq f_2{}^{\ddagger}$ but $f_3 \subseteq f_1{}^{*}$.

Terms fitting the model are 'man' 'white' 'man', and as our fourth term 'horse'. The model supposes that there exist a non-white man, something that is necessarily white (say a swan), and a white horse. The man, swan and horse are necessarily of their own kind and cannot be of any other kind. The man and the horse are contingently white.

QaLaXi-1

28.6		*	¤	‡
f_1	d_1, d_2	d_1	–	d_2, d_3
f_2	d_3	d_3	d_1, d_2	–
f_3	d_3	d_3	d_1, d_2	–
f_4	d_1	d_1	d_2, d_3	–
f_5	d_2	d_2	d_1, d_3	–

$f_2 \subseteq f_1{}^{\ddagger}, f_3 \subseteq f_2{}^{*}$ but $f_3 \cap f_1 = \varnothing$.

Aristotle's terms 'animal' 'white' 'man' and 'animal' 'white' 'cloak', used at A15, 35b11-19 against First Figure moods with Qi-majors, cannot be used against the present moods, since animal does not belong *contingently* to everything white. Terms adapted to the present model are 'white' 'man' 'man'. The model supposes that there are three beings - a white non-man for whom being white is necessary and being a man impossible (say a swan, f_4 in our model), a white non-man for whom being white is contingent and being a man impossible (say a horse, f_5 in our model), and a non-white man who is necessarily a man and contingently white.

QiXaMi-1

28.7		*	¤	‡
f_1	d_1	d_1	d_2	–
f_2	d_1, d_2	d_2	–	d_1
f_3	d_2	d_2	d_1	–

$f_1 \cap f_2{}^{\ddagger} \neq \varnothing, f_3 \subseteq f_2$ but $f_3 \subseteq f_1{}^{¤}, f_1 \subseteq f_3{}^{¤}$

Terms fitting the model are 'swan' 'white' 'man'. The model supposes that a contingently white man and a necessarily white swan exist. The swan cannot be a man, nor the man a swan.

QiXaMo-1

28.8		*	¤	‡
f_1	d_1	d_1	–	–
f_2	d_1	–	–	d_1
f_3	d_1	–	–	d_1

$f_1 \cap f_2^{\ddagger} \neq \varnothing, f_3 \subseteq f_2$ but $f_3 \subseteq f_1^{*}$.

The model supposes that a single thing exists, say a sleeping koala. Let the terms be 'koala' 'sleeping' 'sleeping'. Then the koala is necessarily a koala but contingently asleep.

QaLeMi-1

28.9		*	¤	‡
f_1	d_1	–	d_2	d_1
f_2	d_1	d_1	d_2	–
f_3	d_2	d_2	d_1	–

$f_2 \subseteq f_1^{\ddagger}, f_3 \subseteq f_2^{\text{¤}}$ but $f_3 \subseteq f_1^{\text{¤}}, f_1 \subseteq f_3^{\text{¤}}$.

Aristotle's terms 'white' 'animal' 'pitch' (A16, 36a27-31) fit our model as follows. The model supposes that there are two beings - a white animal and some pitch. For the white animal being an animal is taken to be necessary, being white contingent, and being pitch impossible. For the pitch, being pitch is necessary, and being an animal or white impossible.

LaQaQi-1

28.10		*	¤	‡
f_1	d_1	d_1	–	–
f_2	d_1	–	–	d_1
f_3	d_1	d_1	–	–

$f_2 \subseteq f_1^{*}, f_3 \subseteq f_2^{\ddagger}$ but $f_3 \cap f_1^{\ddagger} = \varnothing, f_1 \cap f_3^{\ddagger} = \varnothing$.

Terms fitting the model are 'man' 'white' 'man'. The model supposes that a solitary white man exists, necessarily human but contingently white.

LaQiXi-1

28.11		*	¤	‡
f_1	d_1	d_1	d_2	–
f_2	d_1	–	–	d_1, d_2
f_3	d_2	d_2	d_1	–

$f_2 \subseteq f_1{}^*, f_3 \cap f_2{}^{\ddagger} \neq \varnothing$ but $f_3 \cap f_1 = \varnothing$.

Terms fitting the model are 'man' 'sleeping' 'horse'. The model supposes there to be just a man and a horse. The man is sleeping, but the horse is not. Both man and horse are contingently sleeping. The man and the horse are each necessarily of their own kind and cannot be of the other kind.

LaQiXo-1

28.12		*	¤	‡
f_1	d_1, d_2	d_1, d_2	–	–
f_2	d_1	d_1	d_2	–
f_3	d_1, d_2	–	–	d_1, d_2
f_4	d_2	d_2	d_1	–

$f_2 \subseteq f_1{}^*, f_2 \cap f_3{}^{\ddagger} \neq \varnothing$ but $f_3 \subseteq f_1$.

Terms fitting the model are 'animal' 'man' 'sleeping' 'horse'. The model supposes there to be just two animals, a man and a horse, both of them asleep. The man must be a man and an animal, and cannot be a horse. The horse must be a horse and an animal, and cannot be a man. Each one is contingently sleeping.

LiQaMi-1

28.13		*	¤	‡
f_1	d_1	d_1	d_2	–
f_2	d_1, d_2	–	–	d_1, d_2
f_3	d_2	d_2	d_1	–

$f_2 \cap f_1{}^* \neq \varnothing, f_3 \subseteq f_2{}^{\ddagger}$ but $f_3 \subseteq f_1{}^{¤}$ and $f_1 \subseteq f_3{}^{¤}$.

Aristotle's terms 'animal' 'white' 'cloak' (A16, 36b3-7) fit this model. The model supposes that there exists an animal and a cloak, each necessarily of its own kind and necessarily not of the other kind. Each is white but contingently so.

LiQaMo-1

28.14		*	¤	‡
f_1	d_1, d_2	d_1, d_2	–	–
f_2	d_1, d_2	–	–	d_1, d_2
f_3	d_1	d_1	d_2	–
f_4	d_2	d_2	d_1	–

$f_2 \cap f_1^* \neq \varnothing, f_3 \subseteq f_2^‡$ but $f_3 \subseteq f_1^*$.

Aristotle's terms 'animal' 'white' 'man' (A16, 36b3-7) fit the model. The model supposes that a man and another animal (say a horse, f_4 in the model) exist. Both are necessarily animals, both white, and both contingently white. Each must be of its own kind and cannot be of the other's.

LeQaQi-1

28.15		*	¤	‡
f_1	d_1	d_1	d_2, d_3, d_4, d_5	–
f_2	d_2, d_3	d_2	d_1	d_3, d_4, d_5
f_3	d_4	d_4	d_1, d_2, d_3, d_5	–
f_4	d_2, d_3, d_4, d_5	d_2, d_3, d_4, d_5	d_1	–
f_5	d_3	d_3	d_1, d_2, d_4, d_5	–
f_6	d_5	d_5	d_1, d_2, d_3, d_4	–

$f_2 \subseteq f_1^¤, f_1 \subseteq f_2^¤, f_3 \subseteq f_2^‡$ but $f_3 \cap f_1^‡ = \varnothing, f_1 \cap f_3^‡ = \varnothing$.

Terms fitting the model are 'pitch' 'white' 'man', and as f_4, f_5 and f_6, 'animal', 'horse' and 'koala'. The model supposes that five things exist:

d_1 is some pitch, which is necessarily pitch, and cannot be white or an animal, or a man;

d_2 is a white thing that is necessarily white and necessarily an animal (say a swan), but cannot be man or pitch;

d_3 is a white thing that is contingently white and necessarily an animal (say a white horse), but cannot be a man or pitch;

d_4 is a non-white man who is contingently white, necessarily a man and necessarily an animal, but cannot be pitch;

d_5 is a non-white animal, not a man or swan, that is necessarily an animal but cannot be a man or swan or pitch. This is the koala.

LoQaMi-1

28.16		*	¤	‡
f_1	d_1	d_1	d_2, d_3	–
f_2	d_1, d_2, d_3	–	–	d_1, d_2, d_3
f_3	d_3	d_3	d_1, d_2	–
f_4	d_2	d_2	d_1, d_3	–

$f_2 \cap f_1^{¤} \neq \varnothing, f_3 \subseteq f_2^{‡}$ but $f_3 \subseteq f_1^{¤}, f_1 \subseteq f_3^{¤}$.

Aristotle's terms 'animal 'white' 'cloak' (A16, 36b3-7) fit our model. The model supposes that there are three beings - two animals and a white cloak. For the animals, being an animal is necessary, being white contingent, and being a cloak impossible. For the cloak, being white is contingent, being a cloak necessary, and being an animal impossible.

LoQaMo-1

28.17		*	¤	‡
f_1	d_1	d_1	d_2	–
f_2	d_1, d_2	–	–	d_1, d_2
f_3	d_1	d_1	d_2	–
f_4	d_2	d_2	d_1	–

$f_2 \cap f_1^{¤} \neq \varnothing, f_3 \subseteq f_2^{‡}$ but $f_3 \subseteq f_1^{*}$.

Aristotle's terms 'animal 'white' 'man' (A16, 36b3-7) fit this model. The model supposes that there are two beings - a white man and a white non-human animal (say a horse). It supposes that both are necessarily animals, and contingently white. And it supposes that the horse is necessarily a horse and that it cannot be a man.

§28.3.2 Second Figure

QaLaMi-2

28.18		*	¤	‡
f_1	d_1, d_2	d_2	–	d_1
f_2	d_1	d_1	d_2	–
f_3	d_2	d_2	d_1	–

$$f_2 \subseteq f_1{}^{\ddagger}, f_3 \subseteq f_1{}^{*} \text{ but } f_3 \subseteq f_2{}^{¤}, f_2 \subseteq f_3{}^{¤}.$$

Terms fitting the model are 'white' 'man' 'swan'. A white man and a white swan are supposed to exist. Each must be of its own kind and cannot be of the other's. The swan is necessarily white, the man contingently white.

QaLaLo-2

28.19		*	¤	‡
f_1	d_1, d_2	d_1	–	d_2, d_3
f_2	d_2, d_3	–	–	d_1, d_2, d_3
f_3	d_1	d_1	d_2, d_3	–
f_4	d_2	d_2	d_1, d_3	–
f_5	d_3	d_3	d_1, d_2	–

$$f_2 \subseteq f_1{}^{\ddagger}, f_3 \subseteq f_1{}^{*} \text{ but } f_3 \cap f_2{}^{¤} = \varnothing.$$

Terms fitting this model are 'white' 'sleeping' 'swan', and as f_4 and f_5 'man' and 'horse'. The model supposes that a white swan, a white man and a non-white horse exist. Each is necessarily of its own kind and cannot be of the others'. The swan is necessarily white, the man contingently white. All three are contingently asleep, but only the man and the horse are actually asleep.

QaLaQi-2

28.20		*	¤	‡
f_1	d_1, d_2	d_1	–	d_2
f_2	d_2	d_2	d_1	–
f_3	d_1	d_1	d_2	–

$$f_2 \subseteq f_1{}^{\ddagger}, f_3 \subseteq f_1{}^{*} \text{ but } f_3 \cap f_2{}^{\ddagger} = \varnothing, f_2 \cap f_3{}^{\ddagger} = \varnothing.$$

Terms fitting the model are 'white' 'swan' 'man'. The model supposes that a swan and a man exist, each necessarily of its own kind and only that kind. Both are white, the swan necessarily and the man contingently.

QaXaMo-2
XaQaMo-2

28.21		*	¤	‡
f_1	d_1	–	–	d_1
f_2	d_1	d_1	–	–
f_3	d_1	d_1	–	–

$$\begin{cases}(i) f_2 \subseteq f_1^{\ddagger}, f_3 \subseteq f_1 \\ (ii) f_2 \subseteq f_1, f_3 \subseteq f_1^{\ddagger}\end{cases} \text{but } f_3 \subseteq f_2^{*}.$$

Aristotle's terms 'healthy' 'animal' 'man' (A18, 37b35-38) fit the model. The model supposes that a healthy man is the sole being. He is necessarily a man and necessarily an animal but contingently healthy.

LaQaLo-2

28.22		*	¤	‡
f_1	d_1	d_1	–	d_2
f_2	d_1	d_1	–	d_2
f_3	d_2	–	–	d_1, d_2

$$f_2 \subseteq f_1^{*}, f_3 \subseteq f_1^{\ddagger} \text{ but } f_3 \cap f_2^{¤} = \varnothing.$$

This is a development of model 22.7, simply making explicit what is taken as contingent there. As there, the terms 'animal' 'man' 'white', which Aristotle deploys against Baroco LXL, can be adapted to fit the model on either of two metaphysical assumptions - that (1) an aggregate, such as a white-man-on-a-white-horse, or (2) a purely potential member of some species, such as a seed, might be not necessarily anything, and therefore contingently an animal and a man.

Model 28.22 can be extended, as we earlier extended 22.7:

28.23		*	¤	‡
f_1	d_1	d_1	–	d_2
f_2	d_1	d_1	–	d_2
f_3	d_2	–	–	d_1, d_2
f_4	d_1, d_2	d_1, d_2	–	–

As before, this extension re-classifies d_2 as being necessarily in the genus f_4 to which f_1 belongs but not necessarily in any species of that genus.

LaQaQi-2

28.24		*	¤	‡
f_1	d_1, d_2	d_2	–	d_1, d_3
f_2	d_2	d_2	d_1, d_3	–
f_3	d_3	d_3	d_1, d_2	–
f_4	d_1	d_1	d_2, d_3	–

$f_2 \subseteq f_1^{*}, f_3 \subseteq f_1^{\ddagger}$ but $f_3 \cap f_2^{\ddagger} = \varnothing, f_2 \cap f_3^{\ddagger} = \varnothing$.

Aristotle's terms 'white' 'swan' 'man' (38b17-22) fit our model. The model supposes that there are three beings - a white being that is neither swan nor man (say a horse), a (white) swan, and a non-white man. For the first of these, being a horse is necessary, being white contingent, and being either a swan or a man impossible. For the swan, being a swan and being white are necessary, and being a man or a horse impossible. For the man, being a man is necessary, being white contingent, and being a swan or a horse impossible.

XeQaMo-2

28.25		*	¤	‡
f_1	d_1, d_2	–	–	d_1, d_2, d_3
f_2	d_3	d_3	d_1, d_2	–
f_3	d_3	d_3	d_1, d_2	–
f_4	d_1	d_1	d_2, d_3	–
f_5	d_2	d_2	d_1, d_3	–

$f_2 \cap f_1 = \varnothing, f_3 \subseteq f_1^{\ddagger}$ but $f_3 \subseteq f_2^{*}$.

Terms fitting the model are 'approaching' 'man' 'man', with 'cow' and 'hen' as f_4 and f_5. The model supposes that a cow, a hen and a man exist. Each must be of its own kind and cannot be of any other kind. The man is not approaching but the cow and the hen are. All three are contingently approaching.

LoQaMo-2

28.26		*	¤	‡
f_1	d_1	–	d_2	d_1
f_2	d_1, d_2	d_1, d_2	–	–
f_3	d_1	d_1	d_2	–
f_4	d_2	d_2	d_1	–

$f_2 \cap f_1^{¤} \neq \varnothing, f_3 \subseteq f_1^{\ddagger}$ but $f_3 \subseteq f_2^{*}$.

Terms fitting the model are 'white', 'animal', 'man'. Assuming that there exists a raven d_2 and a man d_1, some animal (d_2) is necessarily not white, and every man is contingently white, but every man is necessarily an animal.

§28.3.3 Third Figure

QiLaMo-3

28.27		*	¤	‡
f_1	d_1, d_2	d_1, d_2	–	–
f_2	d_1, d_2	d_1, d_2	–	–
f_3	d_1, d_2	d_1	–	d_2
f_4	d_1	d_1	d_2	–
f_5	d_2	d_2	d_1	–

$f_1 \cap f_3^{\ddagger} \neq \varnothing, f_3 \subseteq f_2^{*}$ but $f_2 \subseteq f_1^{*}$.

Terms fitting the model are 'animal' 'animal' 'white', and as f_4 and f_5, 'swan' and 'horse'. The model supposes that a white swan and a white horse exist. Each is necessarily of its own kind and cannot be of the other's. Each is necessarily an animal. The swan is necessarily white, the man contingently so.

QaLeMi-3

28.28		*	¤	‡
f_1	d_1	–	d_2	d_1
f_2	d_2	d_2	d_1	–
f_3	d_1	d_1	d_2	–

$f_3 \subseteq f_1^{\ddagger}, f_3 \subseteq f_2^{\text{¤}}, f_3 \subseteq f_3^{\text{¤}}$ but $f_1 \subseteq f_2^{\text{¤}}, f_2 \subseteq f_1^{\text{¤}}$.

At A15,34b31-35a2 Aristotle uses the terms 'raven' 'reasoning' 'man' to show that premisses X'eQa-1 yield no Q-conclusion and no L-conclusion. If we change the order to 'reasoning' 'raven' 'man', the terms fit our present model. The model supposes that a raven and a man exist. For the raven, it is necessary to be a raven, but it is impossible to be a man or to reason. The man is necessarily a man, cannot be a raven, and - like us, dear reader - is reasoning though contingently.

QaLeMo-3

28.29		*	¤	‡
f_1	d_1, d_2	d_2	–	d_1
f_2	d_1	d_1	d_2	–
f_3	d_2	d_2	d_1	–

$f_2 \subseteq f_1{}^{\ddagger}, f_2 \subseteq f_3{}^{¤}$ and $f_3 \subseteq f_2{}^{¤}$ but $f_3 \subseteq f_1{}^{*}$.

Aristotle's terms 'white' 'animal' 'snow' (A15, 35a20-24), designed to refute QaXe-1 moods with negative or Q-conclusions, fit our model as follows. The model supposes that there are two white beings - an animal which is contingently white and some snow which is necessarily white. For the animal, being an animal is necessary, being snow impossible. For the snow, it is the other way round. Neither being is contingently either animal or snow.

§28.4 Corollaries

As foreshadowed in §20.2, our semantics does not accept the opposition $La \parallel Qi$ because, even though Lab^a is incompatible with the first disjunct of Qab^i, it is compatible with the second, as Figure 28.2 shows.

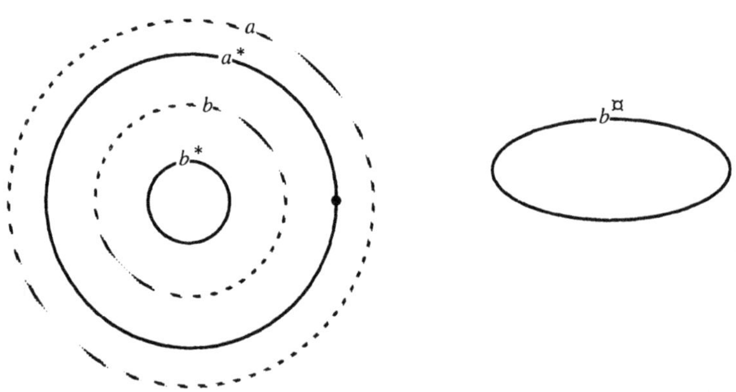

Fig. 28.2

The dot represents an a which is contingently b, when every b is necessarily a. Under these circumstances, Qab^i and Lab^a are both true.

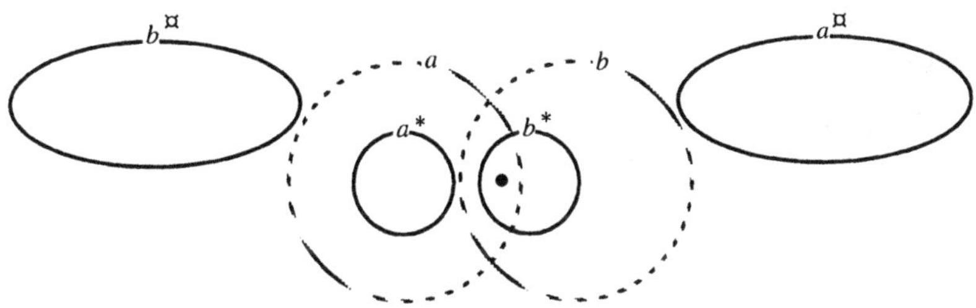

Fig. 28.3

The dot represents an a which is necessarily b, when every b is contingently a. Under these circumstances, Lab^i and Qab^a are both true.

However, the opposition $Le \parallel Qi$ does hold:

Proof If either $f_2 \cap f_1^{\ddagger} \neq \varnothing$ or $f_1 \cap f_2^{\ddagger} \neq \varnothing$
then either $f_2 \not\subset f_1^{¤}$ or $f_1 \not\subset f_2^{¤}$
$\because$ if $f_2 \cap f_1^{\ddagger} \neq \varnothing$ then $f_2 \not\subset f_1^{¤}$ ($\because f_1^{\ddagger} \cap f_1^{¤} = \varnothing$)
and if $f_1 \cap f_2^{\ddagger} \neq \varnothing$ then $f_1 \not\subset f_2^{¤}$ ($\because f_2^{\ddagger} \cap f_2^{¤} = \varnothing$).

The oppositions which are sanctioned by our interpretation are represented in Figure 28.4.

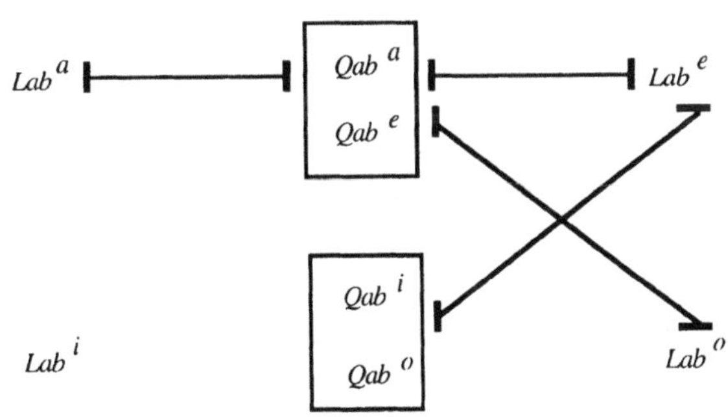

Fig. 28.4

This diagram should be compared with Figure 2.2. We noted in §2 that Aristotle thinks that Lab^i and Lab^o are both contrary to Qab^e; the diagram shows that on our interpretation he is not right.

Figure 28.4 makes it clear that the following Q/M subordination principles (and only these) are valid:

$$\frac{Qab^a}{Mab^a} \quad \frac{Qab^i}{Mab^i} \quad \frac{Qab^a}{Mab^i} \quad \frac{Qab^e}{Mab^o}$$

These are derivable by rule K from Lo-rep and Le-rep. Their validity, along with that of Barbara QXQ and Darii QXQ, implies that of Barbara QXM and Darii QXM. Of the other Figure 1 QXM moods, Ferio is equivalent by rule C to QaLaXe-2, but Celarent is invalid:

28.30		*	‡
f_1	d_1	–	d_1
f_2	d_1	d_1	–
f_3	d_1	d_1	–

$f_2 \subseteq f_1^{\ddagger}, f_3 \subseteq f_2$ but $f_1 \cap f_3^{*} \neq \varnothing$.

(This model also shows the invalidity of Celarent QLM.)

As indicated in §10 Bocardo QXQ is invalid. Its invalidity follows from that of Bocardo QLM (see model 28.25).

Barbara and Darii XQM are valid, so long as we assume the Sun set Theorem.

Barbara XQM If $f_2 \subseteq f_1$ and $f_3 \subseteq f_2^{\ddagger}$ then $f_3 \subseteq f_1^{\dagger}$
$\because f_2^{\dagger} \subseteq f_1^{\dagger}$ (**22.2.1**) and $f_2^{\ddagger} \subseteq f_2^{\dagger}$.

Darii XQM If $f_2 \subseteq f_1$ and either $f_3 \cap f_2^{\ddagger} \neq \varnothing$ or $f_2 \cap f_3^{\ddagger} \neq \varnothing$
then either $f_3 \cap f_1^{\dagger} \neq \varnothing$ or $f_1 \cap f_3^{\dagger} \neq \varnothing$
$\because \begin{cases} f_2^{\ddagger} \subseteq f_2^{\dagger}, f_2^{\dagger} \subseteq f_1^{\dagger} \text{(22.2.1)} \\ f_3^{\ddagger} \subseteq f_3^{\dagger}. \end{cases}$

These moods are provable by K from Baroco and Camestres XLL:

$$\text{Baroco XLL}\ \frac{ab^a \quad Lac^o}{Lbc^o} \ \xrightarrow[Lo \,\|\, Qa]{K} \ \frac{ab^a \quad Qbc^a}{Mac^a}$$

$$\text{Camestres XLL}\ \frac{ab^a \quad Lac^e}{Lbc^e} \ \xrightarrow[Le \,\|\, Qi]{K} \ \frac{ab^a \quad Qbc^i}{Mac^i}$$

However, Celarent and Ferio XQM are invalid, as is shown by model 28.5.

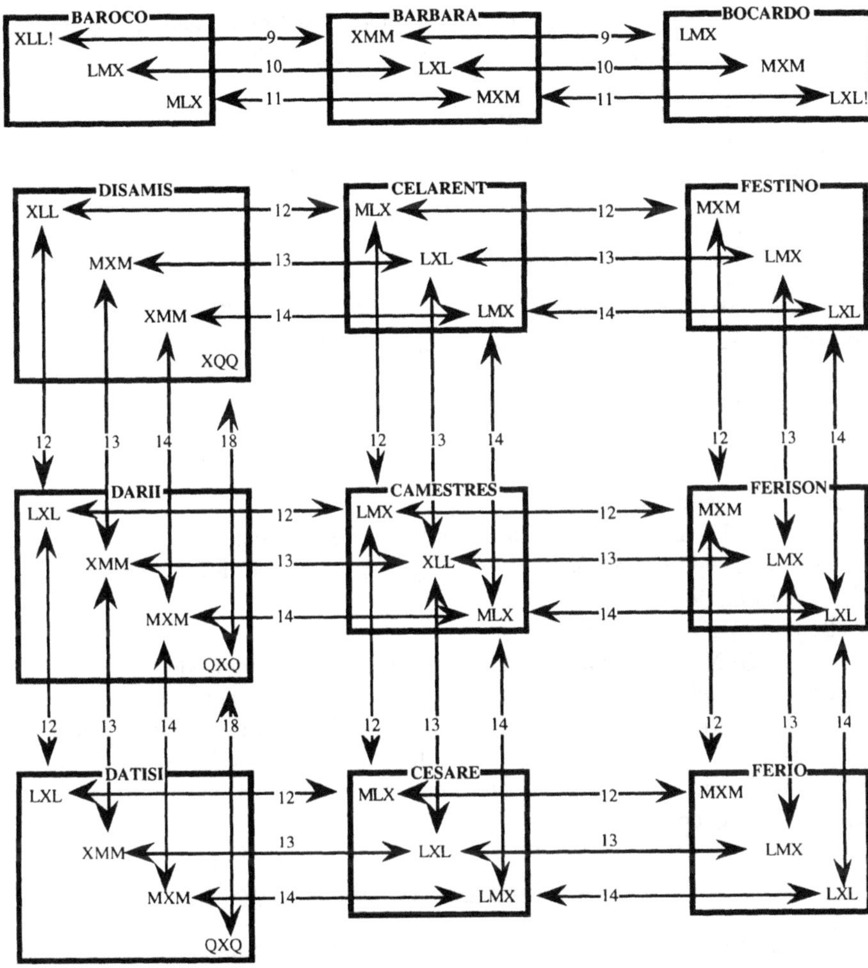
BAROCO
XLL!
LMX
MLX
BARBARA
XMM
LXL
MXM
BOCARDO
LMX
MXM
LXL!
9
10
11
DISAMIS
XLL
MXM
XMM
XQQ
CELARENT
MLX
LXL
LMX
FESTINO
MXM
LMX
LXL
DARII
LXL
XMM
MXM
QXQ
CAMESTRES
LMX
XLL
MLX
FERISON
MXM
LMX
LXL
DATISI
LXL
XMM
MXM
QXQ
CESARE
MLX
LXL
LMX
FERIO
MXM
LMX
LXL
12
13
14
18

Fig. 28.5

Invalid are Barbara (and hence Celarent) QQQ, Darii (and hence Ferio, Datisi, Disamis, Ferison) QQQ, Darapti (and hence Felapton) QQQ.

28.31		*	¤	‡
f_1	d_1	–	–	d_1
f_2	d_1	–	–	d_2
f_3	d_2	d_2		

$f_2 \subseteq f_1{}^{\ddagger}$ and $f_3 \subseteq f_2{}^{\ddagger}$ but $f_3 \not\subset f_1{}^{\ddagger}$.

Any axiomatisation of *AQ* based on rules *U, M, C, K, T,* will have to include, in addition to a basis for assertoric syllogistic, one axiom from each of the following groups:
(1) La-rep or Mo-rep;
(2) Lo-rep or Ma-rep;
(3) Li-conv or Me-conv;
(4) Le-conv or Mi-conv;
(5-8) La-, Le-, Li- and Lo-subordination, or their M-equivalents;
plus one modal syllogism from each of the groups (9-14, 18) in Figure 28.5;
(15) Qi-conv or Qo-conv;
(16) Barbara QXQ or Celarent QXQ;
(17) Cesare QLX or Ferio QXM or Camestres LQX or Ferison QXM.

Figure 28.5 displays the salient relations. The Figure does not show subaltern or subordinate syllogisms, or moods containing negative Q-forms. Moods connected by unbroken vertical lines are inter-derivable by Direct Reduction, ones connected by unbroken horizontal lines by Indirect Reduction. Broken lines indicate derivability by Q-conversion.

§29. AMPLIATED CONTINGENCY-FORMS

§1 introduced the idea that some contingency-forms are ampliated, *i.e.* that their logical subject adds the qualification "possible" to their grammatical subject. That section also mentioned that "possible" can be taken to mean either M or Q. It is time now to face the question, Do Aristotle's ampliated contingency-forms ampliate to the possible (M) or to the contingent? The two interpretations, in the case of universal contingency-forms, are shown in Figure 29.1. Angelelli comments:

> I do not see any reason to choose one-sided rather than two-sided possibility as an interpretation of this ενδεχεται.[1]

---

[1] ANGELELLI p.205.

Notwithstanding, he goes on to use one-sided possibility.

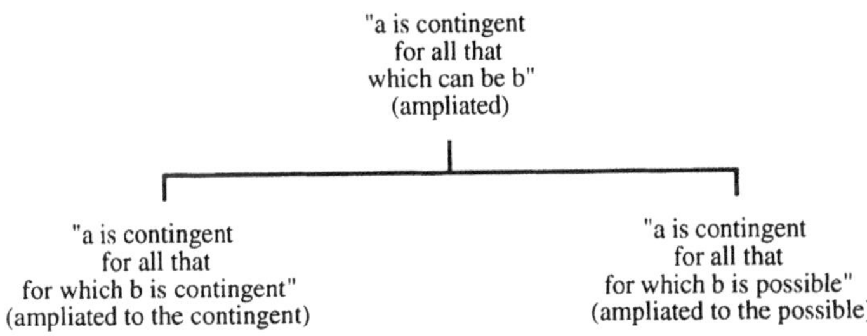

Fig. 29.1

Angelelli's position can be justified by reference to Aristotle's examples of contingency-propositions in the chapters we are interpreting as dealing with ampliated forms, *i.e.* A14, A17, A20. 33b3-17 claims as true the proposition $Q'(\text{white})(\text{man})^a$; but, if that proposition is ampliated to the contingent, its truth would imply the existence of subjects which are contingently men. 37a38-b10 claims as true both the above proposition and $Q'(\text{white})(\text{horse})^a$; so, the implication of ampliation to the contingent would be the existence of subjects which are contingently horses. A20, 39b2-6 claims as true the propositions $Q'(\text{man})(\text{white})^i$, $Q'(\text{horse})(\text{white})^i$ and $Q'(\text{animal})(\text{white})^i$. If we assume that nothing is contingently man, horse or animal, then we must take it that these contingency-particulars are true by virtue of conversion from the propositions

$$Q'(\text{white})(\text{man})^i, Q'(\text{white})(\text{horse})^i, Q'(\text{white})(\text{animal})^i.$$

If ampliation is to the contingent, however, then the implication again is that there are contingent men, contingent horses, and contingent animals. The only way to escape these implications is by supposing that ampliated contingency-forms ampliate to the possible, not to the contingent. Then the above propositions imply only that there are *possible* men, horses and animals. This is the approach I shall follow. The alternative approach - ampliating to the contingent - will be investigated later.

We define a system AQ', which adds Q'-forms to the language of AQ. We shall also add for each Q'-form a contradictory:

$$\tilde{Q}'ab^a,\ \tilde{Q}'ab^e,\ \tilde{Q}'ab^i,\ \tilde{Q}'ab^o.$$

As axioms we take the axioms of *AQ* plus

Q'-subord $\dfrac{Q'ab^a}{Qab^a}$ *Q'-subord* $\dfrac{Qab^i}{Q'ab^i}$

Q'i-conv $\dfrac{Q'ab^i}{Q'ba^i}$

Q'-conv $\dfrac{Q'ab^a}{Q'ab^e}$ *Q'-conv* $\dfrac{Q'ab^e}{Q'ab^a}$

Q'-conv $\dfrac{Q'ab^i}{Q'ab^o}$ *Q'-conv* $\dfrac{Q'ab^o}{Q'ab^i}$

Barbara Q'Q'Q' $\dfrac{Q'ab^a \quad Q'bc^a}{Q'ac^a}$ *Darii Q'Q'Q'* $\dfrac{Q'ab^a \quad Q'bc^i}{Q'ac^i}$

A semantics for this system ought to make the axioms valid. (There is no need to add Q'aLaXe-2 as an axiom, since it is implied by QaLaXe-2 and Q'-subordination. And we do not need partial Q'a-conversion as axiomatic, since it is provable:

$$\dfrac{\dfrac{\dfrac{Q'ab^a}{Qab^a}\,Q'-subord}{Qba^i}\,Qa-conv}{Q'ba^i}\,Q'-subord$$

from Q'-subord and Qa-conv.) A semantics should make invalid the formulae: Q'a-conv, Q'iQ'aQ'i-1, Q'iQ'aXi-1, Q'iQ'aXo-1, Q'aQ'aQ'i-2, Q'aQ'aXi-2, Q'aQ'aXo-2, Q'iQ'iQ'i-3, Q'iQ'iXi-3, Q'iQ'iXo-3.

§29.1 Formal semantics

Such a semantics can be constructed by extending our semantics for AQ. AQ' models and AQ' validity are defined analogously to AQ models and AQ validity. The distinctive features of AQ' semantics are as follows.

29.1.1 $Q'ab^a$ is true iff $f_2^\dagger \subseteq f_1^\ddagger$, *i.e.* iff the *b*s' dagger set is included in the *a*s' double-dagger set.

29.1.2 $Q'ab^i$ is true iff $f_2^\dagger \cap f_1^\ddagger \neq \varnothing \vee f_1^\dagger \cap f_2^\ddagger \neq \varnothing$, *i.e.* iff EITHER the *b*s' dagger set intersects the *a*s' double-dagger set OR the *a*s' dagger set intersects the *b*s' double-dagger set.

Q'e- and Q'o-forms will again be taken as equivalent to their affirmative counterparts.

§29.2 Validity of the axioms

The following formulae are valid.

Q'-subord If $f_2^\dagger \subseteq f_1^\ddagger$ then $f_2 \subseteq f_1^\ddagger$
$\because f_2 \subseteq f_2^\dagger$.
If either $f_2 \cap f_1^\ddagger \neq \varnothing$ or $f_1 \cap f_2^\ddagger \neq \varnothing$
then either $f_2^\dagger \cap f_1^\ddagger \neq \varnothing$ or $f_1^\dagger \cap f_2^\ddagger \neq \varnothing$
$\because f_2 \subseteq f_2^\dagger, f_1 \subseteq f_1^\dagger$.

Q'-conv By definition.

Q'i-conv If either $f_2^\dagger \cap f_1^\ddagger \neq \varnothing$ or $f_1^\dagger \cap f_2^\ddagger \neq \varnothing$
then either $f_1^\dagger \cap f_2^\ddagger \neq \varnothing$ or $f_2^\dagger \cap f_1^\ddagger \neq \varnothing$.

Barbara Q'Q'Q' If $f_2^\dagger \subseteq f_1^\ddagger$ and $f_3^\dagger \subseteq f_2^\ddagger$ then $f_3^\dagger \subseteq f_1^\ddagger$
$\because f_2^\ddagger \subseteq f_2^\dagger$.

Darii Q'Q'Q' If $f_2{}^{\dagger} \subseteq f_1{}^{\ddagger}$ and either $f_3{}^{\dagger} \cap f_2{}^{\ddagger} \neq \varnothing$ or $f_2{}^{\dagger} \cap f_3{}^{\ddagger} \neq \varnothing$
then either $f_3{}^{\dagger} \cap f_1{}^{\ddagger} \neq \varnothing$ or $f_1{}^{\dagger} \cap f_3{}^{\ddagger} \neq \varnothing$
$\because f_2{}^{\ddagger} \subseteq f_2{}^{\dagger}, f_1{}^{\ddagger} \subseteq f_1{}^{\dagger}$.

§29.3 *Invalidity of the axiomatic rejections*

Invalid are

Q'a conv

29.1		*	¤	‡	†
f_1	d_1, d_2	d_1	–	d_2, d_3	d_1, d_2, d_3
f_2	d_3	d_3	d_1, d_2	–	d_3
f_3	d_2	d_2	d_1, d_3	–	d_2
f_4	d_1	d_1	d_2, d_3	–	d_1

$f_2{}^{\dagger} \subseteq f_1{}^{\ddagger}$ but $f_1{}^{\dagger} \not\subset f_2{}^{\ddagger}$.

Terms fitting this model are 'white' 'man' 'horse'. The model then supposes the existence of three things - something white which is neither man nor horse (say a swan), a non-white man and a white horse. The man must be a man, the horse a horse and the swan a swan. The man cannot be a horse or swan, the horse cannot be a man or swan, and the swan cannot be a man or horse. The swan is necessarily white, the man and horse contingently so. All possible men, horses and swans are actual. But one possible white - the man - is not actually white.

Q'iQ'aQ'i-1
Q'iQ'aXi-1

29.2		*	¤	‡	†
f_1	d_1	d_1	d_2	–	d_1
f_2	d_1, d_2	–	–	d_1, d_2	d_1, d_2
f_3	d_2	d_2	d_1	–	d_2

$$f_1{}^{\dagger} \cap f_2{}^{\ddagger} \neq \varnothing, f_3{}^{\dagger} \subseteq f_2{}^{\ddagger} \text{ but } \begin{cases} (i) f_3{}^{\dagger} \cap f_1{}^{\ddagger} = \varnothing, f_1{}^{\dagger} \cap f_3{}^{\ddagger} = \varnothing \\ (ii) f_3 \cap f_1 = \varnothing. \end{cases}$$

In §16.1 we briefly discussed Aristotle's disproof of Q'iQ'aQ'i-1. The above model gives a semantic justification for the procedure he uses there. Following Aristotle's specifications, the *c*s are those *b*s that are not *a*s. They are also the *b*s for which *a* is impossible. *a* is then not contingent for *c*, even though *a* is contingent for some *b* (because the sole *a* is a contingent *b*).

Aristotle's terms 'animal' 'white' 'cloak' (33b3-17) can be adapted to our model. The model supposes that an animal and a cloak exist. Each is necessarily of its own kind and incapable of belonging to the other kind. Both are white, though contingently so. Each term actually applies to everything for which it is possible.

Q'iQ'aXo-1

29.3		*	¤	‡	†
f_1	d_1, d_2	d_1, d_2	–	–	d_1, d_2
f_2	d_1, d_2	–	–	d_1, d_2	d_1, d_2
f_3	d_2	d_2	d_1	–	d_2
f_4	d_1	d_1	d_2	–	d_1

$f_1{}^{\dagger} \cap f_2{}^{\ddagger} \neq \varnothing, f_3{}^{\dagger} \subseteq f_2{}^{\ddagger}$ but $f_3 \subseteq f_1$.

Aristotle's terms 'animal' 'white' 'man' (33b3-17) can be adapted to our model. The model supposes that a man and a non-human animal exist, say a horse. Each must belong to its own kind and cannot belong to the other kind, though both are necessarily animals. Both are white though contingently so. In every case, the actual coincides with the possible.

Q'aQ'aQ'i-2
Q'aQ'aXi-2

29.4		*	¤	‡	†
f_1	d_1, d_2	–	–	d_1, d_2	d_1, d_2
f_2	d_1	d_1	d_2	–	d_1
f_3	d_2	d_2	d_1	–	d_2

$$f_2{}^{\dagger} \subseteq f_1{}^{\ddagger}, f_3{}^{\dagger} \subseteq f_1{}^{\ddagger} \text{ but } \begin{cases} (i) f_2{}^{\dagger} \cap f_3{}^{\ddagger} = \varnothing, f_3{}^{\dagger} \cap f_2{}^{\ddagger} = \varnothing \\ (ii) f_2 \cap f_3 = \varnothing. \end{cases}$$

Aristotle's terms 'white' 'man' 'horse' (37a38-b10) fit our model. The model supposes that a man and a horse exist, each belonging necessarily to its own kind and impossibly to the other. Both are supposed to be white but contingently so. In the case of all terms, the possible coincides with the actual.

Q'aQ'aXo-2

29.5		*	¤	‡	†
f_1	d_1	–	–	d_1	d_1
f_2	d_1	d_1	–	–	d_1
f_3	d_1	d_1	–	–	d_1

$f_2{}^{\dagger} \subseteq f_1{}^{\ddagger}, f_3{}^{\dagger} \subseteq f_1{}^{\ddagger}$ but $f_3 \subseteq f_2$.

The terms 'white' 'man' 'man' fit our model, which supposes that only a white man exists, necessarily human but contingently white. Again the possible is the same as the actual in this model.

Q'iQ'iQ'i-3
Q'iQ'iXi-3

29.6		*	¤	‡	†
f_1	d_1	d_1	d_2	–	d_1
f_2	d_2	d_2	d_1	–	d_2
f_3	d_1, d_2	–	–	d_1, d_2	d_1, d_2

$$f_1^{\dagger} \cap f_3^{\ddagger} \neq \varnothing, f_2^{\dagger} \cap f_3^{\ddagger} \neq \varnothing \text{ but } \begin{cases} (i) f_1^{\dagger} \cap f_2^{\ddagger} = \varnothing, f_2^{\dagger} \cap f_1^{\ddagger} = \varnothing \\ (ii) f_2 \cap f_1 = \varnothing. \end{cases}$$

Aristotle's terms 'horse' 'man' 'white' (39b2-6) fit our model. The model supposes that a horse and a man exist, each necessarily of its own kind and impossibly of the other. Both are white though contingently so. In all cases what is possible is also actual.

Q'iQ'iXo-3

29.7		*	¤	‡	†
f_1	d_1	d_1	–	–	d_1
f_2	d_1	d_1	–	–	d_1
f_3	d_1	–	–	d_1	d_1

$$f_1^{\dagger} \cap f_3^{\ddagger} \neq \varnothing, f_2^{\dagger} \cap f_3^{\ddagger} \neq \varnothing \text{ but } f_2 \subseteq f_1.$$

Aristotle's terms 'animal' 'man' 'white' (39b2-6) fit the model, which supposes that a white man exists, necessarily human but contingently white. Again, what is possible is actual in this model.

§29.4 Corollaries

The oppositions endorsed by this semantics are not quite the same as for our semantics for unampliated contingency-forms. (i) $Q'i \parallel La$ is rejected as with unampliated Q-forms: Lab^a is compatible with either one of the disjuncts of $Q'ab^i$. It is compatible with some possible *a*'s being a contingent *b*and with some possible *b*'s being a contingent *a*, as Figure 29.2 shows. The dot in Figure 29.2 can indifferently represent a possible *a* which is contingently *b*, or a possible *b* which is contingently *a*, when every *b* is necessarily *a*. (ii) Also not accepted (again as with unampliated forms) is the opposition $Li \parallel Q'a$, as Figure 29.3 shows.

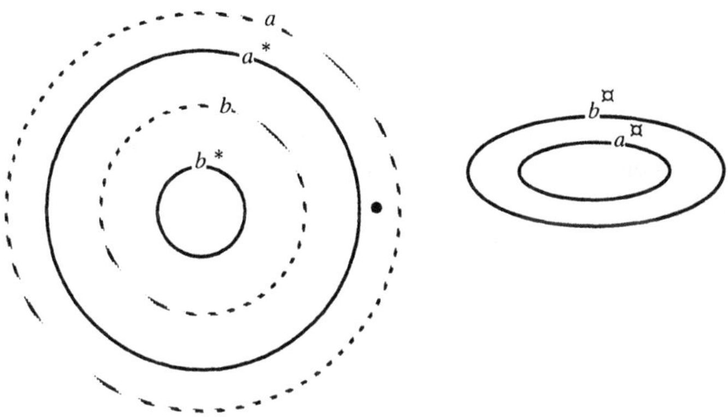

Fig. 29.2

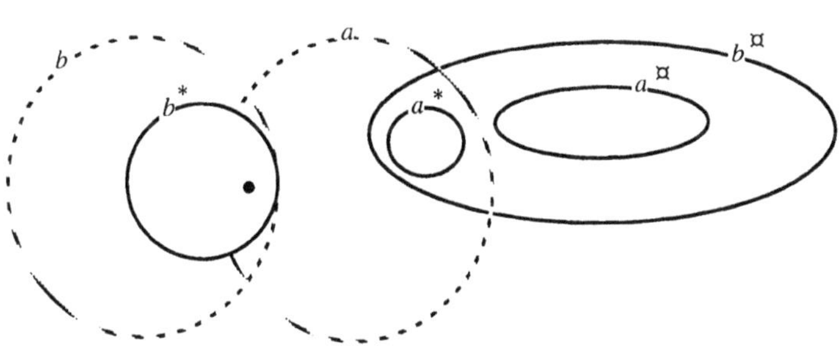

Fig. 29.3

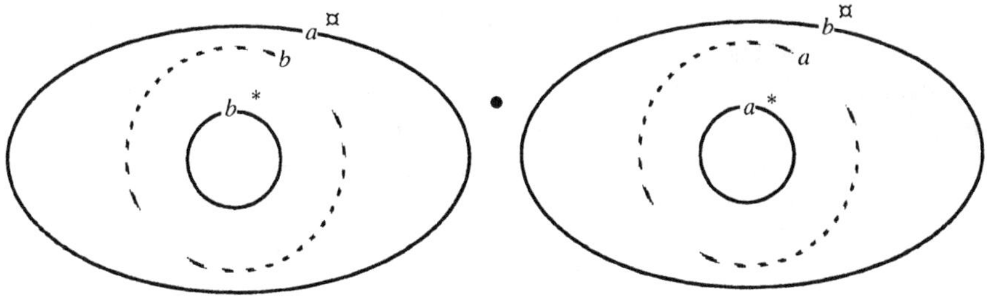

Fig. 29.4

The dot in Figure 29.3 represents an a which is necessarily b, when every possible b is contingently a. Under these circumstances, Lab^i and $Q'ab^a$ are both true. (iii) Finally, in contrast with the unampliated case, the opposition $Le \parallel Q'i$ does *not* hold, as Figure 29.4 shows. The dot in Figure 29.4 represents a possible a which is a contingent b, when no b is a possible a and no a is a possible b. Under this circumstance, both $Q'ab^i$ and Lab^e are true. The oppositions which are sanctioned by our interpretation are shown in Figure 29.5.

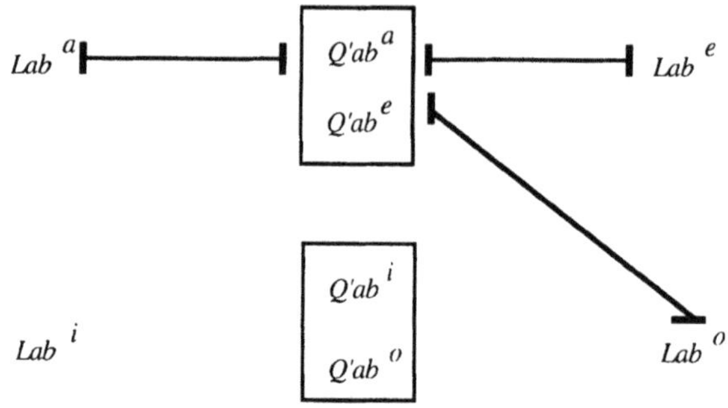

Fig. 29.5

§30. THE RELATION OF AMPLIATED TO UNAMPLIATED SYSTEMS

I argued in §7 (i) that Aristotle intends the mixed contingency-moods to involve unampliated forms but the uniform moods to involve ampliated contingency forms, and (ii) that the two types of contingency form are semantically distinct. We have seen that the two types of contingency-form stand in slightly different relations of opposition with apodeictic forms. We are about to see that, so far as Aristotle's axioms go, we will not fall into error by treating them all as ampliated, provided we accept the Sun set Theorem. Nonetheless, when we consider derived theses and formulae rejected by him, semantic differences are evident between ampliated and unampliated forms. Further, we shall see that, given strong enough base assumptions point (ii) does not hold in general, because universal Q- and Q'-forms are equivalent to one another.

Given the Sun set Theorem, the ampliated versions of the axioms of AQ are all valid. This has already been demonstrated for the one-premissed axioms. For Barbara (and Celarent) Q'XQ' and Darii (and Ferio) Q'XQ' we have:

If $f_2{}^{\dagger} \subseteq f_1{}^{\ddagger}$ and $f_3 \subseteq f_2$ then $f_3{}^{\dagger} \subseteq f_1{}^{\ddagger}$
$\because f_3{}^{\dagger} \subseteq f_2{}^{\dagger}$ - by **22.2.1**.

If $f_2{}^{\dagger} \subseteq f_1{}^{\ddagger}$ and $f_3 \cap f_2 \neq \varnothing$ then $f_3{}^{\dagger} \cap f_1{}^{\ddagger} \neq \varnothing$
$\because f_3{}^{\dagger} \cap f_2{}^{\dagger} \neq \varnothing$ - $\because f_3 \subseteq f_3{}^{\dagger}, f_2 \subseteq f_2{}^{\dagger}$.

Relative to the Sun set Theorem, we can therefore agree with Hintikka's[2] claim that the definition **29.1.1** "suffices to explain everything that Aristotle says and does" in the contingency-syllogistic, if we take Aristotle's sayings and doings to include just his axioms. We get all of his Q-axioms (mixed as well as pure) if we treat *all* contingency-forms as ampliating to the possible.

However, if we assume S0 rather than S3, and ampliate all contingency-forms to the possible, then Barbara Q'XQ' is invalid:

30.1		*	¤	‡	†
f_1	d_1	d_1	–	d_2	d_1, d_2
$\overline{f_1}$	d_2	–	d_1	d_2	d_2
f_2	d_2	–	d_1	d_2	d_2
$\overline{f_2}$	d_1	d_1	–	d_2	d_1, d_2
f_3	d_2	–	–	d_1, d_2	d_1, d_2
$\overline{f_3}$	d_1	–	–	d_1, d_2	d_1, d_2

$f_2{}^{\dagger} \subseteq f_1{}^{\ddagger}$ and $f_3 \subseteq f_2$ but $f_3{}^{\dagger} \not\subseteq f_1{}^{\ddagger}$.

but Darii Q'XQ' is valid:

Proof If $f_2{}^{\dagger} \subseteq f_1{}^{\ddagger}$ and $f_3 \cap f_2 \neq \varnothing$ then $f_3{}^{\dagger} \cap f_1{}^{\ddagger} = \varnothing$
$\because f_2 \subseteq f_2{}^{\dagger}, f_2 \subseteq f_2{}^{\dagger}$.

When we look to derived theses, we find differences resulting from the differences in opposition. For instance, Disamis QLM is valid and is provable from Celarent LLL by rule *K*, relying on the opposition $Le \parallel Qi$. But Disamis Q'LM is invalid, as the following model shows.

[2] HINTIKKA p.39.

30.2		*	¤	‡	†
f_1	d_1	d_1	d_2	d_3	d_1, d_3
f_2	d_2	d_2	d_1	d_3	d_2, d_3
f_3	d_2	d_2	d_1	d_3	d_2, d_3

$f_3^{\dagger} \cap f_1^{\ddagger} \neq \varnothing$ and $f_3 \subseteq f_2^{*}$ but $f_2 \subseteq f_1^{¤}$ and $f_1 \subseteq f_2^{¤}$.

Further differences exist among rejected formulae. For instance, QaLaLe-2 is invalid and is rejected by Aristotle, because he rejects QaLaLo-2. But Q'aLaLe-2 is valid:

Proof If $f_2^{\dagger} \subseteq f_1^{\ddagger}$ and $f_3 \subseteq f_1^{*}$ then $f_3 \subseteq f_2^{¤}$
$\because f_1^{\ddagger} \cap f_1^{*} = \varnothing$.

Let us now note an important difference between two of our earlier proposals regarding the definition of sun sets. According to definition **25.4**, a set's sun set is the union of all star sets which it excludes. According to **25.5**, a set's sun set is the union of all star sets which exclude a star set including the given set. It turns out that the first of these definitions, but not the second, annihilates the semantic distinction between universal ampliated and unampliated forms.

On **25.4** it follows that a given set's double-dagger set itself has a sun set, and that this sun set is the union of the given set's star- and sun sets.

30.1 $\left(f^{\ddagger}\right)^{¤} = \bigcup_{h^*}\left(h^* \cap f^{\ddagger} = \varnothing\right) = \bigcup_{h^*}\left(h^* \subseteq \left(f^* \cup f^{¤}\right)\right) = \left(f^* \cup f^{¤}\right)$

We also saw earlier that any set has a dagger set, namely the complement of its sun set. Hence a given set's double-dagger set has a dagger set. This dagger set turns out to be the double-dagger set itself.

30.2 $\left(f^{\ddagger}\right)^{\dagger} = \overline{\left(f^{\ddagger}\right)^{¤}} = \overline{\left(f^{*} \cup f^{¤}\right)} = f^{\ddagger}$.

What is possibly contingently a is the same as what is contingently a.

Hence, we have the **Ampliation Theorem**

30.3 $\left(f_2 \subseteq f_1^{\ddagger}\right) \supset \left(f_2^{\dagger} \subseteq f_1^{\ddagger}\right)$

Proof $\left(f_2 \subseteq f_1^{\ddagger}\right) \supset \left(f_2^{\dagger} \subseteq \left(f_1^{\ddagger}\right)^{\dagger}\right)$ - **22.2.1**

and $\left(f_2^{\dagger} \subseteq \left(f_1^{\ddagger}\right)^{\dagger}\right) \supset \left(f_2^{\dagger} \subseteq f_1^{\ddagger}\right)$ - **30.2.**

Thus, if Qab^a then $Q'ab^a$. Since the reverse implication obviously holds, the ampliated and unampliated universal forms are mutually equivalent, given **25.4**. This annihilates any semantic distinction between ampliated and unampliated universal forms.

Certainly, it seems that Barbara QQQ, as against Barbara Q'Q'Q', is invalid because the premisses imply only that every actual (not every possible)*c* is an *a*. And certainly, it seems that Barbara Q'XQ', as against Barbara QXQ, is invalid for want of a genuine middle term: the major premiss is about all *b*s, but the minor relates to contingent *b*s. In fact, however, Barbara QQQ is valid given **30.3**:

Proof If $f_2 \subseteq f_1^{\ddagger}$ and $f_3 \subseteq f_2^{\ddagger}$ then $f_3 \subseteq f_1^{\ddagger}$
$\because f_2^{\ddagger} \subseteq f_2^{\dagger}$
and $f_2^{\dagger} \subseteq f_1^{\ddagger}$ - by **22.2.1** and **30.3**.

Not only that, but the stronger Barbara QMQ is valid, as is clear from the above proof.

On the other hand, if we adopt the definition **25.5**, according to which a set's sun set is the union of all star sets which exclude a star set including the given set, then **30.2** cannot be proved. A distinction exists between what is contingently *a* and what is possibly contingently *a*. For, suppose something is necessarily *a* and something is contingently *a* but no star set includes all that is contingently *a*. Then nothing is impossibly contingently *a* (because no star set includes the contingently *a*); but something is not contingently *a* (because something is necessarily *a*). Thus, what is not contingently *a* is not included in the impossibly contingently *a*. Consequently, the possibly contingently *a* is not included in the contingently *a*. This is shown in the following model:

30.3		*	¤	‡	†
f_1	d_1	d_1	–	d_2	d_1, d_2
$(f_1)_Q$	d_2	–	–	d_1, d_2	d_1, d_2

$(f_1)_Q{}^{\dagger} \not\subset f_1^{\ddagger}$.

Neither definition, however, does away with the distinction between *particular* ampliated and unampliated forms. For the existence of possible *b*s that are contingent *a*s does not imply the existence of actual *b*s that are contingent *a*s, given either definition **25.4** or definition **25.5** - as the following model shows.

30.4		*	¤	‡	†
f_1	d_1	d_1	d_2	d_3	d_1, d_3
f_2	d_2	d_2	d_1	d_3	d_2, d_3

$f_2^{\dagger} \cap f_1^{\ddagger} \neq \varnothing$ but $f_2 \cap f_1^{\ddagger} = \varnothing$ and $f_1 \cap f_2^{\ddagger} = \varnothing$.

Darii QQQ is invalid on either definition **25.4** or definition **25.5**:

30.5		*	¤	‡	†
f_1	d_1	d_1	d_2	d_3	d_1, d_3
f_2	d_3	–	d_1, d_2	d_3	d_3
f_3	d_2	d_2	d_1	d_3	d_2, d_3

$f_2 \subseteq f_1^{\ddagger}$ and $f_3^{\ddagger} \cap f_2 \neq \varnothing$ but $f_3 \cap f_1^{\ddagger} = \varnothing$ and $f_1 \cap f_3^{\ddagger} = \varnothing$.

30.6		*	¤	‡	†
f_1	d_1	d_1	d_2	d_3	d_1, d_3
f_2	d_3	–	–	d_1, d_2, d_3	d_1, d_2, d_3
f_3	d_2	d_2	d_1	d_3	d_2, d_3

$f_2 \subseteq f_1^{\ddagger}$ and $f_3^{\ddagger} \cap f_2 \neq \varnothing$ but $f_3 \cap f_1^{\ddagger} = \varnothing$ and $f_1 \cap f_3^{\ddagger} = \varnothing$.

§31. REDUCTION TO ASSERTORIC SYLLOGISTIC

As with system *AL* we can define an assertoric translation for each wff in *AQ*. We expand the system A_L as follows. Corresponding to any term v we introduce the term v_Q. The resulting language will contain, in addition to ordinary assertoric forms and the modalised assertoric forms of A_L, assertoric forms with modalised terms of the type v_Q.

We define a system A_Q by adding to A_L the contingency-rules

R_{QM} $$\frac{R \quad a_M a_Q{}^a}{p} \longrightarrow \frac{R}{p}$$

provided R contains some member $va_Q{}^a, va_Q{}^i, a_Q v^a$ or $a_Q v^i$

R_{QL} $$\frac{R \quad a_L a_Q{}^e}{p} \longrightarrow \frac{R}{p}$$

The semantic interpretation of assertorics with modalised terms of the type v_Q will be as for ordinary assertorics, with the additional rule

31.1 If f is assigned to v then $f^{\ddagger}$ is assigned to v_Q.

For every form *f* of AQ, there is a form g - or a pair of forms g_1, g_2 - of A_Q such that either (i) under any valuation *f* is true iff g is true, or (ii) under any valuation *f* is true iff both g_1 and g_2 are true, or (iii) under any valuation *f* is true iff either g_1 or g_2 is true. The correspondences are as in A_L, with the following additions.

Qab^a is true in AQ iff $a_Q b^a$ is true in A_Q

$Q'ab^a$ is true in AQ iff $a_Q b_M{}^a$ is true in A_Q

Qab^i is true in AQ iff either $a_Q b^i$ or $b_Q a^i$ is true in A_Q

$Q'ab^i$ is true in AQ iff either $a_Q b_M{}^i$ or $b_Q a_M{}^i$ is true in A_Q

AQ validity statements correspond to truth-functions of A_Q validity statements as AL validity statements to truth-functions of A_L validity statements, with obvious additions.

The decision-procedure is as for A_L, with two extra Extension rules

Extension rule QM If a term v_Q in one member is matched by v_M in another, then add $v_M v_Q{}^a$ to the sequence (provided there is some member $wv_Q{}^a, wv_Q{}^i, v_Q w^a$ or $v_Q w^i$).

Extension rule QL If a term v_Q in one member is matched by v_L in another, then add $v_L v_Q{}^e$ to the sequence.

Here are two examples:

(1) Baroco Q'Q'Q'

$$\frac{Q'ab^a \quad Q'ac^o}{Q'bc^o}$$

is AQ-valid iff both $\left\{\text{either } 1\ \frac{a_Q b_M{}^a \quad a_Q c_M{}^i}{b_Q c_M{}^i} \text{ or } 2\ \frac{a_Q b_M{}^a \quad a_Q c_M{}^i}{b_M c_Q{}^i}\right\}$ and $\left\{\text{either } 3\ \frac{a_Q b_M{}^a \quad a_M c_Q{}^i}{b_Q c_M{}^i} \text{ or } 4\ \frac{a_Q b_M{}^a \quad a_M c_Q{}^i}{b_M c_Q{}^i}\right\}$ are A_Q-valid. In fact, *1-4* are all A_Q invalid since

$\langle a_Q b_M{}^a, a_Q c_M{}^i, b_Q c_M{}^e, b_M b_Q{}^a, \rangle$

$\langle a_Q b_M{}^a, a_Q c_M{}^i, c_M c_Q{}^a, b_M c_Q{}^e \rangle$

$\langle a_Q b_M{}^a, a_M a_Q{}^a, a_M c_Q{}^i, c_M c_Q{}^a, b_Q c_M{}^e, b_M b_Q{}^a \rangle$

$\langle a_Q b_M{}^a, a_M a_Q{}^a, a_M c_Q{}^i, b_M c_Q{}^e \rangle$

are none of them Aristotelian chains (a_Q, c_M, a_M and a_M respectively are twice undistributed).

(2) Cesare QLX

$$\frac{Qab^e \quad Lac^a}{bc^e}$$

is AQ-valid iff $\dfrac{a_Q b^a \quad a_L c^a}{bc^e}$ is A_Q-valid. It is, since it is provable in A_Q:

$$\text{Camestres}\,\frac{a_Q b^a \qquad \text{Celarent}\,\dfrac{a_Q a_L{}^e \quad a_L c^a}{a_Q c^e}}{bc^e} \quad \xrightarrow{R_{QL}} \quad \frac{a_Q b^a \quad a_L c^a}{bc^e}$$

§32. ALTERNATIVE *DE RE* APPROACHES

We have seen how, by giving distinct semantic analyses for contingency-forms in pure and in mixed moods, we can provide an A-style treatment that renders such forms convertible but not opposable. All of Aristotle's contingency-axioms are then valid, along with Cesare QLX. Let us now see whether this result can be altered if we change our base assumptions. We shall also see what results can be achieved within other superstructures.

§32.1 Superstructure A

When in §28.2 we proved the validity of AQ's axioms, we made no use of any E- or S-assumption. So adopting any of those assumptions is not going to make Cesare QLX or any Aristotelian perfect modal syllogism invalid. However, the Ampliation Theorem (**30.3**) implies that if definition **25.4** is added to $A0$-semantics then Barbara QQQ becomes valid. Darii QQQ remains invalid even with that addition (see model 30.5). However, Darii QQQ is validated by the addition of the assumption

E6 $f_1{}^{\dagger} \cap f_2{}^{\ddagger} \neq \varnothing \supset f_1 \cap f_2{}^{\ddagger} \neq \varnothing$.

Proof If $f_2 \subseteq f_1{}^{\ddagger}$ and $f_3 \cap f_2{}^{\ddagger} \neq \varnothing$ then $f_3 \cap f_1{}^{\ddagger} \neq \varnothing$
$\because f_2{}^{\ddagger} \subseteq f_2{}^{\dagger}$ – **22.2.1, 30.3**;
and if $f_2 \subseteq f_1{}^{\ddagger}$ and $f_2 \cap f_3{}^{\ddagger} \neq \varnothing$ then $f_3 \cap f_1{}^{\ddagger} \neq \varnothing$
$\because f_3{}^{\ddagger} \subseteq f_3{}^{\dagger}$ – E6.

In fact, the stronger Darii QMQ is also valid, as the above proof shows.

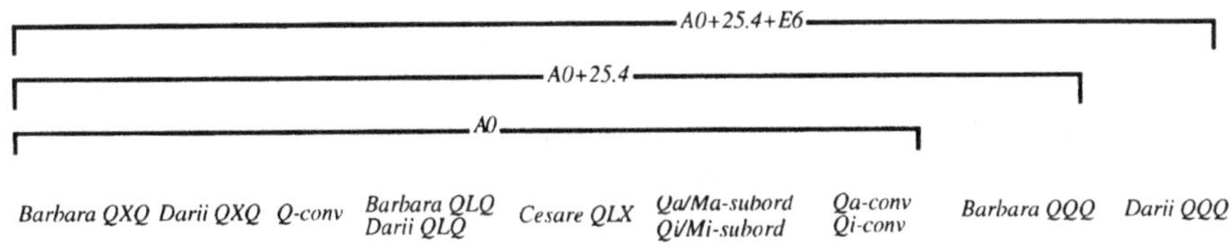

Fig. 32.1

Figure 32.1 shows that under definition **25.4** *A*-semantics can account for all of Aristotle's contingency-syllogisms, given E6. Thus, given this assumption and this definition of sun sets, we need not resort to the Alexandrian ploy of reading mixed contingency-moods unampliated and pure ones ampliated.

Recall that in §26.1 we showed that the addition of S3 to undoctored *A* semantics validates all of Aristotle's laws of apodeictic conversion and perfect apodeictic syllogisms. It also validates the otherwise invalid Barbara XQM and LQM, Bocardo LQX and LQM, Festino LQX and LQM, and Disamis QXM and QLM. These moods' invalidity without S3 can be shown by adapting matrices 26.2 and 26.1 respectively. But given S3, Baroco XLL and Celarent LXL are valid; and the first four of these moods are deducible from Baroco XLL, the last four from Celarent LXL:

$$\textit{Baroco XLL}\ \frac{ab^a \quad Lac^o}{Lbc^o} \xrightarrow[Lo \,\|\, Qa]{K} \frac{\dfrac{Lab^a}{ab^a}\,La-sub \qquad Qbc^a}{Mac^a}$$

$$\textit{Baroco XLL}\ \frac{ab^a \quad Lac^o}{Lbc^o} \xrightarrow[Lo \,\|\, Qa]{K} \frac{Lac^o \quad Qbc^a}{\dfrac{ab^o}{Mab^o}\,Mo-subord}$$

$$\textit{Celarent LXL}\ \frac{Lab^e \quad bc^a}{Lac^e} \xrightarrow[Le \,\|\, Qi]{K} \frac{Lab^e \quad Qac^i}{\dfrac{bc^o}{Mbc^o}\,Mo-subord}$$

$$\textit{Celarent LXL}\ \frac{Lab^e \quad bc^a}{Lac^e} \xrightarrow[Le \,\|\, Qi]{K} \frac{Qac^i \qquad \dfrac{Lbc^a}{bc^a}\,La-subord}{Mab^i}$$

Combining all these results, we can construct a unified semantics for the whole modal syllogistic by incorporating **25.4**+E6 in *A* semantics. (S3 is implicit in **25.4**.) In this way we can achieve a higher degree of Aristotelicity.

Table 32.1 shows which moods are accepted, which rejected, on *A* semantics. "!" indicates a conflict with Aristotle's explicit statements (as given in Table 17.1). A second entry in a cell indicates a result that arises only given S3 or **25.4**. Thus 'raw' *A* semantics delivers 104/133=78.2% Aristotelian results. If we add the favourable assumptions mentioned, this rises to 124/133=93.2%.

A formal system corresponding to these semantics (call it *AQ*+) will differ from *AQ* in having Barbara QMQ, Celarent QMQ, Darii QMQ, Ferio QMQ as theses, and in reflecting definition **25.4**.

TABLE 32.1

	XLL	LLL	LXL	QLQ	QXQ	QLM	QXM	QLX	QQQ	LQQ	XQQ	LQM	XQM	LQX
Barbara	-	+	+	+	+	+	+	-	-!+	-	-	-!+	-+!	-
Celarent	-	-!+	-!+	+	+	-	-	-	-!+	-	-	+	-	+
Darii	-	+	+	+	+	+	+	-	-!+	-	-	+	+	-
Ferio	-	+	+	+	+	+	+	-	-!+	-	-	+	-	+
Cesare	-	-!+	-!+	-	-	+!	-	+!	-	-	-	+	-	+
Camestres	-!+	-!+	-	-	-	+	-	+	-	-	-	+!	-	+!
Festino	-	+	+	-	-	-	-	-	-	-	-	-!+	-	-!+
Baroco	-+!	-!+	-	-	-	+	-	+	-	-	-	-	-	-
Darapti	+	+	+	+	+	+	+	-	-!+	+	+	+	+	-
Felapton	-	+	+	+	+	+	+	-	-!+	-	-	+	-	+
Disamis	+	+	-	-	-	-!+	-!+	-	-!+	+	+	+	+	-
Datisi	-	+	+	+	+	+	+	-	-!+	-	-	+	+	-
Bocardo	-	+	+!	-	-	-!	-!	-	-!+	-	-	-+	-	+
Ferison	-	+	+	+	+	+	+	-	-!+	-	-	+	-	+

Any axiomatisation of *AQ*+ based on rules *U, M, C, K* and *T* will have to include, in addition to a basis for assertoric syllogistic, one axiom from each of the following groups:

(1) La-rep or Mo-rep;
(2) Lo-rep or Ma-rep;
(3) Li-conv or Me-conv;
(4) Le-conv or Mi-conv;
(5-8) La-, Le-, Li- and Lo-subordination, or their M-equivalents;
one modal syllogism from each of the groups (9-14, 19) in Figure 32.2;
(15) Qi-conv or Qo-conv;
(17) Cesare QLX or Ferio QXM or Camestres LQX or Ferison QXM;
(20) Barbara QMQ or Celarent QMQ.

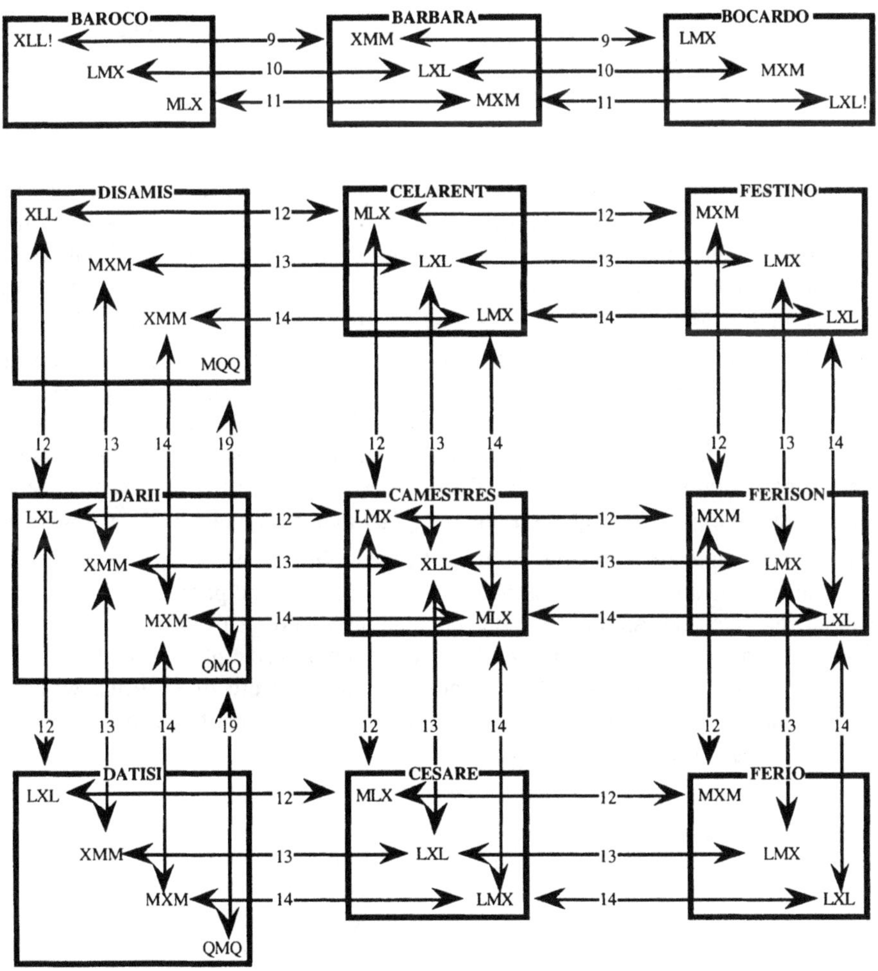

Fig. 32.2

(For the validity of group (19) see above. Consequential on this group are Darii QXQ *etc.* as well as Darii QQQ *etc.* For the validity of group (20), see §30. Consequential on this group are group (16) - Barbara and Celarent QXQ - as well as Barbara and Celarent QQQ.)

If we use rules of ecthesis, the number of axioms can be reduced. Baroco XLL and Bocardo LXL are provable in ways that should by now be familiar. Lo-rep follows from Felapton LXL, which in turns follows from Ferio LXL and a-conv. Li-subord and Lo-subord follow from e-rep and a-rep. Li-conv and Qi-conv follow by ecthesis from Darapti LLL and Darapti QQQ, which in turn follow from Darii LXL (group 12) and Darii QMQ (group 19).

AQ, in contrast to *AQ+*, will have Darii QXQ *etc.* and Barbara QXQ *etc.* in place of groups (19) and (20).

§32.2 *Superstructure* AA

So far no *AA*-semantics for Q-forms has been published; but its character can be predicted. *AA*-semantics restricts the subject-terms of particular L-forms to the necessary; so universal M-forms (if they contradict particular L-forms) should have their subjects restricted to the necessary. Thus if universal Q-forms are to entail the corresponding M-forms (*i.e.* if we accept the oppositions $Qa \parallel Li$ and $Qa \parallel Lo$) then they must at least imply Q-forms whose subjects are restricted to the necessary. On the other hand, *AA*-semantics neither restricts nor ampliates the subject-terms of universal L-forms; so particular M-forms (if they are contradictory to universal L-forms) should have their subjects neither restricted nor ampliated. Thus if particular Q-forms are to entail the corresponding particular M-forms (*i.e.* if we accept the oppositions $Qi \parallel La$ and $Qi \parallel Le$) then they must at least imply Q-forms whose subjects are neither restricted nor ampliated. All of this suggests that we read universal Q-forms as restricted to the necessary and particular Q-forms as neither restricted nor ampliated. But this reading would result in the invalidity of Qa-subalternation. In order to save that subalternation-principle we need to find a way of maintaining our necessary conditions for the truth of Q-forms without making these conditions sufficient. The simplest way of doing this is as follows:[3]

32.2.1 Qab^a is true iff $f_2 \subseteq f_1^{\ddagger}$

32.2.2 Qab^i is true iff $f_2 \cap f_1^{\ddagger} \neq \varnothing$

[3]This analysis is given by BRENNER p.13.

This analysis of Q-forms makes them opposable but not convertible. The modal oppositions are then as in Figure 32.3. All the classical oppositions hold. Because the opposition *Qo* ‖ *La* is accepted, we have all the Q/M subordination principles.

However, partial Qa-conv and Qi-conv are invalid:

32.1		*	¤	‡	†
f_1	d_1	–	–	d_1	d_1
f_2	d_1	d_1	–	–	d_1

$$\left.\begin{array}{l}(i) f_2 \subseteq f_1{}^{\ddagger} \\ (ii) f_2 \cap f_1{}^{\ddagger} \neq \varnothing\end{array}\right\} \text{but } f_1 \cap f_2{}^{\ddagger} = \varnothing.$$

Barbara QXQ and Darii QXQ are obviously valid. Cesare QLX is valid *malgré Aristote*:

Proof If $f_2 \subseteq f_1{}^{\ddagger}$ and $f_3 \subseteq f_1{}^{*}$ then $f_3 \cap f_2 = \varnothing$
$\because f_1{}^{\ddagger} \cap f_1{}^{*} = \varnothing$.

And its validity is independent of all E- and S-assumptions.

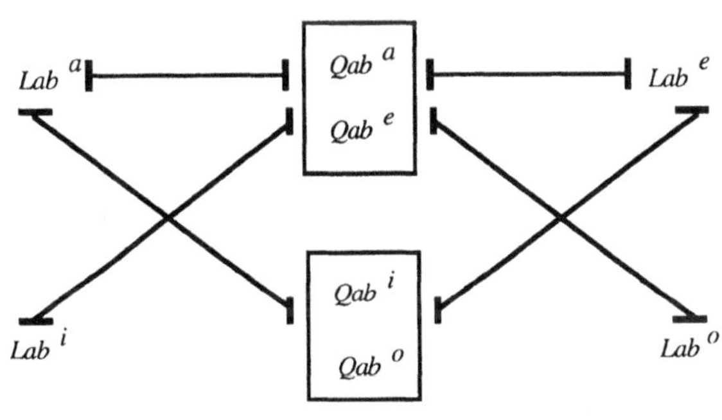

Fig. 32.3

A minimal system with *AA* semantics would contain L-subord, Li-conv, Barbara LXL, Celarent LXL, Darii LLL, Ferio LLL and Bocardo LLL, along with Barbara QXQ, Darii QXQ and Cesare QLX.[4]

Bocardo QXM (and thus QLM) is provable in Aristotelian fashion in this system:

[4] See §26.2.

$$\text{Barbara } LXL\frac{Lab^a \quad bc^a}{Lac^a}\xrightarrow[La\,\|\,Qo]{K}\frac{Qac^o \quad bc^a}{Mab^o}$$

Nonetheless, the failure of the laws of conversion makes a simple *AA* semantics unattractive for Aristotelian purposes. However, if we adopt the assumption

E7 $f_1 \cap f_2{}^{\ddagger} \neq \varnothing \supset f_2 \cap f_1{}^{\ddagger} \neq \varnothing$

the conversion-principles become valid.

Barbara QQQ and Darii QQQ are invalid:

32.2		*	¤	‡
f_1	d_1	d_1	–	d_2
f_2	d_2	–	–	d_1, d_2
f_3	d_1	d_1	–	d_2

$$f_2 \subseteq f_1{}^{\ddagger}, \begin{Bmatrix} (i) f_3 \subseteq f_2{}^{\ddagger} \\ (ii) f_3 \cap f_2{}^{\ddagger} \neq \varnothing \end{Bmatrix} \text{ but } \begin{Bmatrix} (i) f_3 \not\subset f_1{}^{\ddagger} \\ (ii) f_3 \cap f_1{}^{\ddagger} = \varnothing. \end{Bmatrix}$$

However, if we define ampliated contingency-forms:

32.2.3 $Q'ab^a$ is true iff $f_2{}^{\dagger} \subseteq f_1{}^{\ddagger}$

32.2.4 $Q'ab^i$ is true iff $f_2{}^{\dagger} \cap f_1{}^{\ddagger} \neq \varnothing$

then Barbara Q'Q'Q' and Darii Q'Q'Q' are valid:[5]

Proof

$$\text{If } f_2{}^{\dagger} \subseteq f_1{}^{\ddagger} \text{ and } \begin{Bmatrix} f_3{}^{\dagger} \subseteq f_2{}^{\ddagger} \\ f_3{}^{\dagger} \cap f_2{}^{\ddagger} \neq \varnothing \end{Bmatrix} \text{ then } \begin{Bmatrix} f_3{}^{\dagger} \subseteq f_1{}^{\ddagger} \\ f_3{}^{\dagger} \cap f_1{}^{\ddagger} \neq \varnothing \end{Bmatrix}$$

$\because f_2{}^{\ddagger} \subseteq f_2{}^{\dagger}$.

Partial Q'a-conv and Q'i-conv are invalid, as model 32.1 shows. But if we adopt the assumption

E8 $f_1{}^{\dagger} \cap f_2{}^{\ddagger} \neq \varnothing \supset f_2{}^{\dagger} \cap f_1{}^{\ddagger} \neq \varnothing$

then those conversions become valid.

Darii Q'XQ' is valid:

5 There seems to be a mistake in BURIDAN 4.3.14: "In the first ... figure, with a contingent major ... there follows a ... conclusion which is ... *de possibili* if it [the minor] is contingent." The truth is that if the minor is *de possibili* (or stronger) then the conclusion is contingent.

Proof If $f_2{}^\dagger \subseteq f_1{}^\ddagger$ and $f_3 \cap f_2 \neq \varnothing$ then $f_3{}^\dagger \cap f_1{}^\ddagger \neq \varnothing$
$\because f_2 \subseteq f_2{}^\dagger, f_3 \subseteq f_3{}^\dagger$.

But Barbara Q'XQ' is not:

32.3		*	¤	‡	†
f_1	d_1, d_2	d_1	d_3	d_2	d_1, d_2
f_2	d_2	d_2	d_1, d_3	–	d_2
f_3	d_2	d_2	d_1	d_3	d_2, d_3
f_4	d_3	d_3	d_1, d_2	–	d_3

$f_2{}^\dagger \subseteq f_1{}^\ddagger$ and $f_3 \subseteq f_2$ but $f_3{}^\dagger \not\subset f_1{}^\ddagger$.

Thus, an *AA* interpretation that treats the mixed contingency-syllogistic as requiring Q-forms and the pure contingency-syllogistic as requiring Q'-forms is defensible.

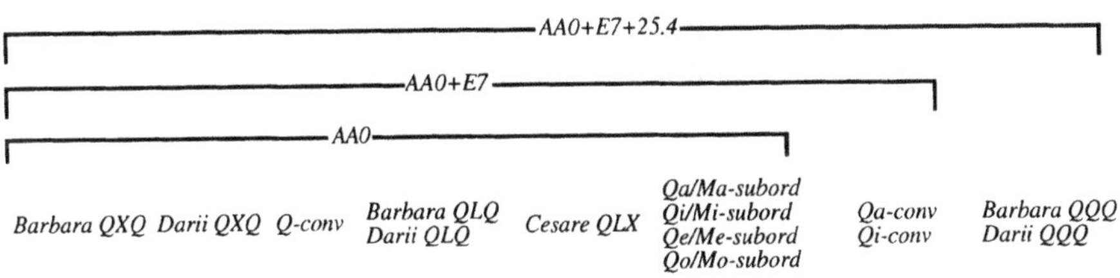

Fig. 32.4

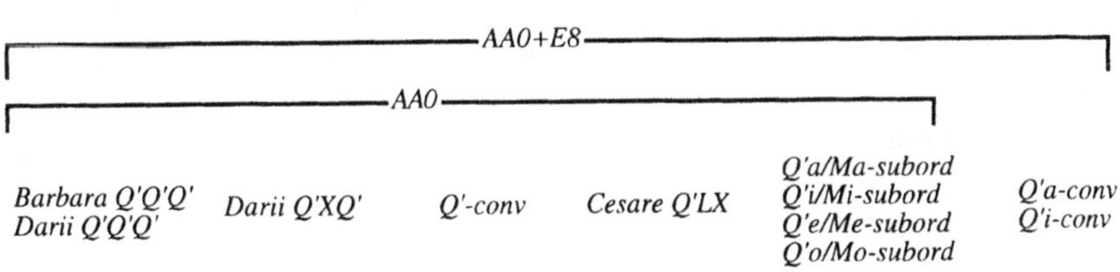

Fig. 32.5

Alternatively, a unified interpretation is possible if we adopt definition **25.4**: Barbara QQQ and Darii QQQ are then valid as in *A* semantics. The price to be paid for invoking **25.4**, however, is acceptance of S3 and thus of Barbara XQM and

Baroco XLL. The relations of various *AA*-semantics to *AQ*''s axioms are shown in Figures 32.4 and 32.5.

Recall that in §26.2 we showed that the addition of E1+E3+E4+S2 to undoctored *A* semantics validated all of Aristotle's laws of apodeictic conversion and perfect apodeictic syllogisms. Combining that result with what we have just shown, we can construct a unified semantics for the whole modal syllogistic by incorporating **25.4**+E7 along with E1+E3+E4 in *AA* semantics. (S2 is implicit in **25.4**.)

The Aristotelicity of this semantics is 88/133 or 66.2% without extra assumptions, and 121/133 or 91% with favourable assumptions. See Table 32.2.

AA semantics, even in its weakest (*AA0*) form, validates some moods that are not *A*-valid but which Aristotle considers valid. These include Bocardo QXM (QiXaMo-3) and thus Bocardo QLM:

Proof If $f_3 \cap f_1^{\ddagger} \neq \varnothing$ and $f_3 \subseteq f_2$ then $f_2 \not\subset f_1^{*}$
$\because f_1^{\ddagger} \cap f_1^{*} = \varnothing$.

AA semantics, even in its weakest form, validates moods that are not valid according to either Aristotle or *A*-semantics. These include Baroco LQX (LaQiXo-2):

Proof If $f_2 \subseteq f_1^{*}$ and $f_3 \cap f_1^{\ddagger} \neq \varnothing$ then $f_3 \not\subset f_2$
$\because f_1^{\ddagger} \cap f_1^{*} = \varnothing$.

This syllogism follows from Barbara LXL by rule *K* relying on the opposition *La* ‖ *Qi* which is endorsed by *AA*- but not by *A*-semantics.

Any axiomatisation of an *AA* system based on rules *U, M, C, K, T,*- if it is to include Le-conv, Darii LXL, Ferio LXL, Baroco LLL, Qi-conv - will have to include, in addition to a basis for assertoric syllogistic, one axiom from each of the following groups:
(1) La-rep or Mo-rep;
(2) Lo-rep or Ma-rep;
(3) Li-conv or Me-conv;
(4) Le-conv or Mi-conv;
(5-8) La-, Le-, Li- and Lo-subordination, or their M-equivalents;
one modal syllogism from each of the groups (11-14, 19, 21-22) in Figure 32.4;
(15) Qi-conv or Qo-conv;
(20) Barbara QMQ or Celarent QMQ.

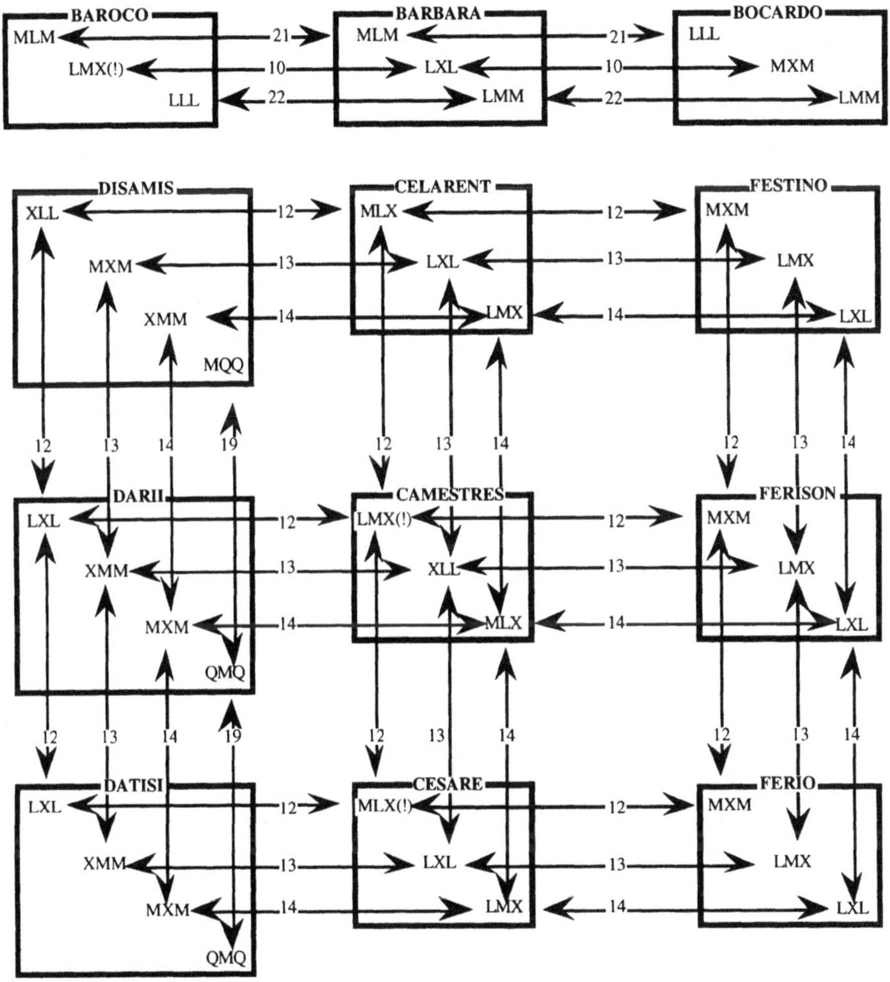

Fig. 32.6

(Group (17) - Cesare QLX *etc.* - are derivable from group (12) given Qe/Me-subord.) The axiomatisation can be simplified by the use of rules of ecthesis:

Li-ecthesis AA $\frac{R \quad Lad^a \quad Lbd^a}{p} \longrightarrow \frac{R \quad Lab^i}{p}$

Lo-ecthesis AA $$\frac{R \quad Lad^e \quad Lbd^a}{p} \longrightarrow \frac{R \quad Lab^o}{p}$$

Qi-ecthesis AA $$\frac{R \quad Qad^a \quad bd^a}{p} \longrightarrow \frac{R \quad Qab^i}{p}$$

TABLE 32.2

	XLL	LLL	LXL	QLQ	QXQ	QLM	QXM	QLX	QQQ	LQQ	XQQ	LQM	XQM	LQX
Barbara	-	+	+	+	+	+	+	-	-!+	-	-	-!+	-+!	-
Celarent	-	+	+	+	+	+	+	-+!	-!+	-	-	-!+	-	-!+
Darii	-	+	-!+	+	+	+	+	-	-!+	-	-	-!+	-+	-
Ferio	-	+	-!+	+	+	+	+	-	-!+	-	-	-+	-	-!+
Cesare	-	-!+	-!+	-	-	+!	-	+!	-	-	-	+	-	+
Camestres	-!+	-!+	-	-	-	+	-	+	-	-	-	+!	-	+!
Festino	-	-!+	-!+	-	-	+!	-+!	+!	-	-	-	+	-	+
Baroco	-+!	-!+	-	-	-	+	-	+	-	-	-	+!	-	+!
Darapti	-!+	+	-!+	+	+	+	+	-	-!+	-+	-+	-!+	-+	-
Felapton	-	+	-!+	+	+	+	+	-	-!+	-	-	-!+	-	-!+
Disamis	-!+	+	-	+	+	+	+	-	-!+	-!+	-!+	-+	-+	-
Datisi	-	+	-!+	+	+	+	+	-	-!+	-	-	-!+	-+	-
Bocardo	-	+	-	+	+	+	+	-	-!+	-	-	-+	-	-+
Ferison	-	+	-!+	+	+	+	+	-	-!+	-	-	-+	-	-!+

Lo-rep (group 2) is provable by Lo-ecthesis and rule K from Barbara MLM (group 21):

$$\text{Barbara } MLM\,\frac{Mab^a \quad Lbd^a}{Mad^a} \xrightarrow[Ma \,\|\, Le]{K} \frac{Lad^e \quad Lbd^a}{Lab^o} \xrightarrow{\text{Lo-ecthesis AA}} \frac{Lab^o}{Lab^o}$$

Li-conv (group 3) follows from Darii LXL (group 12) and La-subord (5):

$$\text{Darii } LXL\,\cfrac{Lad^a \qquad a-conv\,\cfrac{La-subord\,\cfrac{Lbd^a}{bd^a}}{bd^i}}{Lba^i} \xrightarrow{\text{Li-ecthesis AA}} \frac{Lab^i}{Lba^i}$$

Li-subord (7) is provable by Li-ecthesis, given La-subord (5):

$$\text{Darapti}\,\cfrac{La-subord\,\cfrac{Lad^a}{ad^a} \qquad La-subord\,\cfrac{Lbd^a}{bd^a}}{ab^i} \xrightarrow{\text{Li-ecthesis AA}} \frac{Lab^i}{ab^i}$$

Lo-subord (8) is provable by Lo-ecthesis, given Le-subord (6):

$$\text{Felapton}\,\cfrac{Le-subord\,\cfrac{Lad^e}{ad^e} \qquad La-subord\,\cfrac{Lbd^a}{bd^a}}{ab^o} \xrightarrow{\text{Lo-ecthesis AA}} \frac{Lab^o}{ab^o}$$

Qi-conv (group 15) follows by Qi-ecthesis from Darii QMQ (group 19) and Le-subord (6):

$$Le-subord\,\frac{Ldb^e}{db^e} \xrightarrow{C} \text{Darii } QMQ\,\cfrac{Qad^a \qquad \cfrac{a-conv\,\cfrac{bd^a}{db^i}}{Mdb^i}}{Qab^i} \xrightarrow{\text{Qi-ecthesis AA}} \frac{Qab^i}{Qba^i}$$

Bocardo LLL (group 21) follows from Barbara LXL (group 11), Ferio LXL (group 14) and La-subord (5):

$$\text{Ferio } LXL\,\cfrac{a-conv\,\cfrac{La-subord\,\cfrac{\text{Barbara } LXL\,\cfrac{Lbc^a \qquad La-subord\,\cfrac{Lcd^a}{cd^a}}{Lbd^a}}{bd^a}}{db^i} \qquad Lad^e}{Lab^o} \xrightarrow{\text{Lo-ecthesis AA}} \frac{Lac^o \quad Lbc^a}{Lab^o}$$

Baroco LLL (group 22) follows from Celarent LXL (group 13), Ferio LXL (group 14) and La-subord (5):

$$\text{Ferio LXL}\ \dfrac{\dfrac{\text{Celarent LXL}\ \dfrac{La\text{-subord}\ \dfrac{Lab^a}{ab^a}\quad Le\text{-conv}\ \dfrac{Lad^e}{Lda^e}}{Ldb^e}}{Lbd^e}\qquad La\text{-subord}\ \dfrac{\dfrac{Lcd^a}{cd^a}}{dc^i}}{Lbc^o}\qquad \xrightarrow{Lo\text{-ecthesis}\ \wedge\wedge}\ \dfrac{Lab^a \quad Lac^o}{Lbc^o}$$

Thus given our three rules of L-ecthesis we can reduce our axiom-set from 17 to 10 - one each from groups (1), (4)-(6), (11)-(14), (19), (20).

§32.3 Superstructure C

Within *C* semantics as advocated by Buridan, contingency-forms are normally understood as ampliating to the possible, though this normal implicature can be cancelled by the insertion of a 'what is' phrase before the subject. Thus there are two types of contingency-propositions, one having truth-conditions:

32.3.1 Qab^a is true iff $f_2{}^{\dagger} \neq \varnothing \wedge f_2{}^{\dagger} \subseteq f_1{}^{\ddagger}$

32.3.2 Qab^i is true iff $f_2{}^{\dagger} \cap f_1{}^{\ddagger} \neq \varnothing$,

the other having truth-conditions **32.2.1-2**. Their logic has therefore been covered to some extent in §32.2. Buridan, however, uses *C*- rather than *A*-semantics for assertorics, and this makes a difference to the principles of L-subordination: as noted in §26.3, only Le-subord is valid. Apart from this, the above results for ampliated forms transfer to *C* semantics. Thus, Le-conv, Barbara LLL, Celarent LLL, Darii LLL, Ferio LLL, Baroco LLL, Bocardo LLL, Darii LXL, Ferio LXL, Barbara QQQ, Darii QQQ, Barbara QLQ and Darii QLQ are valid; and while Darii QXQ (in the normal divisive sense) is valid, Barbara QXQ is not.[6] In addition to these, Cesare QLL is valid (not just Cesare QLX); the proof is given below.

A minimal system with *C* semantics would therefore contain Le-subord, Le-conv, Barbara LLL, Celarent LLL, Darii LLL, Ferio LLL, Baroco LLL, Bocardo LLL, Darii LXL, Ferio LXL, along with Cesare QLL, Barbara QQQ, Darii QQQ, Barbara QLQ, Darii QLQ and Darii QXQ.

Bocardo QXM is invalid:[7]

[6] BURIDAN 4.3.17: "In the first figure, a syllogism from a contingent major and an assertoric minor to a particular contingent conclusion is valid, but not to a universal contingent conclusion."

[7] BURIDAN 4.3.21: "... although someone running is contingently riding and everything running is a horse, nevertheless no horse is contingently riding. Similarly if the major is taken as negative, since it is equivalent to an affirmative."

32.4		*	¤	‡	†
f_1	d_1	–	–	d_1	d_1
f_2	d_1	–	–	d_1	d_1
f_3	d_1	d_1	–	d_1	d_1

$f_3^{\dagger} \cap f_1^{\ddagger} \neq \varnothing$ and $f_3 \neq \varnothing \wedge f_3 \subseteq f_2$ but $f_2 \neq \varnothing \wedge f_2 \subseteq f_3^{*}$.

But Bocardo QLM is provable in Aristotelian fashion:[8]

$$\text{Barbara LLL}\frac{Lab^a \quad Lbc^a}{Lac^a}\xrightarrow[La\,\|\,Qo]{K}\frac{Qac^o \quad Lbc^a}{Mab^o}$$

As shown in §30, the validity of Barbara QXQ (on the present semantics) depends on what S-assumptions we make: this mood is valid given S3, but is invalid if we assume any definition of sun sets that does not imply S3 (*i.e.* **25.1**).

However Barbara QLQ and Darii QLQ are both valid:[9]

Proof
$$\text{If } f_2^{\dagger} \subseteq f_1^{\ddagger} \text{ and } \begin{Bmatrix} f_3^{\dagger} \subseteq f_2^{*} \\ f_3^{\dagger} \cap f_2^{*} \neq \varnothing \end{Bmatrix} \text{ then } \begin{Bmatrix} f_3^{\dagger} \subseteq f_1^{\ddagger} \\ f_3^{\dagger} \cap f_1^{\ddagger} \neq \varnothing \end{Bmatrix}$$
$$\because f_2^{*} \subseteq f_2^{\dagger}.$$

Here we should note Mignucci's purported proof of Barbara QLQ:[10] His ingenious reasoning (See p.240) comes close to being sanctioned by the present semantics. Rule z is acceptable, requiring as it does only that if Qac^a is false then either Lac^i or Lac^o be true. Step x is valid, depending on the opposition $Lo \,\|\, Qa$. And step y is valid, depending on the opposition $La \,\|\, Qi$. But Qi-conversion, also used by Mignucci, is invalid on the present semantics.

Qa-conv and Qi-conv are invalid on *C0*-semantics. However they are validated by the addition of E8. Figure 32.7 shows the effect of different base assumptions.

8 BURIDAN Theorem IV-6 (c): ""In the third figure ... from one premiss *de necessario* and the other *de possibili* there always follows a conclusion with the same mode as the mode of the major." 4.2.13: " ... the whole <of Theorem IV-6> is proved *per impossibile*, since from the minor and the opposite of the conclusion the opposite of the major is deduced by the first figure."

9 BURIDAN 4.3.14: "In the first ... figure, with a contingent major ... there follows a similar conclusion which is contingent, if the minor is *de necessario* ...."

10 MIGNUCCI *ad* 35b32.

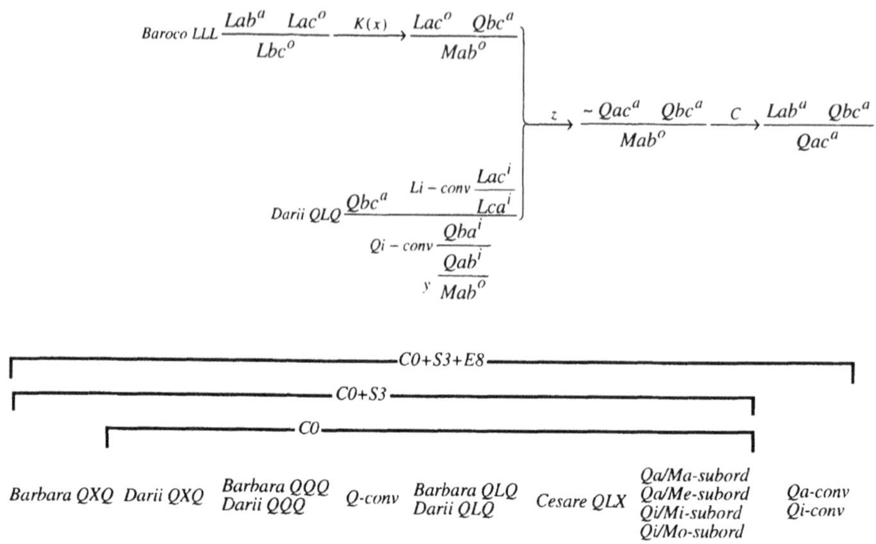

Fig. 32.7

Recall that in §26.3 we showed that the addition of E3+E4+E5+S1.1+S3 to undoctored *C* semantics validated all of Aristotle's laws of apodeictic conversion and perfect apodeictic syllogisms. Combining that result with what we have just shown, we can construct a unified semantics for the whole modal syllogistic by incorporating E3+E4+E5+E8+S1.1+S3 into *C* semantics.

The Aristotelicity of *C* semantics is 111/133, 83.5%. If we adopt assumptions the figure rises to 120/133, 90.2%. See Table 32.3.

Some moods that are invalid according to both Aristotle and *A*-semantics are *C*-valid, *e.g.* Cesare QLL (QaLaLe-2):[11]

Proof If $f_2{}^{\dagger} \subseteq f_1{}^{\ddagger}$ and $f_3{}^{\dagger} \subseteq f_1{}^{*}$ then $f_3{}^{\dagger} \cap f_2{}^{\dagger} = \varnothing$ $\because f_1{}^{\ddagger} \cap f_1{}^{*} = \varnothing$.

[11] BURIDAN 4.2.5: "In the second figure ... there is always a valid syllogism from a pair of premisses ... one of which is *de necessario* and the other *de possibili* to a conclusion which is *de necessario* ...." 4.3.10: "In any syllogism to some given conclusion which has some premiss *de possibili* with the mode affirmed ... if a contingent sentence with the mode affirmed (whether affirmative or negative) is put in the place of that premiss, then the same conclusion follows ...."

Buridan reduces this syllogism to Figure 1 thus:

$$\text{Ferio } MMM\,\frac{Mab^e \quad Mbc^i}{Mac^o} \xrightarrow{C} \frac{\dfrac{Qab^e}{Mab^e}\,Q/M-subord \qquad Lac^a}{Lbc^e}$$

He accepts as "perfect (or quasi-perfect)"[12] all direct first figure MMM, LLL, LML and MLM syllogisms. The moods valid relative to *C0+E3+E4+E5+S1.1+S3* are shown in Table 32.3.

TABLE 32.3

	XLL	LLL	LXL	QLQ	QXQ	QLM	QXM	QLX	QQQ	LQQ	XQQ	LQM	XQM	LQX
Barbara		+		+		+		-		-		+		-
	-		-!+		-!+		-		+		-		-+!	
Celarent		+		+		+		-		-		+		+
	-		-!+		-!+		-		+		-		-	
Darii		+		+		+		-		-		+		-
	-		+		+		+		+		-		-	
Ferio		+		+		+		-		-		+		-!+
	-		+		+		+		+		-		-	
Cesare		+		-		+!		+!		-		+		+
	-		-!+		-		-		-		-		-	
Camestres		+		-		+		+		-		+!		+!
	-!+		-				-		-		-		-	
Festino		+		-		+!		-		-		+		-!+
	-		+		-		-		-		-		-	
Baroco		+		-		+		-		-		+!		-
	-+!		-		-		-		-		-		-+!	
Darapti		+		+		+		-		-		+		-
	-!+		+		+		+		+		-		+	
Felapton		+		+		+		-		-		+		-!+
	-		+		+		+		+		-		+	
Disamis		+		+		+		-		-!		+		-
	-!+		-		-		-!+		+		-!		+	
Datisi		+		+		+		-		-		+		-
	-		+		+		+		+		-		-	
Bocardo		+		+		+		-		-		+		-+
	-		-+!		-		-!		+		-		+	
Ferison		+		+		+		-		-		+		-!+
	-		+		+		+		+		-		-	

Any axiomatisation based on rules *U*, *M*, *C*, *K* and *T* will have to include, in addition to a basis for assertoric syllogistic, one axiom from each of the following groups:

[12] BURIDAN 4.2.4.

(1) La-rep or Mo-rep;
(2) Lo-rep or Ma-rep;
(3) Li-conv or Me-conv;
(4) Le-conv or Mi-conv;
(5-8) La-, Le-, Li- and Lo-subordination, or their M-equivalents;
one modal syllogism from each of the groups (19, 23-27) in Figure 32.8; and
(15) Qi-conv or Qo-conv;
(20) Barbara QMQ or Celarent QMQ.

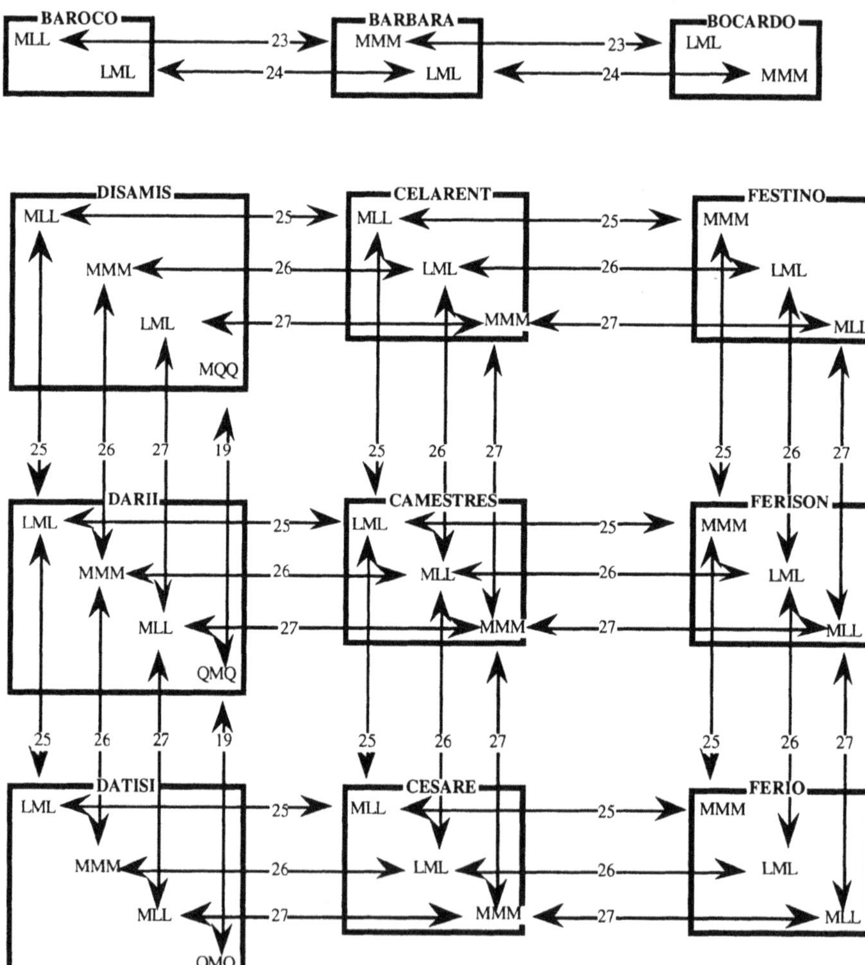

Fig. 32.8

Axiomatisation can be simplified by the use of ecthetic rules:

Li-ecthesis C $$\frac{R \quad Lad^a \quad Mbd^a}{p} \longrightarrow \frac{R \quad Lab^i}{p}$$

Lo-ecthesis C $$\frac{R \quad Lad^e \quad Mbd^a}{p} \longrightarrow \frac{R \quad Lab^o}{p}$$

Lo-rep (group 2) is derivable from Ferio LML (group 26)

$$\text{Ferio LML}\,\frac{Lad^e \qquad \text{Ma}-\text{conv}\,\dfrac{Mbd^a}{Mdb^i}}{Lab^o} \xrightarrow{\text{Lo}-\text{ecthesis C}} \frac{Lab^o}{Lab^o}$$

Li-conv (group 3) is derivable from Darii MLL (group 27)

$$\text{Darii MLL}\,\frac{\text{La}-\text{conv}\,\dfrac{Lad^a}{Lda^i} \qquad Mbd^a}{Lba^i} \xrightarrow{\text{Li}-\text{ecthesis C}} \frac{Lab^i}{Lba^i}$$

Baroco MLL (group 23) is derivable from Camestres MLL and Ferio LML (both in group 26)

$$\text{Ferio LML}\,\frac{\text{Camestres MLL}\,\dfrac{Mab^a \quad Lad^e}{Lbd^e} \qquad \text{Ma}-\text{conv}\,\dfrac{Mcd^a}{Mdc^i}}{Lbc^o} \xrightarrow{\text{Lo}-\text{ecthesis C}} \frac{Mab^a \quad Lac^o}{Lbc^o}$$

Ferio MLL (group 27) is derivable from Celarent MLL (group 25) and Ferio LML (group 26)

$$\text{Ferio LML}\,\frac{\text{Celarent MLL}\,\dfrac{Mab^e \quad Lbd^a}{Lad^e} \qquad \text{Ma}-\text{conv}\,\dfrac{Mcd^a}{Mdc^i}}{Lac^o} \xrightarrow{\text{Li}-\text{ecthesis C}} \frac{Mab^e \quad Lbc^i}{Lac^o}$$

Thus given the rules of ecthesis we can reduce our axiom-set from 16 to 12 - one from each of groups (1), (4)-(8), (15), (19)-(20), (24)-(26).

§32.4 Superstructure CC

Ockham thinks that the subject-terms of contingency-propositions may be understood either as unampliated, or as ampliated to the contingent, or else as ampliated to what either is actual or is contingent.[13]

For unampliated Q-forms his truth-conditions are as in **32.2.1-2**. Modal oppositions will again be as in Figure 32.1. And again, Barbara QXQ and Darii QXQ will be valid (along with Cesare QLX),[14] but partial Qa-conv and Qi-conv will be invalid, along with Barbara QQQ and Darii QQQ.[15]

A minimal system with *CC* semantics of the above type would contain La-subord, Le-subord, Li-subord, Lo-subord, Barbara LXL, Celarent LXL, Darii LXL and Ferio LXL, along with Cesare QLX, Barbara QXQ and Darii QXQ.

Bocardo QXM (and thus QLM) is valid,[16] and is provable in Aristotelian fashion:

$$\textit{Barbara LXL}\frac{Lab^a \quad bc^a}{Lac^a} \xrightarrow[La\,\|\,Qo]{K} \frac{Qac^o \quad bc^a}{Mab^o}$$

The Aristotelicity is 92/133 (69.2%) without assumptions, or 121/133 (91%) with assumptions. See Table 32.4.

If ampliated forms ampliate to the contingent, their semantic analysis will be

32.4.1 $Q''ab^a$ is true iff $f_2{}^{\ddagger} \subseteq f_1{}^{\ddagger}$

32.4.2 $Q''ab^i$ is true iff $f_2{}^{\ddagger} \cap f_1{}^{\ddagger} \neq \varnothing$

This approach delivers the minimum number of oppositions between contingency- and necessity-forms, namely none:

[13] OCKHAM IIIa.37.24-27.

[14] OCKHAM IIIa.37.27-35; IIIa.48.24-25. (Ockham does not bother to state explicitly that the Q-form in Cesare QLX must be unampliated.)

[15] OCKHAM II.27.34-44. IIIa.26 does not state explicitly that Barbara/Celarent QQQ is invalid with unampliated Q-forms, but this was clearly what he thought: see his rejection (IIIa.28) of third figure QQQ unampliated moods.

[16] OCKHAM IIIa.39.14.

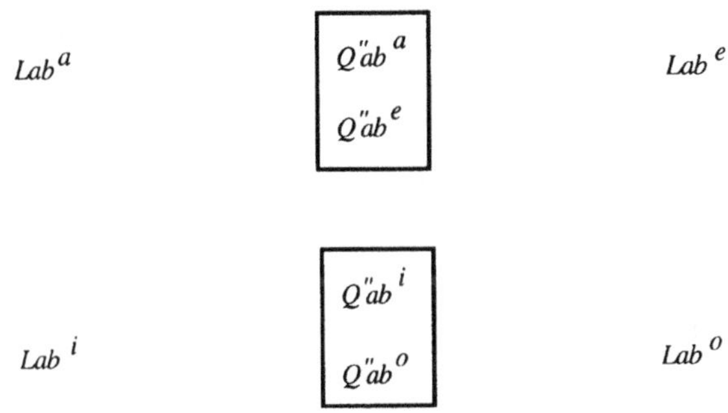

Fig. 32.9

TABLE 32.4

	XLL	LLL	LXL	QLQ	QXQ	QLM	QXM	QLX	QQQ	LQQ	XQQ	LQM	XQM	LQX
Barbara	-	+	+	+	+	+	+	-	-!+	-	-	-!+	-+!	-
Celarent	-	+	+	+	+	+	+	-	-!+	-	-	-!+	-	-!+
Darii	-	+	+	+	+	+	+	-	-!+	-	-	-!+	-+	-
Ferio	-	+	+	+	+	+	+	-	-!+	-	-	-+	-	-!+
Cesare	-	-!+	-!+	-	-	+!	-	+!	-	-	-	+	-	+
Camestres	-!+	-!+	-			+	-	+	-	-	-	+!	-	+!
Festino	-	-!+	-!+	-	- -	+!	-	+!	-	-	-	+	-	+
Baroco	-+!	-!+	-		-	+	-	+	-	-	-	+!	-	+!
Darapti	-!+	+	+	+	+	+	+	-	-!+	-	-	-!+	-+	-
Felapton	-	+	+	+	+	+	+	-	-!+	-	-	-!+	-	-!+
Disamis	-!+	+	+!		+	+	+	-	-!+	-!+	-!+	+	-	-
Datisi	-	+	+	+	+	+	+	-	-!+	-	-+	-!+	-+	-
Bocardo	-	+	+!		+	+	+	-	-!+	-	-	-+	-	-+
Ferison	-	+	+	+	+	+	+	-	-!+	-	-	-+	-	-!+

To see this, it will be enough to show that $Q''ab^a$ is compatible with Lab^a and also with Lab^e. This is done bymatrices 32.5 and 32.6.

32.5		*	¤	‡
f_1	d_1	d_1	–	d_2
f_2	d_1	d_1	–	d_2

$f_2 \subseteq f_1{}^*$ but $f_2{}^{\ddagger} \subseteq f_1{}^{\ddagger}$.

32.6		*	¤	‡
f_1	d_1	d_1	d_2	d_3
f_2	d_2	d_2	d_1	d_3

$f_2 \subseteq f_1{}^{¤}$ but $f_2{}^{\ddagger} \subseteq f_1{}^{\ddagger}$.

But Barbara Q"Q"Q" and Darii Q"Q"Q" are obviously valid, as are partial Q"-a conv and Q"i-conv.

The ecthesis rule for Q"i-forms will be

$$Q''i\text{-}ecthesis \quad \frac{R \quad Q''ad^a \quad Q''bd^a}{p} \longrightarrow \frac{R \quad Q''ab^i}{p}$$

Mignucci proposes an ecthetic proof for Q"i-conversion which rests on this rule:[17]

$$Darapti\ Q''Q''Q''\ \frac{Q''ac^a \quad Q''bc^a}{Q''ba^i} \longrightarrow \frac{Q''ab^i}{Q''ba^i}$$

(This proof tacitly uses rule *C*; and Darapti Q"Q"Q" is provable from Darii Q"Q"Q" and partial Q"a-conv.)

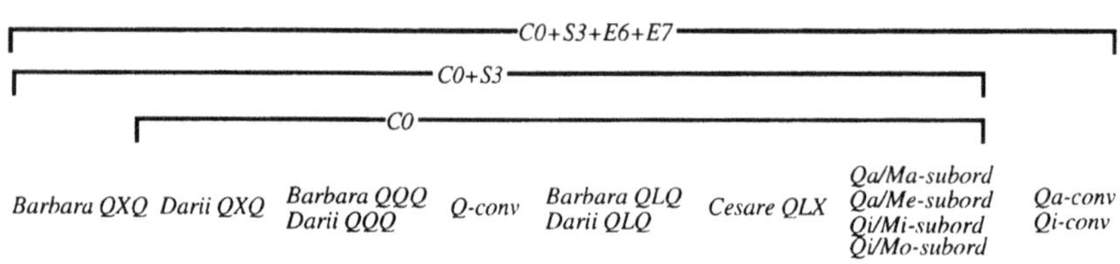

Fig. 32.10

Aristotle's contingency-syllogistic can thus be interpreted, in the spirit of *CC*-semantics, as falling into two parts. In the mixed moods contingency-propositions are read as unampliated, in the pure moods as ampliated to the contingent. This interpretation delivers all the required axioms, provided assumption E7 is adopted.

[17] MIGNUCCI *ad* 25b2.

But there is no obvious way of constructing a unified interpretation of L- and Q-moods based on ampliating Q-forms to the contingent.

Ockham's other idea - of ampliating to what either is or is contingent - amounts to ampliating to the possible. The truth-conditions for universal and particular Q'-forms are then as in **32.1.3-4**, and the oppositions are as in Figure 29.5. The result of combining an unampliated reading for mixed contingency-moods with an ampliated one for pure moods would be exactly the same as on an *A A* interpretation, and would require the same extra assumptions if it were to be defensible.

§32.5 Comparing superstructures

If we eschew all E- and S-assumptions, then of all superstructures *C* is perhaps the most intuitive, giving the truest representation of Aristotle's logic with the minimum of complication. But if we add to each superstructure the assumptions most favourable to it then *A* semantics gives the best results - though there isn't much in it. Figure 32.11 shows the Aristotelicity of the four different superstructures, with and without additional assumptions.

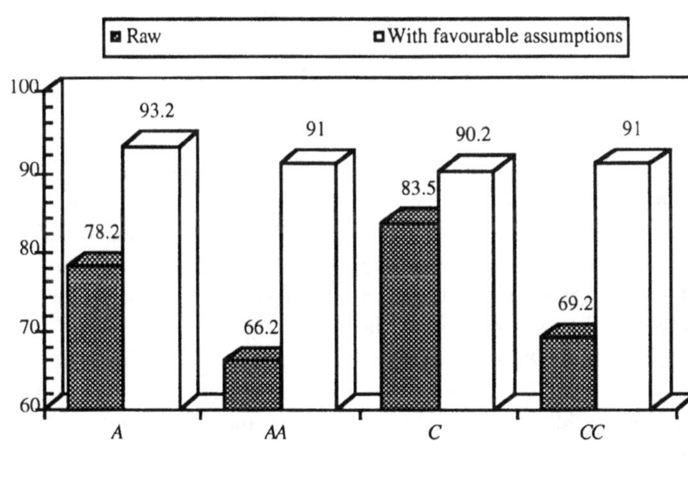

Fig. 32.11

§33. ALTERNATIVE APPROACHES: *DE DICTO* THEORIES

The simplest *de dicto* analysis of contingency-forms is as follows.

Qab^a $\quad \nabla ab^a$, *i.e.* $\Diamond ab^a \wedge \Diamond ab^o$

Qab^e $\quad \nabla ab^e$, *i.e.* $\Diamond ab^e \wedge \Diamond ab^i$

Qab^i $\quad \nabla ab^i$, *i.e.* $\Diamond ab^i \wedge \Diamond ab^e$

Qab^o $\quad \nabla ab^i$, *i.e.* $\Diamond ab^o \wedge \Diamond ab^a$

L.A. Kosman has argued that this analysis is used by Brogan.[18] On this analysis, Qi-conversion is valid, but so is the unAristotelian $\dfrac{Qab^e}{Qba^e}$. Qa-subalternation (and therefore partial Qa-conversion) is invalid. It does not follow from the antecedent's being contingent that the consequent is contingent. And a counter-example can be constructed. Let "the set" mean some non-null set of numbers, and let "lowest" apply to any member of the set than which none is lower. Then it is not contingent (but necessary) that some member of the set be lowest, even though (i) it is possible that every member of the set be lowest and (ii) it is possible that not every member of the set be lowest.

Q-conversion is not valid. Brogan himself notes[19] that on his understanding of Q-forms, Aristotle's principles of Q-conversion have to be replaced by

$$\frac{Qab^a}{Qab^o} \qquad \frac{Qab^e}{Qab^i} \qquad \frac{Qab^i}{Qab^e} \qquad \frac{Qab^o}{Qab^a}$$

Invalid too is full Qa-conversion (as with Aristotle), but so is the Aristotelian Qo-conversion.

Barbara QXQ is invalid: even supposing that (i) in the actual world all koalas are asleep, (ii) in some possible world all sleepers are non-koalas, and (iii) in some possible world not all sleepers are non-koalas, it does not follow that in some possible world all koalas are non-koalas.

Equally, Barbara QQQ is invalid: even supposing that (i) in some possible world all koalas are asleep, (ii) in some possible world not all koalas are asleep, (iii) in some possible world all sleepers are non-koalas, and (iv) in some possible world not all sleepers are non-koalas, it does not follow that in some possible world all koalas are non-koalas.

All in all, this interpretation has little in its favour.

---

[18] KOSMAN.

[19] BROGAN, p.60.

Angelelli's two versions of the internal reading of Qa-forms are $\forall x(\Diamond bx \supset \nabla ax)$ and $\forall x(\nabla bx \supset \nabla ax)$. Qa-conversion and Qi-conversion are invalid on either reading; Angelelli proposes to save them by making them depend on extra assumptions concerning the metaphysical status of the terms involved. We shall examine assumptions of this type in Chapter Ten. Further moods validated by Angelelli's interpretations but rejected by Aristotle are: Darapti XQX, Darapti QLL, Datisi XQX, Datisi QLL, Disamis XQX, Disamis QLL, Felapton XQX, Felapton QLL, Ferison XQX, Ferison QLL.[20]

In Łukasiewicz's contingency-syllogistic, which is a further development of his apodeictic system (§27.1), *all* MMM, MXM, XMM, MLL and LML analogues of assertoric syllogisms are provable.[21] Thus all QLM, QXM, QLX, LQM, XQM, LQX analogues of assertoric syllogisms are provable, a total of 84 cases; Aristotle is committed to accepting only 29 of these (see Table 17.1).

The best of the *de dicto* interpretations is that of Nortmann, who analyses universal Q-modals against an S5 background thus:[22]

$Qab^{a,e}$ $\qquad \forall x\Box(bx \supset \nabla ax)$

To date he has not published an analysis of Qi-forms, but on the evidence we have we can see that Q-conv is valid. Barbara QQQ is also valid:

Suppose $\Box(r \supset \nabla q), \Box(q \supset \nabla p)$
then $\Box(\Diamond r \supset (\Diamond\Diamond q \wedge \Diamond\Diamond \sim q)), \Box(\Diamond q \supset (\Diamond\Diamond p \wedge \Diamond\Diamond \sim p))$
so $\Box(\Diamond r \supset (\Diamond\Diamond p \wedge \Diamond\Diamond \sim p))$
so $\Box(\Diamond r \supset \nabla p)$
$\therefore \Box(r \supset \nabla p)$

Barbara QXQ is valid given Nortmann's analysis $\forall x\Box(bx \supset ax)$ of assertoric a-forms. In §27 we saw that that analysis faces problems as a textual interpretation. Another problem with Nortmann's analysis is that Celarent QLL becomes valid.[23] Aristotle asserts Celarent QLX (A16, 36a8-15). We saw in §10.2 that Alexander's purported proof of the stronger QLL must be rejected by Aristotle. Celarent QLL can be shown to be invalid (relative to *AQL* semantics) by reference to model 33.1.

---

[20] ANGELELLI p.211.

[21] ŁUKASIEWICZ pp.192-193. Łukasiewicz opines that Aristotle would have accepted Barbara MMM and Barbara MXM; but in fact Aristotle explicitly rejects Bocardo LXL and is thus committed to rejecting the equivalent Barbara MXM and hence also the weaker Barbara MMM.

[22] NORTMANN 1990 p.79.

[23] NORTMANN 1994, p.123.

33.1		*	¤	‡
f_1	d_1	d_1	d_2	d_3
f_2	d_2	d_2	d_1	d_3
f_3	d_3	–	–	d_1, d_2, d_3

$f_2 \subseteq f_1{}^{¤}, f_1 \subseteq f_2{}^{¤}, f_3 \subseteq f_2{}^{‡}$ but $f_3 \not\subset f_1{}^{¤}$.

Terms fitting this model are 'pitch' 'white' 'man'. Nothing white is pitch (necessarily so), and all men are contingently white. But it is not the case that all men are necessarily pitch.

CHAPTER EIGHT. SYSTEM *AX'*

The last class of modal syllogisms in Aristotle's system are those that contain unqualified assertoric forms. In this Chapter we propose a semantics, based on Angelelli's analysis of unqualifieds as ampliated assertorics, which distinguishes these forms both from ordinary assertorics and from apodeictics. Angelelli's analysis is shown to be superior to the analyses of Ockham and Buridan which treat unqualifieds as assertorics that are always or necessarily true; and it is shown to be superior to Pseudo-Scotus's analysis of unqualifieds as ampliated *problematics*. Our semantics has the consequence that Barbara and Darii XQM (with ordinary assertoric major) are valid despite Aristotle's argument to the contrary. Our semantics also shows that Aristotle's procedure of modal *reductio*, though unsound as a general principle, is sound for the particular case in which he applies it.

§34. UNQUALIFIED ASSERTORICS

What sense can be given to those unqualified assertoric universal affirmatives that Aristotle (A15,34b6) calls *haplōs* universals? What did he have in mind in contrasting them with assertorics true merely *kata to nun*? Their interpretation is disputed. Some take Aristotle to mean any proposition which is true at all times.[1] This is a reasonable interpretation, because if an X'a proposition means that the predicate attaches to the subject at all times, and if in addition we take possibility to be truth at some time, then Barbara X'QM will be valid as Aristotle says it is: if at all times every *b* is an *a* and at some time every *c* is a *b* (and at some time none) then at some time (*viz.*, at that time when every *c* is a *b*) every *c* is an *a*. Others take unqualified assertorics to be necessary propositions.[2] This interpretation also is reasonable, since it is plausible that when *a* necessarily belongs to all *b*, *a* possibly belongs to anything to which *b* possibly belongs; and given this, Barbara X'QM is valid. Yet I do not think either of these interpretations should be accepted.

I shall mention two difficulties, one of them specific to the temporal interpretation, the other common to both interpretations. One problem with the temporal interpretation is that there is no obvious reason why possibility should be identified with truth at a time. What prevents a possibility remaining forever unrealised? Aristotle is happy to suppose that at some time only men move. What then prevents the whole of time being such that only non-men move? But if that is admitted then Aristotle's own counter-example to Barbara XQM can be turned against Barbara X'QM: all movers might eternally be men, and movement might

---

[1] OCKHAM IIIa.31.60-63: "... an assertoric *simpliciter* is one in which the predicate cannot belong to the subject at one time and be denied at another, but always stands uniformly, so that it is always or never truly predicated." According to Manekin (GERSONIDES p. 194) Averroës is one such, but Averroës does not give an example.

[2] BURIDAN 4.1.3: "... Aristotle means by 'simply assertoric sentence' an assertoric sentence which is necessary ...." Some indeed (HINTIKKA) see them as necessary *because* true at all times.

be contingent for all horses, but for horses it is not contingent that they be men. Or if the eternal non-movement of non-human animals strains metaphysical credence, then try a feature like cooking one's own food. Eternally it might be that only men do this. But it might be contingent for some non-human animal. Dogs might be trainable to do it, but never get so trained in all of time. Yet dogs cannot be men.[3]

Another problem affects both interpretations in that they make Barbara X'LL valid, contrary (so it seems) to Aristotle's view.[4] Ockham attempts to deflect this problem:[5]

And if it were said that the Philosopher denies one [XLL] and grants the other [LXL], so that it is false what is said here [*sc.* by Ockham] that both are valid, it must be said that the Philosopher speaks of necessity-propositions in the divided sense; and then it is clearly evident that from a necessity-major and an assertoric minor a necessity-conclusion follows, and this whether the minor is assertoric *simpliciter* or *ut nunc*. However it is not thus that such a discourse is valid if the major is assertoric, and therefore he denies one mixture and grants the other.

What Ockham says is true; but what he does not say is that, given that Aristotle took the trouble to distinguish Barbara X'QM from Barbara XQM, claiming that only the former is valid, his failure to mention Barbara X'LL suggests that he did not think it valid.[6]

Underlying the medieval idea that Barbara X'LL is valid, is the principle that if *a* belongs without qualification to all *b* then anything that is necessarily *b* is necessarily *a*. Ockham states this explicitly.[7] Now, it is quite true that if ab^a is necessary *de dicto* and bc^a (or bc^i) is necessary *de dicto* then ac^a (or ac^i) is necessary *de dicto*. But there is no reason to think that if ab^a is necessary *de dicto* then anything in the *b*s' star set must be in the *a*s' star set. Our account of star sets does not require that the *a*s' star set be non-empty, given that ab^a is necessarily true and the *b*s' star set is non-empty. And this seems intuitively right. Necessarily every ape is either an ape or a lion, but this does not entail that what is

---

[3] See PATTERSON 1995 pp.180-181 for a critique of temporal analyses of modal notioins.

[4] OCKHAM IIIa.31.104-108: "Note that if the major is assertoric *simpliciter*, such a mixture holds, because when something cannot be predicated of something at one time and be denied at another, but is predicated either always or never, then the subject cannot be predicated of something with the mode of necessity unless the predicate is predicated of the same with the same mode".. BURIDAN Theorem IV-19: "In the first and second figure, from one assertoric and the other *de necessario* there follows a conclusion *de necessario* under the hypothesis that the assertoric is necessary".

[5] OCKHAM IIIa.31.109-116.

[6] A further philosophical difficulty is merely apparent. OCKHAM IIIa.37.52-56 asserts without proof that Barbara X'QQ is valid. He was mistaken here. Suppose all horses are contingently walking horses, and suppose all walking horses are horses. The major is according to Ockham assertoric *simpliciter*, but we cannot infer that all horses are contingently horses.

[7] OCKHAM IIIa.31.105-108.

essentially an ape is an essential ape-or-lion: nothing, one might hold, is essentially that.

Angelelli offers a distinct interpretive approach:

> Without introducing the idea of time, we may formulate Aristotle's thought in a fruitful way.[8]

Angelelli's way, as we noted earlier, is to define unqualifieds as ampliated assertorics, so that an unqualified universal affirmative states that every possible b is an a. This gives the right logical results, validating X'-subordination and X'e-conversion, as well as First Figure X'QM moods, without validating Barbara X'LL. (See below for proofs.)

Furthermore, the Angelelli analysis is clearly better than the alternative analysis put forward by Pseudo-Scotus in the following passage:[9]

> The third conclusion is this, that if the major is an unqualified assertoric, the syllogism [Barbara X'QM] is valid, because from that unqualified assertoric follows a possibility-proposition about what can be [*una de possibili, pro his, quae possunt esse*]. For instance, it follows *Necessarily every man is an animal*, so *Everything that can be a man can be an animal*. Now, a syllogism from a possibility-major about what can be and a possibility-minor is valid, as was clear above; so a syllogism from an unqualified assertoric major and a possibility-minor is valid, because whatever follows from the consequent follows from the antecedent. And this same conclusion can be posited about two-way contingency, but then a possibility-conclusion, not a two-way contingency-conclusion, is inferred.

Barbara X'QM is valid, according to Scotus, provided the major implies that every possible b is a possible a. Thus, Scotus's condition for the validity of the said Barbara is that the 'unqualified' major be read as an M'-form. He is clearly right about this: if every c is a contingent (and thus a possible) b and every possible b is a possible a then every c is a possible a.

But how can Scotus's idea be extended to the other First Figure moods with 'unqualified' majors? Consider Darii X'QM: its premisses can be true in Scotus's sense while the conclusion is false, as model 34.1 shows.

34.1		*	¤	†
f_1	d_1	d_1	d_3	d_1, d_2
f_2	d_2	–	d_1, d_3	d_2
f_3	d_3	d_3	d_1	d_2, d_3

$$f_2^{\dagger} \subseteq f_1^{\dagger}, f_3^{\dagger} \cap f_2 \neq \varnothing \text{ but } f_3^{\dagger} \cap f_1 = \varnothing, f_1^{\dagger} \cap f_3 = \varnothing.$$

And this can happen consistent with the Sun set Theorem, which is satisfied by the model. Another difficulty with the Scotus analysis is that it does not validate X'-subordination: every possible b might be a possible a even if not every b is an a:

8 ANGELELLI p.201.

9 SCOTUS Q.32 Part I conclusion 3.

34.2		*	¤	†
f_1	d_1	–	–	d_1, d_2
f_2	d_2	–	–	d_1, d_2

$f_2^{\dagger} \subseteq f_1^{\dagger}$ but $f_2 \not\subset f_1$.

Thus we shall follow Angelelli in taking $X'ab^a$ to mean that all possible *b*s are *a*s. $X'ab^e$ we will take to mean that no possible *b*s are *a*s and no possible *a*s are *b*s. There is then no need to appeal to *de dicto* notions of necessity or truth at a time in order to explain the validity of Barbara X'QM.

34.1 $X'ab^a$ is true iff $f_2^{\dagger} \subseteq f_1$, *i.e.* the possible *b*s are among the *a*s.

34.2 $X'ab^e$ is true iff $f_2^{\dagger} \cap f_1 = \varnothing \wedge f_1^{\dagger} \cap f_2 = \varnothing$, *i.e.* the possible *b*s exclude the *a*s and the possible *a*s exclude the *b*s.

Affirmative unqualified assertorics, so understood, neither imply nor are implied by the corresponding apodeictics, as is clear from the following counter-models.

34.3		*	¤	†
f_1	d_1	d_1	–	d_1, d_2
f_2	d_1	d_1	–	d_1, d_2
f_3	d_2	–	–	d_1, d_2

$f_1 \subseteq f_2^{*}$ but $f_1^{\dagger} \not\subset f_2$.

34.4		*	¤	†
f_1	d_1, d_2	d_1	–	d_1, d_2
f_2	d_1, d_2	d_2	–	d_1, d_2

$f_1^{\dagger} \subseteq f_2$ but $f_1 \not\subset f_2^{*}$.

By contrast, Le-forms are equivalent to X'e-forms:[10]

$$f_1 \subseteq f_2^{¤} \wedge f_2 \subseteq f_1^{¤} \text{ iff } f_1^{\dagger} \subseteq \overline{f_2} \wedge f_2^{\dagger} \subseteq \overline{f_1}$$

$$\therefore \begin{cases} f_1 \subseteq f_2^{¤} \text{ iff } f_2^{\dagger} \subseteq \overline{f_1} \\ f_2 \subseteq f_1^{¤} \text{ iff } f_1^{\dagger} \subseteq \overline{f_2}. \end{cases}$$

Using these definitions we can extend our semantics for *AQ'* to provide a semantics for a system *AX'* whose axioms will be given in the next section. Definitions of *AX'* model and *AX'* validity follow the usual pattern.

Valid are the following:

[10] In §13.1 three interpretations were suggested for Aristotle's demonstrations of the validity of First Figure X'QM moods. The first required X' forms to be incompatible with M forms of the opposite quantity and quality; the second and third required an incompatibility between X' and either L or X forms. The above models show that our present semantics conflicts with the first of those interpretations, but not with the second or third: if X' forms do not entail the corresponding L-forms then they are not incompatible with the M-forms of opposite quantity and quality.

X'a-subord If $f_2^{\dagger} \subseteq f_1$ then $f_2 \subseteq f_1$
$\because f_2 \subseteq f_2^{\dagger}$.

X'e-subord If $f_2^{\dagger} \subseteq \overline{f_1}$ and $f_1^{\dagger} \subseteq \overline{f_2}$ then $f_2 \subseteq \overline{f_1}$
$\because f_2 \subseteq f_2^{\dagger}$.

X'-e conv If $f_2^{\dagger} \cap f_1 = \varnothing \wedge f_1^{\dagger} \cap f_2 = \varnothing$
then $f_1^{\dagger} \cap f_2 = \varnothing \wedge f_2^{\dagger} \cap f_1 = \varnothing$.

Barbara X'QM If $f_2^{\dagger} \subseteq f_1$ and $f_3 \subseteq f_2^{\ddagger}$ then $f_3 \subseteq f_1^{\dagger}$
$\because f_2^{\ddagger} \subseteq f_2^{\dagger}, f_1 \subseteq f_1^{\dagger}$.

This last proof relies on our semantics for X'a-forms, but not on the Sun set Theorem. An alternative method first proves the validity of Barbara XQM using the Sun set Theorem, and then appeals to the validity of X'a-subord:

Barbara XQM If $f_2 \subseteq f_1$ and $f_3 \subseteq f_2^{\ddagger}$ then $f_3 \subseteq f_1^{\dagger}$

$$\because \begin{cases} f_2^{\ddagger} \subseteq f_2^{\dagger} \\ f_2^{\dagger} \subseteq f_1^{\dagger} (\mathbf{22.2.1}). \end{cases}$$

Celarent X'QM

$$\text{If } f_2^{\dagger} \subseteq \overline{f_1} \text{ and } f_1^{\dagger} \subseteq \overline{f_2} \text{ and } f_3 \subseteq f_2^{\ddagger} \text{ then } \begin{cases} f_3 \subseteq \overline{f_1^{*}} \\ f_1 \subseteq \overline{f_3^{*}} \end{cases}$$

$$\because \begin{cases} f_2^{\ddagger} \subseteq f_2^{\dagger}, f_1^{*} \subseteq f_1; \\ f_2^{\ddagger} \subseteq f_2^{\dagger}, f_3^{*} \subseteq f_3. \end{cases}$$

This proof, using our semantics for X'e-forms but not the Sun set Theorem, cannot be replaced by a proof of Celarent XQM using the Sun set Theorem. Celarent XQM is invalid (see model 28.5).

Darii X'QM If $f_2^\dagger \subseteq f_1$ and either $f_3 \cap f_2^\ddagger \neq \varnothing$ or $f_2 \cap f_3^\ddagger \neq \varnothing$
then either $f_3 \cap f_1^\dagger \neq \varnothing$ or $f_1 \cap f_3^\dagger \neq \varnothing$

$$\therefore \begin{cases} \text{if } f_3 \cap f_2^\ddagger \neq \varnothing \text{ then } f_3 \cap f_1^\dagger \neq \varnothing \quad \because f_2^\ddagger \subseteq f_2^\dagger, f_1 \subseteq f_1^\dagger; \\ \text{if } f_2 \cap f_3^\ddagger \neq \varnothing \text{ then } f_1 \cap f_3^\dagger \neq \varnothing \quad \because f_3^\ddagger \subseteq f_3^\dagger, f_2 \subseteq f_2^\dagger. \end{cases}$$

We could, instead of relying on our semantics for X'a-forms, invoke the Sun set Theorem to prove Darii XQM and then strengthen the major premiss using X'a-subord:

If $f_2 \subseteq f_1$ and either $f_3 \cap f_2^\ddagger \neq \varnothing$ or $f_2 \cap f_3^\ddagger \neq \varnothing$
then either $f_3 \cap f_1^\dagger \neq \varnothing$ or $f_1 \cap f_3^\dagger \neq \varnothing$

$$\therefore \begin{cases} \text{if } f_3 \cap f_2^\ddagger \neq \varnothing \text{ then } f_3 \cap f_1^\dagger \neq \varnothing \quad \because f_2^\ddagger \subseteq f_2^\dagger, f_2^\dagger \subseteq f_1^\dagger \, \textbf{(22.2.1)} \\ \text{if } f_2 \cap f_3^\ddagger \neq \varnothing \text{ then } f_1 \cap f_3^\dagger \neq \varnothing \quad \because f_3^\ddagger \subseteq f_3^\dagger. \end{cases}$$

Ferio X'QM If $f_2^\dagger \subseteq \overline{f_1}$ and $f_1^\dagger \subseteq \overline{f_2}$ then
if either $f_3 \cap f_2^\ddagger \neq \varnothing$ or $f_2 \cap f_3^\ddagger \neq \varnothing$ then $f_3 \not\subset f_1^*$

$$\because \begin{cases} \text{if } f_3 \cap f_2^\ddagger \neq \varnothing \text{ then } f_3 \not\subset f_1^* \because \begin{cases} f_2^\ddagger \subseteq f_2^\dagger \\ f_2^\dagger \subseteq \overline{f_1} \\ \overline{f_1} \subseteq \overline{f_1^*} \end{cases} \\ \text{if } f_2 \cap f_3^\ddagger \neq \varnothing \text{ then } f_3 \not\subset f_1^* \because \begin{cases} f_3^\ddagger \subseteq f_3^\dagger \\ f_1^\dagger \subseteq \overline{f_2} \text{ so } f_2 \subseteq \overline{f_1^\dagger} \\ \quad \text{if } f_3 \not\subset \overline{f_1^\dagger} \text{ then } f_3 \not\subset f_1 \, \textbf{(22.2.1)} \\ \overline{f_1} \subseteq \overline{f_1^*} \end{cases} \end{cases}$$

In this proof, unlike any of the other proofs of the validity of First Figure X'QM moods, essential use is made of both the semantics for X'-forms and the Sun set Theorem.

An examination of the above proofs will show that our semantics validates Barbara, Celarent and Ferio X'QX as well as X'QM. Darii X'QX, however, is invalid:

Darii X'QX

34.5		*	¤	‡	†
f_1	d_1	d_1	d_2, d_3	–	d_1
f_2	d_1	d_1	d_2, d_3	–	d_1
f_3	d_2, d_3	d_2	–	d_1, d_3	d_1, d_2, d_3
f_4	d_3	d_3	d_1, d_2	–	d_3

$f_2{}^{\dagger} \subseteq f_1$ and $f_2 \cap f_3{}^{\ddagger} \neq \varnothing$ but $f_3 \cap f_1 = \varnothing$.

As terms of the syllogism, take 'horse' 'horse' 'white'. We shall also consider the term 'man' (f_4). Suppose there is a non-white horse (d_1), a swan (d_2) and a white man (d_3). Suppose that the swan is necessarily white, that the horse is necessarily a horse, and that the man is necessarily a man but not necessarily white. Then nothing white is a horse, even though every possible horse is a horse and some horse is contingently white.

Our semantics, however, does not validate Barbara X'LL:

34.6		*	¤	‡	†
f_1	d_1	–	–	d_1	d_1
f_2	d_1	d_1	–	–	d_1
f_3	d_1	d_1	–	–	d_1

$f_2{}^{\dagger} \subseteq f_1$ and $f_3 \subseteq f_2{}^{*}$ but $f_3 \not\subset f_1{}^{*}$.

Take the terms 'sleeping', 'koala' and 'koala'; and suppose the universe contains only a sleeping koala. Then all possible koalas are sleeping, and all koalas are necessarily koalas, but not all koalas are necessarily sleeping.

§35. SYSTEM *AX'*

Let us add the forms $\tilde{X}'ab^a$ and $\tilde{X}'ab^e$ (as contradictories for $X'ab^a$ and $X'ab^e$) to the language of AQ. System AX' adopts this expansion and adds to AQ the axioms

X'-subord $\dfrac{X'ab^a}{ab^a}$ *X'-subord* $\dfrac{X'ab^e}{ab^e}$

X'e-conv $\dfrac{X'ab^e}{X'ba^e}$ *X'/Le-subord* $\dfrac{X'ab^e}{Lab^e}$

L/X'e-subord $\dfrac{Lab^e}{X'ab^e}$ *Barbara X'QX* $\dfrac{X'ab^a \quad Qbc^a}{ac^a}$

Celarent X'QM $\dfrac{X'ab^e \quad Qbc^a}{Mac^e}$ *Ferio X'QM* $\dfrac{X'ab^e \quad Qbc^i}{Mac^o}$

Aristotle does not consider First Figure X'QX moods. He presumably thought they must be invalid because XQX First Figure moods are so. But he was mistaken about Barbara, Celarent and Ferio X'QX if our account of unqualified assertorics is right: this was shown in the previous section. However, we do not have to take Celarent and Ferio X'QX as axiomatic since they are provable from their LQX sisters by X'L-subordination.

Barbara X'QM can be proved from Barbara XQM with the aid of X'-subordination. Darii X'QM can be proved from Camestres XLL:

$$\text{Camestres XLL}\,\frac{ab^a \quad Lac^e}{Lbc^e}\xrightarrow[Le\,\|\,Qi]{K}\frac{\dfrac{X'ab^a}{ab^a}X'-\text{subord} \qquad Qbc^i}{Mac^i}$$

Our semantics comes into conflict with Aristotle's text on these matters. For, both approaches accept Darii XQM. But the text rejects Barbara XQM on the ground that all movers might at some time be men, and all horses contingently move, but some horse is not contingently a man. And this example works equally well against Darii XQM: all movers might at some time be men, and some horses contingently move, but no horse is contingently a man. My view is that Aristotle's counter-example simply does not work. It derives its plausibility from the thought that, timelessly, horses have the capacity for movement. Now, while this can be accepted as a theorem within a plausible metaphysical theory of potentiality, we have to remember that in the context of the modal syllogistic necessities can be ephemeral and contingency is logically opposed to necessity. Within this context there are therefore ephemeral contingencies too. And the movement of horses is one of them. At times when only men move, movement is not contingent for horses.

Further theses of *AX'* are

Barbara X'XM

$$\dfrac{\dfrac{\dfrac{X'ab^a}{ab^a}\,X'\text{-subord} \qquad bc^a}{ac^a}\,Barbara}{Mac^a}\,X/M\text{-subord}$$

Celarent X'XM Similarly.
Darii X'XM
Ferio X'XM

Aristotle considers only a handful of syllogisms containing X'-forms. That there are other X'-syllogisms follows from our adoption of X'-subordination as axiomatic. Any syllogism with an ordinary assertoric premiss can be strengthened by substituting a corresponding unqualified premiss, for example:

Barbara LX'L

$$\frac{Lab^a \quad X'bc^a}{Lac^a}$$

Celarent LX'L

$$\frac{Lab^e \quad X'bc^a}{Lac^e}$$

If the above syllogisms are asserted then so are the following:

Bocardo MX'M

$$\frac{Mac^o \quad X'bc^a}{Mab^o} \xleftarrow{\;C\;} \frac{Lab^a \quad X'bc^a}{Lac^a}\,\text{Barbara } LX'L$$

Disamis MX'M

$$\frac{Mac^i \quad X'bc^a}{Mab^i} \xleftarrow{\;C\;} \frac{Lab^e \quad X'bc^a}{Lac^e}\,\text{Celarent } LX'L$$

§36. REDUCTION TO ASSERTORIC SYLLOGISTIC

System AX', like its forebears, can be reduced to assertoric syllogistic. The reduction is the same as that for system AQ with the following additions.
$X'ab^a$ is true in AX' iff $ab_M{}^a$ is true in A_Q.

$X'ab^e$ is true in AX' iff $ab_M{}^e$ and $ba_M{}^e$ are true in A_Q.

Two examples follow.

(1) Barbara X'QM

$$\frac{X'ab^a \quad Qbc^a}{Mac^a}$$

is AX'-valid iff $\dfrac{ab_M{}^a \quad b_Qc^a}{a_Mc^a}$ is A_Q-valid. It is, and is provable thus:

$$\text{Barbara}\;\dfrac{ab_M{}^a \qquad \text{Barbara}\;\dfrac{b_Mb_Q{}^a \quad b_Qc^a}{b_Mc^a}}{ac^a} \xrightarrow{R_{QM}} \text{Barbara}\;\dfrac{\dfrac{ab_M{}^a \quad b_Qc^a}{ac^a} \qquad a_Ma^a}{a_Mc^a} \xrightarrow{R_M} \dfrac{ab_M{}^a \quad b_Qc^a}{a_Mc^a}$$

(2) Ferio X'QM

$$\frac{X'ab^e \quad Qbc^i}{Mac^o}$$

is AX'-valid iff both either $1\,\dfrac{ab_M{}^e \quad b_Qc^i}{a_Lc^o}$ or $2\,\dfrac{ba_M{}^e \quad b_Qc^i}{a_Lc^o}$ and either $3\,\dfrac{ab_M{}^e \quad c_Qb^i}{a_Lc^o}$ or $4\,\dfrac{ba_M{}^e \quad c_Qb^i}{a_Lc^o}$ are A_Q-valid. Both *1* and *4* are valid and are provable thus:

$$\text{Ferio}\;\dfrac{\text{Darii}\;\dfrac{b_Qc^i \quad b_Mb_Q{}^a}{b_Mc^i} \qquad ab_M{}^e}{ac^o} \xrightarrow{R_{QM}} \text{Baroco}\;\dfrac{\dfrac{ab_M{}^e \quad b_Qc^i}{ac^o} \qquad aa_L{}^a}{a_Lc^o} \xrightarrow{R_L} \dfrac{ab_M{}^e \quad b_Qc^i}{a_Lc^o}\;1$$

$$\text{Frisesomorum}\;\dfrac{ba_M{}^e \qquad \text{Darii}\;\dfrac{c_Mc_Q{}^a \quad c_Qb^i}{c_Mb^i}}{a_Mc_M{}^o} \xrightarrow{R_{QM}} \dfrac{ba_M{}^e \quad c_Qb^i}{a_Mc_M{}^o}$$

$$\dfrac{ac^a}{a_Mc_M{}^a} \xrightarrow{C} \dfrac{a_Mc_M{}^o}{ac^o} \qquad \text{Baroco}\;\dfrac{ac^o \qquad aa_L{}^a}{a_Lc^o} \xrightarrow{R_L} \dfrac{ba_M{}^e \quad c_Qb^i}{a_Lc^o}\;4$$

§37. A REPRIEVE FOR MODAL *REDUCTIO*

We saw in §5 that Aristotle in his proof of Barbara X'QM appears to rely on the unsound rule C'. We are now in a position to see that a restricted version of this rule is sound and can be used to prove that syllogism. The rule is sound for cases where p is an unqualified assertoric, q is an ordinary universal affirmative assertoric form, and s is the corresponding Q-form. I shall use FLANNERY's term "upgrading" to name these rules.

a-upgrading 1 $\dfrac{R \quad X'ab^a \quad bc^a}{p} \longrightarrow \dfrac{R \quad X'ab^a \quad Qbc^a}{p}$

a-upgrading 2 $\dfrac{R \quad X'ab^e \quad bc^a}{p} \longrightarrow \dfrac{R \quad X'ab^e \quad Qbc^a}{p}$

These are indeed consequences of the general rule C' of modal *reductio*, since Qa-forms do not entail their corresponding assertorics. And they can be shown to be sound.

They are sound not for Aristotle's reason, discussed in §5; for that reason mentions nothing specific to unqualified assertorics. The crucial thing about both Barbara and Celarent X'QM is revealed by our semantics for unqualified assertorics. Consider Figure 37.1.

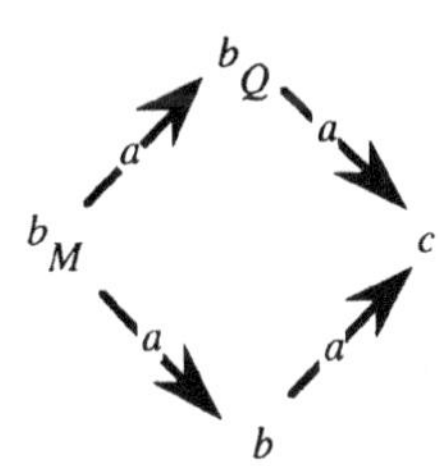

Fig. 37.1

What this diamond shows is that all cs may be possible bs for one of two reasons - either because all cs are contingent bs or because all cs are actual bs. This is so because both the contingent and the actual are possible. So, if (and only if) there is an Aristotelian chain[11] that includes the upper path in Figure 37.2, there is such a chain that includes the lower path.

[11] See §24.

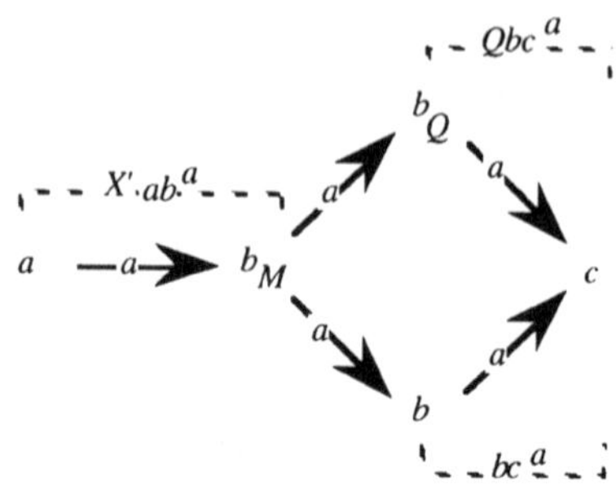

Fig. 37.2

In the case of the upper path, we will have to make use of the extension rule R_{QM} (§31); in the case of the lower path, of the extension rule R_M (§24). The two chains are interchangeable and our first a-upgrading rule is sound.

Now consider the result of combining $X'ab^e$ as major with either Qbc^a or bc^a as minor.

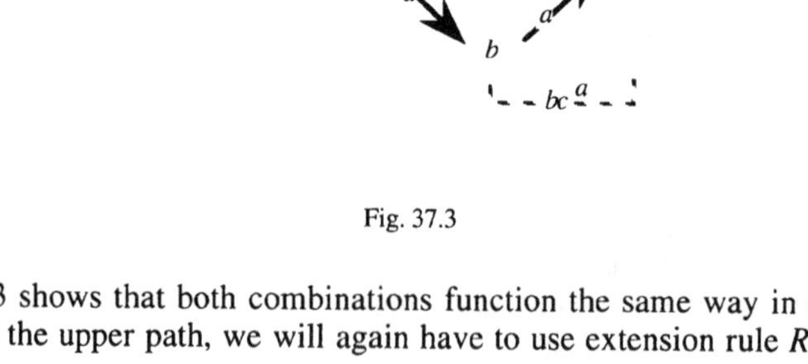

Fig. 37.3

Figure 37.3 shows that both combinations function the same way in a syllogistic chain. For the upper path, we will again have to use extension rule R_{QM}; for the lower path, extension rule R_M. So, all and only conclusions implied by $X'ab^e$ and Qbc^a are implied by $X'ab^e$ and bc^a and the second a-upgrading rule is sound. (Celarent X'QM combines the upper path $a_L - c$ with the upper path $a - c_L$.)

Next consider Figure 37.4:

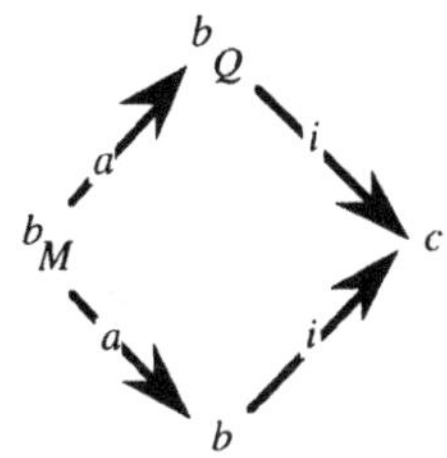

Fig. 37.4

This shows that some c may be a possible b either because some c is a contingent b or because some c is an actual b. It follows that there is an Aristotelian chain that includes the upper path in Figure 37.5 iff there is such a chain including the lower path. So all and only conclusions implied by $X'ab^a$ and b_Qc^i are implied by $X'ab^a$ and bc^a:

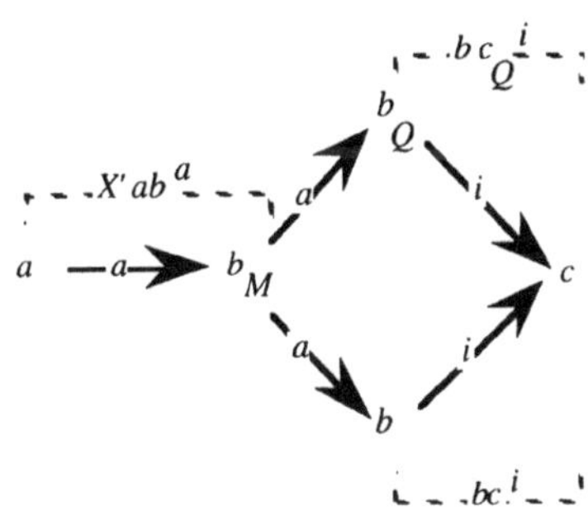

Fig. 37.5

A third diamond is shown in Figure 37.6. This shows that some possible c may be a b either because some contingent c is a b or because some actual c is a b. However, it does not follow that a b - i - c_Q chain can be replaced by a b - i - c chain. These have different end-points. Certainly there is an Aristotelian chain that includes the upper path in Figure 37.7 iff there is such a chain including the lower path. But all this shows is that all and only conclusions implied by $X'ab^a$ and $bc_Q{}^i$ *and* $c_Mc_Q{}^a$ are implied by $X'ab^a$ and bc^a *and* c_Mc^a. In fact, $X'ab^a$ and $bc_Q{}^i$ imply $ac_Q{}^i$ but $X'ab^a$ and bc^a do not, as model 37.1 shows.

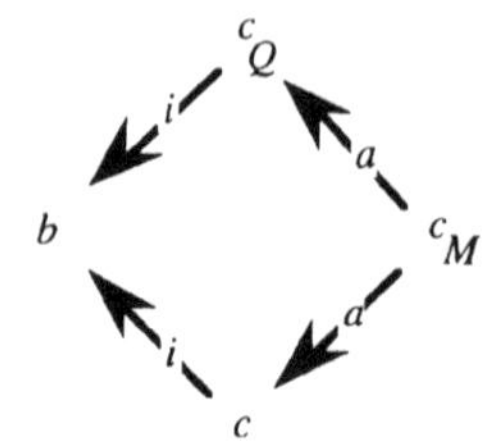

Fig. 37.6

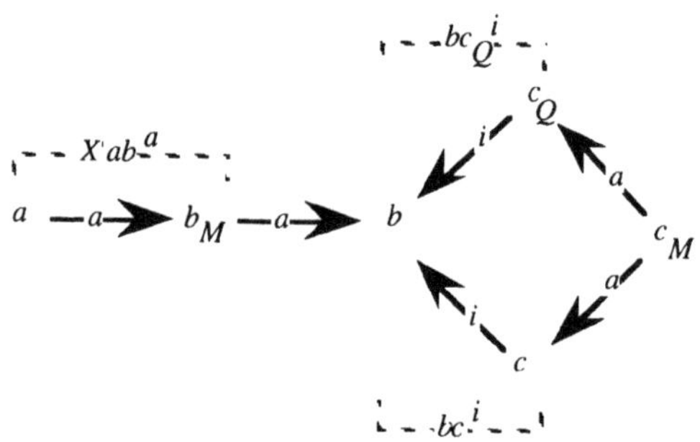

Fig. 37.7

37.1		*	¤	‡	†
f_1	d_1	d_1	–	–	d_1
f_2	d_1	–	–	d_1	d_1
f_3	d_1	d_1	–	–	d_1

$f_2{}^{\dagger} \subseteq f_1, f_3 \cap f_2 \neq \varnothing$ but $f_3{}^{\ddagger} \cap f_1 = \varnothing$.

Further, $X'ab^a$ and bc^a imply ac^i but $X'ab^a$ and Qbc^i do not:

37.2		*	¤	‡	†
f_1	d_1	–	d_2, d_3	d_1	d_1
f_2	d_1	d_1	d_2, d_3	–	d_1
f_3	d_2, d_3	d_3	–	d_1, d_2	d_1, d_2, d_3
f_4	d_2	d_2	d_1, d_3	–	d_2

$f_2{}^{\dagger} \subseteq f_1, f_2 \cap f_3{}^{\ddagger} \neq \varnothing$ but $f_3 \cap f_1 = \varnothing$.

So it is not the case that all and only conclusions implied by $X'ab^a$ and Qbc^i are implied by $X'ab^a$ and bc^a. In other words, the analogous i-upgrading rule is incorrect.

Similar reasoning shows that a Qi-minor cannot always be upgraded to an Xi when the major is an X'e.

In this way a version of modal *reductio* gets a partial reprieve. The upgrading rules stated above can be written into the basis of *AX'* and Barbara and Celarent X'QM dropped. These moods follow from Barbara and Celarent XXX by upgrading:

$$\text{Barbara}\ \dfrac{X'\text{-subord}\ \dfrac{X'ab^a}{ab^a} \qquad bc^a}{ac^a} \quad \xrightarrow{a\text{-upgrading 1}} \quad \dfrac{X'ab^a \quad Qbc^a}{ac^a}$$

$$\text{Celarent}\ \dfrac{X'\text{-subord}\ \dfrac{X'ab^e}{ab^e} \qquad bc^a}{ac^a} \quad \xrightarrow{a\text{-upgrading 2}} \quad \dfrac{X'ab^e \quad Qbc^a}{ac^e}$$

Darii X'QM is provable from Camestres XLL. Ferio X'QM still has to be axiomatic. This leaves us with the axioms and rules of AQ' plus (as axioms) X'-subordination, X'/Le-subordination, L/X'e-subordination, Ferio X'QM; and (as rules) a-upgrading 1, a-upgrading 2.

Our versions of Aristotle's demonstrations of Barbara and Celarent X'QM moods are as follows:

$$X/M\text{-subord}\ \dfrac{\text{Barbara } X'XX\ \dfrac{X'ab^a \quad bc^a}{ac^a}}{Mac^a} \quad \xrightarrow{a\text{-upgrading 1}} \quad \dfrac{X'ab^a \quad Qbc^a}{Mac^a}$$

$$L\text{-subord}\ \dfrac{\text{Festino LLL}\ \dfrac{Lab^e \quad Lac^i}{Lbc^o}}{bc^o} \quad \xrightarrow[C]{X'/Le\text{-subord}} \quad \dfrac{\dfrac{X'ab^e}{Lab^e} \qquad bc^a}{Mac^e} \quad \xrightarrow{a\text{-upgrading 2}} \quad \dfrac{X'ab^e \quad Qbc^a}{Mac^e}$$

There is no need to use *reductio* in proving Barbara X'QM by the upgrading procedure. However, if we want to approximate Aristotle's own proof more closely then we can combine upgrading and *reductio*:

$$\text{Bocardo } LXX\,\frac{Lac^{o} \quad bc^{a}}{ab^{o}} \xrightarrow[C]{X'-subord} \frac{\dfrac{X'ab^{a}}{ab^{a}} \qquad\qquad bc^{a}}{Mac^{a}} \xrightarrow{a-upgrading\ 1} \frac{X'ab^{a} \quad Qbc^{a}}{Mac^{a}}$$

One would be mistaken if one thought that Darii and Ferio X'QM were provable from Ferison and Datisi LXX (or LXL) by means of modal *reductio*. (These proofs were suggested in §13.) Those Third Figure syllogisms are indeed valid, as are Darii and Ferio X'QM; it is just that the validity of the latter does not follow from the validity of the former by reason of upgrading. Upgrading cannot be extended to i-forms.

CHAPTER NINE. SINGULAR MODAL SYLLOGISTIC

Modal syllogistic can be extended to cover syllogisms containing singular terms. The entire modal system can then be based on a system of singular modals. In turn, the system of singular modal syllogisms can be reduced to assertoric singular syllogistic. Many different systems of singular modal syllogistic are possible, corresponding to different existence-assumptions and different definitions of sun sets.

§38. SYSTEM *AEL5*

We open our discussion of singular modal syllogistic by presenting a system *AEL5*, based on a combination of *A*- and *C*-semantics. The reason for this mix will become apparent shortly.

§38.1 Syntactic base

We define an ecthetic system *AEL5*, containing singular term variables

$$x, y, z, \ldots$$

plain general term variables

$$a, b, c, \ldots$$

bold general term variables

$$\boldsymbol{a}, \boldsymbol{b}, \boldsymbol{c}, \ldots$$

and a bar

forming negated terms from plain terms (but not from bold terms). A *term* is then a variable or the result of applying a bar to a term. (We adopt the convention that $\bar{\bar{\tau}}$ is interchangeable with τ.) A *propositional form* is a sequence $\tau\xi$, $\sim \tau\xi$, $\tau\upsilon^a$, $\tau\upsilon^e$, $\tau\upsilon^i$ or $\tau\upsilon^o$, where τ, υ are terms and ξ is an individual variable. Propositional forms are intended to be read as follows:

ax	"x is a"	$\sim$ *ax*	"x is not a"
Lax	"x must be a"	*Uax*	"x cannot be a"
Max	"x may be a"	$\sim$ *Lax*	"x need not be a"

as well as the forms of proposition recognised in system *AL*.

The following pairs are defined as contradictories: $\tau\xi \mid \sim \tau\xi$, $L\tau\xi \mid \sim L\tau\xi$, $M\tau\xi \mid U\tau\xi$, in addition to *AL*'s pairs of contradictories. The sign '$\sim$' is thus ambiguous between its object-language use to form an expression's contradictory, and its meta-language use to form the name of an expression's contradictory. Note that forms $\sim \tau\xi, \bar{\tau}\xi$ are distinct. The following pairs are defined as contraries: $\tau\xi \parallel U\tau\xi$, $\sim \bar{\tau}\xi \parallel U\tau\xi$, $L\tau\xi \parallel U\tau\xi$, $M\tau\xi \parallel L\bar{\tau}\xi$, $\tau\xi \parallel L\bar{\tau}\xi$, $L\tau\xi \parallel L\bar{\tau}\xi$, $M\tau\xi \parallel \bar{\tau}\xi$, $\tau\xi \parallel \bar{\tau}\xi$, $M\tau\xi \parallel \sim \tau\xi$, $L\tau\xi \parallel \sim \tau\xi$, in addition to AL's pairs of contraries.

The non-modal base of the system comprises the axioms:

a-identity $\dfrac{}{aa^a}$ *rep* $\dfrac{ax}{ax}$

a-subalt $\dfrac{ab^a}{ab^i}$ *a-elim* $\dfrac{ab^a \quad bx}{ax}$

e-elim 1 $\dfrac{ab^e \quad bx}{\bar{a}x}$ *e-elim 2* $\dfrac{ab^e \quad ax}{\bar{b}x}$

together with rules of Substitution, Permutation, C, K, and Cut, and elimination rules for i- and o-forms

i-elim $\dfrac{R \quad ax \quad bx}{p} \longrightarrow \dfrac{R \quad ab^i}{p}$

o-elim $\dfrac{R \quad \sim ax \quad bx}{p} \longrightarrow \dfrac{R \quad ab^o}{p}$

provided R has some member $(L)bv^{a(i)}$ or $(L)vb^{a(i)}$

(where R is a possibly null sequence of forms and p is a form and x does not occur in R or p).

The rule of Substitution must be subject to a restriction to block the derivation of formulae whose conclusions go beyond their premisses in implying the non-emptiness of some class. Call the restricted rule U_t. The required restriction is that a term t can be substituted for a variable v in a formula's a- or i-conclusion only if after the substitution there is an a- or i-premiss containing t. This entails that, while bold variables can be substituted for plain, plain terms (including negated terms) cannot be substituted for bold variables. Otherwise we could infer

$$\overline{\quad}\qquad\overline{\quad}$$
$$aa^a \qquad \overline{aa}^a$$

from a-identity; but the conclusions of these formulae imply the existence of *a*s and non-*a*s respectively - something for which we have no general warrant. (See the semantic analysis below.) Still, negated terms can be substituted for plain variables.

Two further introduction rules follow by *C* from i-elim and o-elim:

e-intro $\dfrac{R \quad bx}{\sim ax} \longrightarrow \dfrac{R}{ab^e}$

a-intro $\dfrac{R \quad bx}{ax} \longrightarrow \dfrac{R}{ab^a}$

provided R has some member $(L)bv^{a(i)}$ or $(L)vb^{a(i)}$.

And three further introduction theses follow from a-elim, e-elim 1 and e-elim 2:

o-intro $\dfrac{\sim ax \quad bx}{ab^o}$ *i-intro 1* $\dfrac{\sim \bar{a}x \quad bx}{ab^i}$

i-intro 2 $\dfrac{\sim \bar{b}x \quad ax}{ab^i}$

whence:

i-intro 3 $rep\,\dfrac{\bar{a}x}{\bar{\bar{a}}x} \xrightarrow{K} \dfrac{ax}{\dfrac{\sim \bar{\bar{a}}x \quad bx}{ab^i}\ i-intro\ 1}$

Corresponding to the assumption that the necessary is a sub-class of the actual, we have a thesis of L-subordination, which is provable as follows:

$$rep\,\frac{ax}{ax} \xrightarrow{C} \frac{\sim ax}{\sim ax} \xrightarrow{K} \frac{Lax}{ax}$$

For negative singular forms we have the thesis of U-subordination:

$$rep\frac{ax}{ax}\xrightarrow{K}\frac{Uax}{\sim ax}$$

corresponding to S0 of §25.2. From this follows M-subordination:

$$U-subord\frac{Uax}{\sim ax}\xrightarrow{C}\frac{ax}{Max}$$

As analogues of the axiom of Repetition we have the axioms

L-rep $\frac{Lax}{Lax}$ *U-rep* $\frac{Uax}{Uax}$

which by *C* imply the additional repetition theses

$\frac{\sim Lax}{\sim Lax}$ *M-rep* $\frac{Max}{Max}$

The structural definition of La-forms implicit in *A* semantics is reflected in the axiom

La-elim A $\frac{Lab^a \quad bx}{Lax}$

and the rule

La-intro A $\frac{R \quad bx}{Lax}\longrightarrow\frac{R}{Lab^a}$
provided R has some member $(L)bv^{a(i)}$ or $(L)vb^{a(i)}$

The corresponding structural definition of Le-forms is reflected in the axioms

Le-elim 1 A $\frac{Lab^e \quad bx}{Uax}$ *Le-elim 2* A $\frac{Lab^e \quad ax}{Ubx}$

and the rule

Le-intro A $\left.\begin{array}{c}\dfrac{R \quad ax}{Ubx} \\ \\ \dfrac{R \quad bx}{Uax}\end{array}\right\} \longrightarrow \dfrac{R}{Lab^e}$

A-semantics for Li-forms are reflected in the axioms

Li-intro 1 A $\dfrac{Lax \quad bx}{Lab^i}$ *Li-intro 2* A $\dfrac{ax \quad Lbx}{Lab^i}$

and the rule

Li-elim A $\left.\begin{array}{c}\dfrac{R \quad Lax \quad bx}{p} \\ \\ \dfrac{R \quad Lby \quad ay}{p}\end{array}\right\} \longrightarrow \dfrac{R \quad Lab^i}{p}$

Corresponding to A-semantics for Lo-forms, we have the axiom

Lo-intro A $\dfrac{Uax \quad bx}{Lab^o}$

and the rule

Lo-elim A $\dfrac{R \quad Uax \quad bx}{p} \longrightarrow \dfrac{R \quad Lab^o}{p}$

Finally, corresponding to **25.5** we have

U-intro 5 $\dfrac{Lda^a \quad de^e \quad Lex}{Uax}$

and the rule

U-elim 5 $\dfrac{R \quad Lda^a \quad de^e \quad Lex}{p} \longrightarrow \dfrac{R \quad Uax}{p}$

whence we derive the Sun set Theorem SS:

$$\dfrac{U-intro\ 5\dfrac{Barbara\ LXL\dfrac{Lda^a \quad ab^a}{}\ Ldb^a \qquad de^e \quad Lex}{}}{Ubx} \qquad \xrightarrow{U-elim\ 5} \dfrac{ab^a \quad Uax}{Ubx}$$

Theses include

$$\overline{a\bar{a}^e}$$

Proof $\quad rep\dfrac{ax}{ax} \xrightarrow{K} \dfrac{\bar{a}x}{\sim ax} \xrightarrow{e-intro} \dfrac{}{a\bar{a}^e}$

§38.2 *Formal semantics*

Semantics for *AEL5* depart slightly from our usual pattern. We suppose, as in *A*-style semantics, that some general terms (in this case, bold general terms) are associated with non-empty sets; but we do not make the same supposition about negated general terms or about singular terms. Negated general terms will be associated with non-universal sets; singular terms will be associated either with singletons or with the null set. The general rule is that terms are associated with possibly null sets; and in this sense we have a *C*-style semantics. But if we consider only bold variables we have an *A*-style semantics.

An *AEL5* model is defined as an ordered triple $\langle D, F, V\rangle$ satisfying **21.1**, **21.7** and **21.8**, plus **22.1.1-8**, **24.1-7** and

38.2.1 An assignment *V* of values to the variables in a wff associates with each individual variable in the wff either a member of *D* or the null set, and with each un-negated term a non-null member of *F*.

38.2.2 *ax* is true iff '*x*' stands for d_1 and $d_1 \in f_1$, *i.e.* iff '*x*' stands for some member of the domain that is among the *a*s.

38.2.3 $\bar{a}x$ is true iff '*x*' stands for d_1 and $d_1 \notin f_1$, *i.e.* iff '*x*' stands for some member of the domain that is not among the *a*s.

38.2.4 *Lax* is true iff '*x*' stands for d_1 and $d_1 \in f_1^*$, *i.e.* iff '*x*' stands for some member of the domain that is in the *as*' star-set.

38.2.5 *Uax* is true iff '*x*' stands for d_1 and $d_1 \in f_1^{\text{¤}}$, *i.e.* iff '*x*' stands for some member of the domain that is in the *as*' sun set.

38.2.6 $\sim p$ is true iff p is not true.

System *AEL5* is sound with respect to this semantics. All the axioms are valid:

a-identity $f_1 \neq \varnothing \wedge f_1 \subseteq f_1$ provided $f_1 \neq \varnothing$

rep Trivial

a-subalt If $f_2 \neq \varnothing \wedge f_2 \subseteq f_1$ then $f_2 \cap f_1 \neq \varnothing$

a-elim If $f_2 \neq \varnothing \wedge f_2 \subseteq f_1$ and $d_1 \in f_2$ then $d_1 \in f_1$

e-elim 1 If $f_2 \cap f_1 = \varnothing$ and $d_1 \in f_2$ then $d_1 \in \overline{f_1}$

e-elim 2 If $f_2 \cap f_1 = \varnothing$ and $d_1 \in f_1$ then $d_1 \in \overline{f_2}$

L-rep, U-rep Trivial

La-elim A If $f_2 \neq \varnothing \wedge f_2 \subseteq f_1^*$ and $d_1 \in f_2$ then $d_1 \in f_1^*$

Le-elim 1 A If $f_2 \subseteq f_1^{\text{¤}} \wedge f_1 \subseteq f_2^{\text{¤}}$ and $d_1 \in f_2$ then $d_1 \in f_1^{\text{¤}}$

Le-elim 2 A If $f_2 \subseteq f_1^{\text{¤}} \wedge f_1 \subseteq f_2^{\text{¤}}$ and $d_1 \in f_1$ then $d_1 \in f_2^{\text{¤}}$

Li-intro 1 A If $d_1 \in f_2$ and $d_1 \in f_2^*$ then $f_2 \cap f_1^* \neq \varnothing \vee f_1 \cap f_2^* \neq \varnothing$

Li-intro 2 A If $d_1 \in f_2^* $ and $d_1 \in f_2$ then $f_2 \cap f_1^* \neq \varnothing \vee f_1 \cap f_2^* \neq \varnothing$

Lo-intro A If $d_1 \in f_2$ and $d_1 \in f_1^{¤}$ then $f_2 \cap f_1^{¤} \neq \varnothing$

U-intro 5 If $f_1 \subseteq f_4^*$ and $f_4 \cap f_5 = \varnothing$ and $d_1 \in f_5^*$ then $d_1 \in f_1^{¤}$
– by **22.1.4**

And the rules preserve validity:

Substitution Trivial.

Permutation Trivial

C If $(p \wedge r) \supset q$ is valid then $(\sim q \wedge r) \supset \sim p$ is valid
- by **21.7 - 8**

K If $(p \wedge r) \supset q$ is valid then $(\neg q \wedge r) \supset \sim p$ is valid
$\because \neg q \supset \sim q$ is valid

T If $p \supset q$ and $(q \wedge s) \supset r$ are valid then $(p \wedge s) \supset r$ is valid
- by **21.7 - 8**

i-elim If $(r \wedge d_1 \in f_1 \wedge d_1 \in f_2) \supset p$ is valid
then $(r \wedge f_2 \cap f_1 \neq \varnothing) \supset p$ is valid
$\because (r \wedge f_2 \cap f_1 \neq \varnothing) \supset \exists d[r \wedge d \in f_1 \wedge d \in f_2]$ is valid

o-elim

Let $\dot{R}$ be the conjunction of *R*' s members.
If $(\dot{R} \wedge d_1 \notin f_1 \wedge d_1 \in f_2) \supset p$ is valid
then $(\dot{R} \wedge f_2 \not\subset f_1) \supset p$ is valid
provided *R* has some conjunct
$f \neq \varnothing \wedge f \subseteq f_2, f \neq \varnothing \wedge f \subseteq f_2^*,$
$f \cap f_2 \neq \varnothing, f \cap f_2^* \neq \varnothing \vee f_2 \cap f^* \neq \varnothing,$
$f_2 \neq \varnothing \wedge f_2 \subseteq f, f_2 \neq \varnothing \wedge f_2 \subseteq f^*,$
$f_2 \cap f \neq \varnothing$ or $f_2 \cap f^* \neq \varnothing \vee f \cap f_2^* \neq \varnothing$
$\because (\dot{R} \wedge f_2 \not\subset f_1) \supset \exists d[\dot{R} \wedge d \notin f_1 \wedge d \in f_2]$ is valid
under the proviso

La-intro A

If $(\dot{R} \wedge d_1 \in f_2) \supset d_1 \in f_1^*$ is valid
then $\dot{R} \supset (f_2 \neq \varnothing \wedge f_2 \subseteq f_1^*)$ is valid
provided *R* has some member
$f_2 \neq \varnothing \wedge f_2 \subseteq f, f_2 \neq \varnothing \wedge f_2 \subseteq f^*,$
$f \cap f_2 \neq \varnothing, f \cap f_2^* \neq \varnothing \vee f_2 \cap f^* \neq \varnothing,$
$f \neq \varnothing \wedge f \subseteq f_2, f \neq \varnothing \wedge f \subseteq f_2^*,$
$f_2 \cap f \neq \varnothing$ or $f_2 \cap f^* \neq \varnothing \vee f \cap f_2^* \neq \varnothing$
$\because$ if $\forall d[(\dot{R} \wedge d \in f_2) \supset d \in f_1^*]$ is valid
then $\dot{R} \supset (f_2 \neq \varnothing \wedge f_2 \subseteq f_1^*)$ is valid
under the proviso

Le-intro A

If both $(r \wedge d_1 \in f_1) \supset d_1 \in f_2^{¤}$
and $(r \wedge d_1 \in f_2) \supset d_1 \in f_1^{¤}$ are valid
then $r \supset (f_2 \subseteq f_1^{¤} \wedge f_1 \subseteq f_2^{¤})$ is valid
$\because$ if both $\forall d[(r \wedge d \in f_1) \supset d \in f_2^{¤}]$
and $\forall d[(r \wedge d \in f_2) \supset d \in f_1^{¤}]$ are valid
then $r \supset (f_2 \subseteq f_1^{¤} \wedge f_1 \subseteq f_2^{¤})$ is valid

Li-elim A

If both $\left(r \wedge d_1 \in f_1^{*} \wedge d_1 \in f_2\right) \supset p$
and $\left(r \wedge d_2 \in f_2^{*} \wedge d_2 \in f_1\right) \supset p$ are valid
then $\left(r \wedge \left(f_2 \cap f_1^{*} \neq \varnothing \vee f_1 \cap f_2^{*} \neq \varnothing\right)\right) \supset p$ is valid
$\because \left(r \wedge \left(f_2 \cap f_1^{*} \neq \varnothing \vee f_1 \cap f_2^{*} \neq \varnothing\right)\right) \supset p$ is valid iff
both $(i)\left(r \wedge f_2 \cap f_1^{*} \neq \varnothing\right) \supset p$
and $(ii)\left(r \wedge f_1 \cap f_2^{*} \neq \varnothing\right) \supset p$ are valid; and
(i) if $\exists d\left[r \wedge d \in f_1^{*} \wedge d \in f_2\right] \supset p$ is valid
then $\left(r \wedge f_2 \cap f_1^{*} \neq \varnothing\right) \supset p$ is valid,
(ii) if $\exists d\left[r \wedge d \in f_2^{*} \wedge d \in f_1\right] \supset p$ is valid
then $\left(r \wedge f_1 \cap f_2^{*} \neq \varnothing\right) \supset p$ is valid

Lo-elim A

If $\left(r \wedge d_1 \in f_1^{¤} \wedge d_1 \in f_2\right) \supset p$ is valid
then $\left(r \wedge f_2 \cap f_1^{¤} \neq \varnothing\right) \supset p$ is valid
$\because \left(r \wedge f_2 \cap f_1^{¤} \neq \varnothing\right) \supset \exists d\left[r \wedge d \in f_1^{¤} \wedge d \in f_2\right]$ is valid

U-elim 5

If $\left(r \wedge f_1 \subseteq f_4^{*} \wedge f_5 \cap f_4 = \varnothing \wedge d_1 \in f_5^{*}\right) \supset p$ is valid
then $(r \wedge d_1 \in f_1^{¤}) \supset p$ is valid
$\because (r \wedge d_1 \in f_1^{¤}) \supset \exists d, f, g\left[r \wedge f_1 \subseteq f^{*} \wedge g \cap f = \varnothing \wedge d \in g^{*}\right]$ is valid
- by **22.1.4**

This semantics allows for a distinction between singular denials and singular affirmations with negated predicate. Affirmations may imply denials but denials never imply affirmations. Thus $\dfrac{a\bar{x}}{\sim ax}$ is valid, and is provable as follows:

$$rep\,\frac{ax}{ax} \xrightarrow{\;K\;} \frac{a\bar{x}}{\sim ax}$$

But $\frac{\sim ax}{ax}$ is not valid (since 'x' might not stand for any individual in the domain). This reflects Aristotle's view of the relation between affirmations and denials.[1]

Our semantics possesses another peculiarity, in that two modalities must be taken as undefined - necessity and either possibility or impossibility. These are not inter-definable. $\frac{L\bar{a}x}{Uax}$ is valid (given that a set's star set is in the original set, and that its sun set is in the original set's complement - conditions that obtain in any of S0-S5), and is provable as follows:

$$M-rep\frac{Max}{Max}\xrightarrow{K}\frac{L\bar{a}x}{Uax}$$

But $\frac{Uax}{L\bar{a}x}$ is not valid:

38.1		*	¤
f_1	d_2	d_2	d_1, d_3
$\overline{f_1}$	d_1, d_3	–	–
f_2	d_1	d_1	d_2
f_3	d_3	d_3	d_1

$d_1 \in f_1^{¤}$ but $d_1 \notin \overline{f_1}^{*}$.

The situation is similar to that in A.N. Prior's system Q where necessity and possibility and not inter-definable, 'necessary-not' implying 'not possible' but not being implied by it.[2]

§38.3 Provability of AL*'s base*

AEL5 contains a system isomorphic to *AL*. This system is obtained from *AL* by replacing all its variables with corresponding bold variables. This can be shown by deriving within *AEL5* the basis of that system. a-rep in its bold version is an axiom of *AEL5*. Each of the remaining *AL* axioms is derivable in *AEL5* in its plain version, and the bold version follows from this by Substitution.

---

[1] See THOM 1981 §33.

[2] See HUGHES & CRESSWELL pp.303-305.

e-conv

$$\text{rep}\,\frac{bx}{bx}\xrightarrow{\;K\;}\frac{e-\text{elim }2\,\dfrac{ab^e \quad ax}{\bar{b}x}}{\sim bx}\qquad\xrightarrow{\;e-\text{intro}\;}\frac{ab^e}{ba^e}$$

a-conv (partial)

$$i-\text{intro }3\,\frac{ax \quad bx}{ba^i}\xrightarrow{\;i-\text{elim}\;}\frac{a-\text{subalt}\,\dfrac{ab^a}{ab^i}}{ba^i}$$

i-conv *As above.*

a-rep

$$a-\text{elim}\,\frac{ab^a \quad bx}{ax}\xrightarrow{\;a-\text{intro}\;}\frac{ab^a}{ab^a}$$

Barbara

$$a-\text{elim}\,\frac{ab^a \qquad a-\text{elim}\,\dfrac{bc^a \quad cx}{bx}}{ax}\qquad\xrightarrow{\;a-\text{intro}\;}\frac{ab^a \quad bc^a}{ac^a}$$

Celarent

$$\text{rep}\,\frac{ax}{ax}\xrightarrow{\;K\;}\frac{e-\text{elim }2\,\dfrac{e-\text{conv}\,\dfrac{ab^e}{ba^e}\qquad a-\text{elim}\,\dfrac{bc^a \quad cx}{bx}}{\bar{a}x}}{\sim ax}\qquad\xrightarrow{\;e-\text{intro}\;}\frac{ab^e \quad bc^a}{ac^e}$$

Darii

$$i-\text{intro }3\,\frac{a-\text{elim}\,\dfrac{ab^a \quad bx}{ax}\qquad cx}{ac^i}\qquad\xrightarrow{\;i-\text{elim}\;}\frac{ab^a \quad bc^i}{ac^i}$$

Ferio

$$\text{rep}\ \frac{ax}{ax}\ \xrightarrow{\ K\ }\ \text{o-intro}\ \dfrac{\dfrac{\dfrac{ab^e \quad bx}{\bar{a}x}\ e\text{-elim }1}{\sim ax} \qquad cx}{ac^o} \qquad \xrightarrow{\ i\text{-elim}\ }\ \frac{ab^e \quad bc^i}{ac^o}$$

Le-conv

$$\left.\begin{array}{l} Le\text{-elim }1\ \mathrm{A}\ \dfrac{Lab^e \quad bx}{Uax} \\[2em] Le\text{-elim }2\ \mathrm{A}\ \dfrac{Lab^e \quad ax}{Ubx} \end{array}\right\} \xrightarrow{\ Le\text{-intro A}\ } \frac{Lab^e}{Lba^e}$$

La-conv

$$Li\text{-intro }2\ \mathrm{A}\ \dfrac{\dfrac{Lab^a \quad bx}{Lax}\ La\text{-elim A} \qquad bx}{Lba^i} \qquad \xrightarrow{\ i\text{-intro}\ }\ \frac{Lab^a \quad \dfrac{\dfrac{}{bb^a}\ a\text{-identity}}{bb^i}\ a\text{-subalt}}{Lba^i}$$

Li-conv

$$\left.\begin{array}{l} Li\text{-intro }2\ \mathrm{A}\ \dfrac{Lax \quad bx}{Lba^i} \\[2em] Li\text{-intro }1\ \mathrm{A}\ \dfrac{Lbx \quad ax}{Lba^i} \end{array}\right\} \xrightarrow{\ Li\text{-intro A}\ } \frac{Lab^i}{Lba^i}$$

La-rep

$$La\text{-elim A}\ \frac{Lab^a \quad bx}{Lax}\ \xrightarrow{\ La\text{-intro A}\ }\ \frac{Lab^a}{Lab^a}$$

Lo-rep

$$Lo\text{-intro A}\ \frac{Uax \quad bx}{Lab^o}\ \xrightarrow{\ Lo\text{-elim A}\ }\ \frac{Lab^o}{Lab^o}$$

La-subord

$$\text{L-subord}\;\dfrac{\text{La-elim A}\;\dfrac{Lab^a \quad bx}{Lax}}{ax} \xrightarrow{a\text{-intro}} \dfrac{Lab^a}{ab^a}$$

Le-subord

$$\text{U-subord}\;\dfrac{\text{Le-elim 1 A}\;\dfrac{Lab^e \quad bx}{Uax}}{\sim ax} \xrightarrow{e\text{-intro}} \dfrac{Lab^e}{ab^e}$$

Li-subord

$$\left.\begin{array}{c} \text{i-intro 3}\;\dfrac{\text{L-subord}\;\dfrac{Lax}{ax} \quad bx}{ab^i} \\[3em] \text{i-intro 3}\;\dfrac{ax \quad \text{L-subord}\;\dfrac{Lbx}{bx}}{ab^i} \end{array}\right\} \xrightarrow{Li\text{-elim}} \dfrac{Lab^i}{ab^i}$$

Lo-subord

$$\text{o-intro}\;\dfrac{\text{U-subord}\;\dfrac{Uax}{\sim ax} \qquad bx}{ab^o} \xrightarrow{Lo\text{-elim A}} \dfrac{Lab^o}{ab^o}$$

Barbara LXL

$$\text{La-elim A}\;\dfrac{Lab^a \qquad \text{a-elim}\;\dfrac{bc^a \quad cx}{bx}}{Lax} \xrightarrow{La\text{-intro A}} \dfrac{Lab^a \quad bc^a}{Lac^a}$$

Celarent LXL

$$\text{Le-elim 1 A}\ \dfrac{Lab^e \qquad \text{a-elim}\ \dfrac{bc^a \quad cx}{bx}}{Uax}$$

$$\text{SS}\ \dfrac{\text{Le-elim 2 A}\ \dfrac{Lab^e \quad ax}{Ubx} \qquad bc^a}{Ucx}$$

$$\xrightarrow{\text{Le-intro A}}\ \dfrac{Lab^e \quad bc^a}{Lac^e}$$

Darii LXL

$$\text{Li-intro 1 A}\ \dfrac{\text{La-elim A}\ \dfrac{Lab^a \quad bx}{Lax} \qquad cx}{Lac^i} \quad \xrightarrow{\text{i-elim}}\ \dfrac{Lab^a \quad bc^i}{Lac^i}$$

Ferio LXL

$$\text{Lo-intro A}\ \dfrac{\text{Le-elim 2 A}\ \dfrac{\text{Le-conv}\ \dfrac{Lab^e}{Lba^e} \quad bx}{Uax} \qquad cx}{Lac^o} \quad \xrightarrow{\text{i-elim}}\ \dfrac{Lab^e \quad bc^i}{Lac^o}$$

Aristotle's ecthetic proofs of Bocardo LLL and Baroco LLL, if we take the 'exhibited' terms to be a singular term,[3] are representable in *AEL5* as follows:

$$\text{Lo-intro A}\ \dfrac{Uax \qquad \text{L-subord}\ \dfrac{\text{La-elim A}\ \dfrac{cx \quad Lbc^a}{Lbx}}{bx}}{Lab^o} \quad \xrightarrow{\text{Lo-elim A}}\ \dfrac{Lac^o \quad Lbc^a}{Lab^o}$$

$$\text{Lo-intro A}\ \dfrac{\text{SS}\ \dfrac{\text{La-subord}\ \dfrac{Lab^a}{ab^a} \quad Uax}{Ubx} \qquad cx}{Lbc^o} \quad \xrightarrow{\text{Lo-elim A}}\ \dfrac{Lab^a \quad Lac^o}{Lbc^o}$$

These proofs adapt readily to demonstrate Baroco XLL and Bocardo LXL.

---

[3] As is done by MIGNUCCI 1991, THOM 1993 and JOHNSON 1993. For Johnson's analysis see §40.2 below.

§39. SYSTEMS *AEQ5* AND *AEQ'5*

System *AEQ5* adds to *AEL5* the forms Qax ("x may or may not be a") and $\sim Qax$ ("not: x may or may not be a").

$$Qax \mid \sim Qax$$

are contradictories. The following forms are contraries

$$Qax \parallel Lax \quad Qax \parallel Uax$$

For singular affirmatives, we have the repetition-axiom

Q-rep $\dfrac{Qax}{Qax}$

L-rep and U-rep by *K* imply the principles of Q-elimination:

$$L-rep\frac{Lax}{Lax}\xrightarrow{K}\frac{Qax}{\sim Lax} \qquad U-rep\frac{Uax}{Uax}\xrightarrow{K}\frac{Qax}{Max}$$

But we need to specify Q-introduction axiomatically:

Q-intro $\dfrac{\sim Lax \quad Max}{Qax}$

A semantics for universal and particular Q-forms are reflected in the axioms

Qa-elim A $\dfrac{Qab^{a} \quad bx}{Qax}$

Qi-intro 1 A $\dfrac{Qax \quad bx}{Qab^{i}}$ *Qi-intro 2* A $\dfrac{ax \quad Qbx}{Qab^{i}}$

and the rules

Qa-intro A $\dfrac{R \quad bx}{Qax} \longrightarrow \dfrac{R}{Qab^{a}}$

provided R has some member $(L)ab^{a(i)}$ or $(L)ba^{a(i)}$

Qi-elim A $\left.\begin{array}{c} \dfrac{R \quad Qax \quad bx}{p} \\ \\ \dfrac{R \quad Qby \quad ay}{p} \end{array}\right\} \longrightarrow \dfrac{R \quad Qab^{i}}{p}$

System *AEQ'*5 adds ampliated Q-forms. For ampliated Q-forms we have the rules of Q'a-intro

$$\frac{R \quad Mbx}{Qax} \longrightarrow \frac{R}{Q'ab^{a}}$$

and Q'i-elim

$$\left.\begin{array}{c} \dfrac{R \quad Qax \quad Mbx}{q} \\ \\ \dfrac{R \quad Max \quad Qbx}{q} \end{array}\right\} \longrightarrow \frac{R \quad Q'ab^{i}}{q}$$

plus the axioms

Q'a-elim $\dfrac{Q'ab^{a} \quad Mbx}{Qax}$

Q'i-intro 1 $\dfrac{Qax \quad Mbx}{Q'ab^{i}}$ *Q'i-intro 2* $\dfrac{Max \quad Qbx}{Q'ab^{i}}$

Semantics for the systems is analogous to that for *AEL*5. It also includes the *A* truth-conditions for Qa- and Qi-forms, **28.1.1-3**. Truth-conditions for singular contingency-forms are as follows:

39.1 Qax is true iff neither Lax nor Uax is true

39.2 Qab^a is true iff $f_2 \neq \varnothing \wedge f_2 \subseteq f_1{}^{\ddagger}$

39.3 Qab^i is true iff $f_2 \cap f_1{}^{\ddagger} \neq \varnothing \vee f_1 \cap f_2{}^{\ddagger} \neq \varnothing$

At the level of singulars, truth-conditions for Q-forms reduce to those for L- and M-forms. To this extent McCall was wrong[4] in claiming that contingency cannot be defined in terms of possibility and necessity. All he showed was that Qab^a and Qab^i are not truth-functions of Mab^a, Mab^e, Mab^i, Mab^o or of their sister apodeictics. The mode of contingency plays different rôles in different forms of proposition, these different rôles all being related to its occurrence in singular Q-forms by means of structural principles.

System *AEQ5* is sound with respect to this semantics. The axioms are valid:

Q-intro If $d_1 \notin f_1{}^{*}$ and $d_1 \notin f_1{}^{¤}$ then $d_1 \in f_1{}^{\ddagger}$

Qa-elim A If $f_2 \neq \varnothing \wedge f_2 \subseteq f_1{}^{\ddagger}$ and $d_1 \in f_2$ then $d_1 \in f_2{}^{\ddagger}$

Qi-intro 1 A If $d_1 \in f_1{}^{\ddagger}$ and $d_1 \in f_2$ then $f_2 \cap f_1{}^{\ddagger} \neq \varnothing \vee f_1 \cap f_2{}^{\ddagger} \neq \varnothing$

Qi-intro 2 A If $d_1 \in f_2{}^{\ddagger}$ and $d_1 \in f_1$ then $f_2 \cap f_1{}^{\ddagger} \neq \varnothing \vee f_1 \cap f_2{}^{\ddagger} \neq \varnothing$

And the rules preserve validity:

Qa-intro A If $(\dot{r} \wedge d_1 \in f_2) \supset d_1 \in f_1{}^{\ddagger}$ is valid
then $\dot{r} \supset (f_2 \neq \varnothing \wedge f_2 \subseteq f_1{}^{\ddagger})$ is valid
provided $\dot{r}$ has some conjunct
$f_2 \neq \varnothing \wedge f_2 \subseteq f^{*}, f_2 \neq \varnothing \wedge f_2 \subseteq f,$
$f_2 \cap f^{*} \neq \varnothing \vee f_2{}^{*} \cap f \neq \varnothing, f_2 \cap f \neq \varnothing,$
$f \neq \varnothing \wedge f \subseteq f_2{}^{*}, f \neq \varnothing \wedge f \subseteq f_2,$
$f \cap f_2{}^{*} \neq \varnothing \vee f^{*} \cap f_2 \neq \varnothing$ or $f \cap f_2 \neq \varnothing$
$\because$ if $\forall d[(\dot{r} \wedge d \in f_2) \supset d \in f_1{}^{\ddagger}]$ is valid
then $\dot{r} \supset (f_2 \neq \varnothing \wedge f_2 \subseteq f_1{}^{\ddagger})$ is valid under the proviso.

---

[4] McCALL 1963 §24.

Qi-elim A

If both $\left(r \wedge d_1 \in f_1{}^{\ddagger} \wedge d_1 \in f_2\right) \supset p$
and $\left(r \wedge d_2 \in f_1 \wedge d_2 \in f_2{}^{\ddagger}\right) \supset p$ are valid
then $\left(r \wedge \left(f_1{}^{\ddagger} \cap f_2 \neq \varnothing \vee f_1 \cap f_2{}^{\ddagger} \neq \varnothing\right)\right) \supset p$ is valid
$\because \left(r \wedge \left(f_1{}^{\ddagger} \cap f_2 \neq \varnothing \vee f_1 \cap f_2{}^{\ddagger} \neq \varnothing\right)\right) \supset p$ is valid
iff both $(i)\left(r \wedge f_1{}^{\ddagger} \cap f_2 \neq \varnothing\right) \supset p$
and $(ii)\left(r \wedge f_1 \cap f_2{}^{\ddagger} \neq \varnothing\right) \supset p$ are valid;
and (i) if $\left(r \wedge f_1{}^{\ddagger} \cap f_2 \neq \varnothing\right) \supset p$ is valid
then $\exists d\left[\left(r \wedge d_1 \in f_1{}^{\ddagger} \wedge d_1 \in f_2\right) \supset p\right]$ is valid
and (ii) if $\left(r \wedge f_1 \cap f_2{}^{\ddagger} \neq \varnothing\right) \supset p$ is valid
then $\exists d\left[\left(r \wedge d_1 \in f_1 \wedge d_1 \in f_2{}^{\ddagger}\right) \supset p\right]$ is valid.

*AEQ'*5 contains a system isomorphic to *AQ'*. This system is obtained from *AQ'* by replacing all its variables with corresponding bold variables. This can be shown by deriving within *AEQ'*5 the Q-elements in the basis of that system.

Qi-conv, *Q'i-conv* Proof analogous to that for Li-conv

Q-conv, *Q'-conv* By definition.

Cesare QLX

$$\text{Qa-elim A}\,\frac{Qab^a \quad bx}{Qax} \xrightarrow{\quad C \quad} \frac{\text{Q-rep}\,\frac{Qax}{Qax} \xrightarrow{K} \frac{\text{La-elim}\,\frac{Lac^a \quad cx}{Lax}}{\sim Qax} \qquad \text{Q-conv}\,\frac{Qab^e}{Qab^a}}{\sim bx} \xrightarrow{\text{e-intro}} \frac{Qab^e \quad Lac^a}{bc^e}$$

Darii QXQ, *Q'XQ'* Proof analogous to that for Darii LXL

Barbara QXQ, *Q'XQ'* Proof analogous to that for Barbara LXL

Qa-conv (partial)

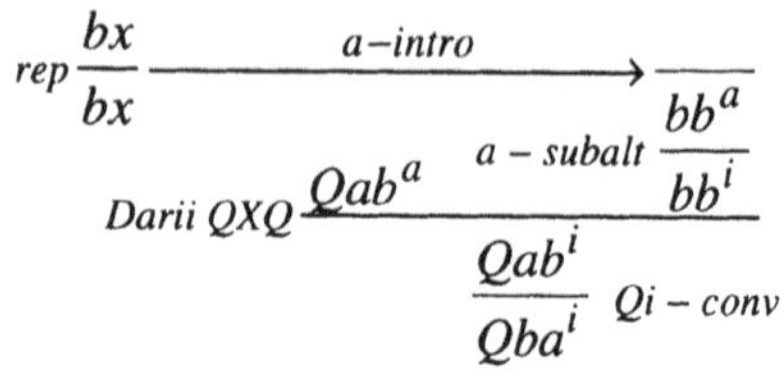

Q'-subord

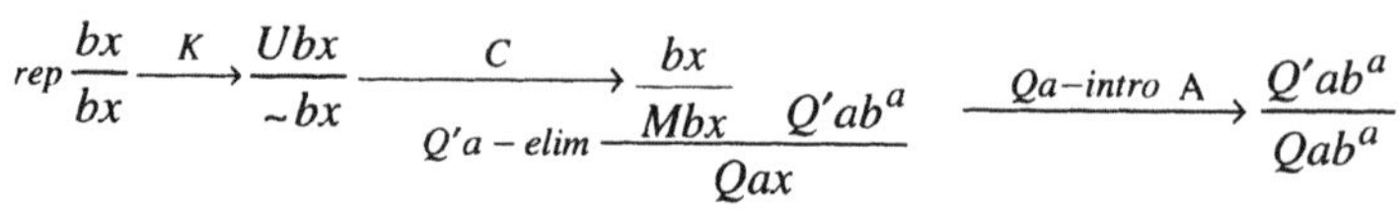

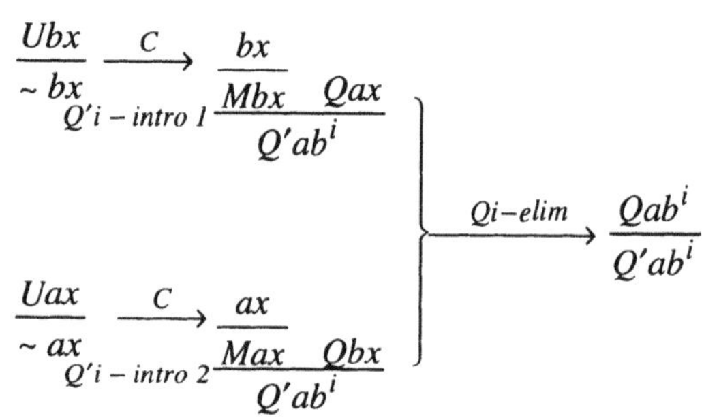

§40. ALTERNATIVE *DE RE* APPROACHES

In this section I present various *de re* approaches to singular syllogistic. *De dicto* approaches to singular syllogistic have already been implicitly considered in §§27 and 33, because the analyses discussed there involved translating modal expressions into modern quantification theory. Here I shall present systems of singular modal syllogistic sound relative to *A*, *AA*, *C* or *CC* superstructures possibly augmented by existential assumptions E1-E5 and sun set definitions **25.1-3**, **25.5**. The axioms and rules of U-elimination and introduction corresponding to these definitions are

U-intro 1 $\dfrac{L\bar{a}x}{Uax}$ *U-elim 1* $\dfrac{Uax}{L\bar{a}x}$

U-intro 2 $\dfrac{L\bar{d}a^{a} \quad Ldx}{Uax}$ *U-elim 2* $\dfrac{R \quad L\bar{d}a^{a} \quad Ldx}{p} \longrightarrow \dfrac{R \quad Uax}{p}$

U-intro 3 $\dfrac{ad^{e} \quad Ldx}{Uax}$ *U-elim 3* $\dfrac{R \quad ad^{e} \quad Ldx}{p} \longrightarrow \dfrac{R \quad Uax}{p}$

(U-intro 5 and U-elim 5 were used in §38.1. We defer discussion of **25.4** to the next section. Our present object-language is incapable of expressing the proposition whose truth-condition is $\exists f\left[f_1 \neq \varnothing \wedge f_1 \subseteq \overline{f^{*}} \wedge d_1 \in f^{*}\right]$.)

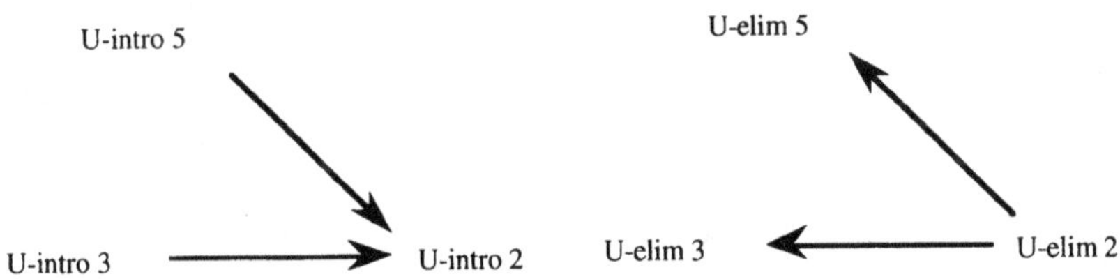

Fig. 40.1

The implication-relations among these axioms and rules are shown in Figure 40.1. These relations hold by virtue of the implications

$$\left.\begin{array}{r} \exists d,e\left[Lea^{a} \wedge ed^{e} \wedge Ldx\right] \\ \exists d\left[da^{e} \wedge Ldx\right] \end{array}\right\} \longleftarrow \exists d\left[L\bar{d}a^{a} \wedge Ldx\right]$$

which are valid if $\bar{d}d^{e}$ and $L\bar{d}a^{a} \supset da^{e}$ are valid; and the latter are valid independently of the interpretation of L-forms. (That U-intro 2 does not imply U-intro 1 - and U-elim 1 does not imply U-elim 2 - is shown by model 25.2.)

§40.1 A-*style singular modal syllogistic*

*AEQ*0 is based on the axiom

a-identity $\dfrac{}{aa}{}^{a}$

plus rep, a-subalt, a-elim, e-elim 1, e-elim 2, L-rep, U-rep, Q-rep, La-elim *A*, Le-elim 1 *A*, Le-elim 2 *A*, Li-intro 1 *A*, Li-intro 2 *A*, Lo-intro *A*, Q-intro, Qa-elim *A*, Qi-intro 1 *A*, Qi-intro 2 *A*, Barbara QXQ, Darii QXQ, Cesare QLX, and the rules of Substitution, Permutation, *C*, *K*, Cut, i-elim, o-elim, La-intro *A*, Le-intro *A*, Li-elim *A*, Lo-elim *A*, Qa-intro *A*, Qi-elim *A*.

*AEQ'*0 adds axioms Q'a-elim,Q'i-intro 1, Q'i-intro 2, and rules Q'a-intro, Q'i-elim.of §39.

System *AEQ*0 is sound with respect to a semantics comprising **21.1**, **21.7** and **21.8**, plus **22.1.1**, **22.1.5-8**, **24.1-7**, **28.1.1-3** and **38.2.1-6**, **39.1**, plus (in place of **22.1.2-4**):

40.1.1 $F^{¤}$ is a non-empty set of possibly empty sets whose members are members of *D*.

40.1.2 An assignment of values to the variables in a wff associates with each term variable v in the wff an ordered *triple* consisting of a member of *F* followed by a member of F^* followed by a member of $F^{¤}$.

These sets I will refer to as the *v*s, the v^*s and the $v^{¤}$s respectively.

40.1.3 If the triple $\langle \alpha_1, \alpha_2, \alpha_3 \rangle$ is assigned to a variable then $\alpha_2 \subseteq \alpha_1, \alpha_3 \subseteq \overline{\alpha_1}$.

Celarent LXL and Baroco LLL are invalid: see models 26.1 and 26.2. But Bocardo LXL is valid and is provable as in §38.3. Not valid is $\dfrac{Uax}{Lax}$ (U-elim 1): see model 38.1. Also not valid are

(i) U-intro 2
(ii) U-intro 3
(iii) U-intro 5

40.1		*	¤
f_1	d_2	d_2	–
f_4	d_1	d_1	–
$\overline{f_4}$	d_2	d_2	–

$(i) f_1 \neq \varnothing \wedge f_1 \subseteq \overline{f_4}^{*}$ and $d_1 \in f_4^{*}$
$(ii) f_1 \neq \varnothing \wedge f_1 \subseteq \overline{f_4}$ and $d_1 \in f_4^{*}$
$(iii) f_1 \neq \varnothing \wedge f_1 \subseteq f_5^{*} \wedge f_5 \subseteq \overline{f_4} \wedge d_1 \in f_4^{*}$
but $d_1 \notin f_1^{¤}$.

Barbara LQM[5]

40.2		*	¤	‡	†
f_1	d_1	d_1	d_2	–	d_1
f_2	d_1	–	–	d_1, d_2	d_1, d_2
f_3	d_2	–	–	d_1, d_2	d_1, d_2

$f_2 \subseteq f_1^{*}$ and $f_3 \subseteq f_2^{‡}$ but $f_3 \not\subseteq f_1^{†}$.

$AEQ1$ is defined as the result of adding U-intro 1 and U-elim 1 to $AEQ0$. Its semantics is the same as $AEQ0$'s with the replacement of **40.1.1-3** by **22.1.2-3** and the addition of **25.1**.

Since $\dfrac{Uax}{L\bar{a}x}$ (U-elim 1) is valid, $AEQ1$ properly contains $AEQ0$.

Adding U-elim 2 and U-intro 2 to $AEQ0$, we get $AEQ2$. Correspondingly, **25.2** should be substituted for **25.1** in $AEQ1$ semantics. Baroco XLL, SS and Celarent LXL are then valid and are provable:

$$\text{Lo-intro A}\ \dfrac{\text{U-intro 2}\ \dfrac{\text{Barbara LXL}\ \dfrac{L\bar{d}a^{a} \quad ab^{a}}{L\bar{d}b^{a}} \qquad Ldx}{Ubx} \qquad cx}{Lbc^{o}} \xrightarrow{\text{U-elim 2}} \dfrac{ab^{a} \quad Uax \quad cx}{Lbc^{o}} \xrightarrow{\text{Lo-elim A}} \dfrac{ab^{a} \quad Lac^{o}}{Lbc^{o}}$$

$$\text{U-intro 2}\ \dfrac{\text{Barbara LXL}\ \dfrac{L\bar{d}a^{a} \quad ab^{a}}{L\bar{d}b^{a}} \qquad Ldx}{Ubx} \xrightarrow{\text{U-elim 2}} \dfrac{ab^{a} \quad Uax}{Ubx}$$

5 The first diagram on p.219 of STAHL 1977 is consistent with our model.

Celarent LXL is provable as in §38.3. Barbara LQM is provable as in §12.1. *AEQ*2 properly contains *AEQ*0.

*AEQ*3 is formed by replacing *AEQ*2's U-intro 2 and U-elim 2 with U-intro 3 and U-elim 3. Semantically, we substitute **25.3** for **25.1** in *AEQ*1 semantics.

*AEQ*3 properly contains *AEQ*0.

Systems *AEQ*1-3, 5 are mutually independent. To show this, given the implication-relations in Figure 40.1, it is enough to show that (i) *AEQ*1 does not contain U-intro 2, (ii) *AEQ*2 does not contain U-intro 3, (iii) *AEQ*2 does not contain U-elim 1, (iv) *AEQ*3 does not contain U-elim 5, (v) *AEQ*3 does not contain U-elim 2, (vi) *AEQ*5 does not contain U-elim 3. This is shown as follows.

(i)

40.3		*	¤
f_1	d_2	–	–
f_4	d_1, d_3	d_1	d_2
$\overline{f_1}$	d_1, d_3	-	–
$\overline{f_4}$	d_2	d_2	d_1

$f_1 \neq \varnothing \wedge f_1 \subseteq \overline{f_4}^*$ and $d_1 \in f_4{}^*$ but $d_1 \notin f_1{}^{¤} - \because d_1 \notin \overline{f_1}^*$.

(ii)

40.4		*	¤
f_1	d_2	–	–
$\overline{f_1}$	d_1	–	–
f_2	d_1	d_1	–
$\overline{f_2}$	d_2	–	–

$f_1 \neq \varnothing \wedge f_1 \cap f_2 = \varnothing \wedge d_1 \in f_2{}^*$ but $d_1 \notin f_1{}^{¤}$
$\because \sim \exists f\left[f_1 \neq \varnothing \wedge f_1 \subseteq \overline{f}^* \wedge d_1 \in f^*\right]$.

(iii)

40.5		*	¤
f_1	d_2	–	d_1
f_2	d_1	d_1	–
$\overline{f_1}$	d_1, d_3	–	–
$\overline{f_2}$	d_2, d_3	d_2	–

$d_1 \in f_1{}^{¤} - \because f_1 \neq \varnothing \wedge f_1 \subseteq \overline{f_2}^* \wedge d_1 \in f_2{}^*$ but $d_1 \notin \overline{f_1}^*$.

(iv)

40.6	*	¤
f_1	d_2 —	d_1
f_2	d_1 d_1	—

$d_1 \in f_1^{¤} - \because f_1 \cap f_2 = \varnothing \wedge d_1 \in f_2^*$

but $\sim \exists f, g\left[f \neq \varnothing \wedge f_1 \subseteq g^* \wedge f \cap g = \varnothing \wedge d_1 \in f^*\right]$.

(v)

40.7	*	¤
f_1	— —	d_1
$\overline{f_1}$	d_1 d_1	—

$d_1 \in f_1^{¤} - \because d_1 \in \overline{f_1}^*, \overline{f_1} \cap f_1 = \varnothing$

but $\sim \exists f\left[f_1 \neq \varnothing \wedge f_1 \subseteq \overline{f}^* \wedge d_1 \in f^*\right]$.

(vi)

40.8	*	¤
f_1	d_2 —	—
f_2	d_1 d_1	d_2

$f_1 \cap f_2 = \varnothing$ and $d_1 \in f_2^*$ but $d_1 \notin f_1^{¤}$

$\because \sim \exists f, g\left[f_1 \neq \varnothing \wedge f_1 \subseteq f^* \wedge f \cap g = \varnothing \wedge d_1 \in g^*\right]$.

AEQ'1-3 are defined analogously to *AEQ*1-3. (See Index of systems.)

§40.2 AA-*style singular modal syllogistic*

*AAEQ*0 is formed from *AEQ*0 by replacing the latter's introduction and elimination axioms and rules for modalised categoricals with

La-intro AA $\dfrac{R \quad bx}{Lax} \longrightarrow \dfrac{R}{Lab^a}$ *La-elim* AA $\dfrac{Lab^a \quad bx}{Lax}$

Le-intro AA $\dfrac{R \quad bx}{Uax} \longrightarrow \dfrac{R}{Lab^e}$ *Le-elim* AA $\dfrac{Lab^e \quad bx}{Uax}$

Li-intro AA $\dfrac{Lax \quad Lbx}{Lab^i}$ *Li-elim* AA $\dfrac{R \quad Lax \quad Lbx}{p} \longrightarrow \dfrac{R \quad Lab^i}{p}$

Lo-intro AA $\dfrac{Uax \quad Lbx}{Lab^o}$ *Lo-elim* AA $\dfrac{R \quad Uax \quad Lbx}{p} \longrightarrow \dfrac{R \quad Lab^o}{p}$

Qa-intro AA $\dfrac{R \quad bx}{Qax} \longrightarrow \dfrac{R}{Qab^a}$ *Qa-elim* AA $\dfrac{Qab^a \quad bx}{Qax}$

Qi-intro AA $\dfrac{Qax \quad bx}{Qab^i}$ *Qi-elim* AA $\dfrac{R \quad Qax \quad bx}{p} \longrightarrow \dfrac{R \quad Qab^i}{p}$

*AAEQ*0's semantics is the same as *AEQ*0's, except that **22.1.5-8** are replaced by **26.2.1-4** and **28.1.2-3** by **32.2.1-2**. In *AAEQ*0 Le-conv, partial La-conv, Darii LXL and Ferio LXL are invalid: see matrices 26.3-6.

We showed in §32.2 that a minimal system with *AA* semantics needs to include as theses L-subord, Barbara LXL, Celarent LXL, Darii LLL, Ferio LLL, Bocardo LLL, Barbara QXQ, Darii QXQ and Cesare QLX. All of these are provable in *AAEQ*0:

La-subord

$$\text{L-subord}\,\dfrac{\text{La-elim AA}\,\dfrac{Lab^a \quad bx}{Lax}}{ax} \quad \xrightarrow{\text{a-intro}} \dfrac{Lab^a}{ab^a}$$

Le-subord

$$\text{rep}\,\dfrac{\bar{ax}}{\bar{ax}} \xrightarrow{C} \dfrac{\sim\bar{ax}}{\sim\bar{ax}} \xrightarrow{K} \dfrac{\text{Le-elim AA}\,\dfrac{Lab^e \quad bx}{Uax}}{\bar{ax}} \quad \xrightarrow{\text{e-intro}} \dfrac{Lab^e}{ab^e}$$

Li-subord

$$\text{i-intro}\,\dfrac{\text{L-subord}\,\dfrac{Lax}{ax} \quad \text{L-subord}\,\dfrac{Lbx}{bx}}{ab^i} \quad \xrightarrow{\text{Li-elim AA}} \dfrac{Lab^i}{ab^i}$$

Lo-subord

$$o\text{-}intro\;\dfrac{U\text{-}subord\,\dfrac{Uax}{\sim ax}\qquad L\text{-}subord\,\dfrac{Lbx}{bx}}{ab^{o}}\qquad \xrightarrow{Lo\text{-}elim\ \mathrm{AA}}\ \dfrac{Lab^{o}}{ab^{o}}$$

Barbara LXL

$$La\text{-}elim\ \mathrm{AA}\,\dfrac{Lab^{a}\qquad a\text{-}elim\,\dfrac{bc^{a}\quad cx}{bx}}{Lax}\qquad \xrightarrow{La\text{-}intro\ \mathrm{AA}}\ \dfrac{Lab^{a}\quad bc^{a}}{Lac^{a}}$$

Celarent LXL

$$Le\text{-}elim\ \mathrm{AA}\,\dfrac{Lab^{e}\qquad a\text{-}elim\,\dfrac{bc^{a}\quad cx}{bx}}{Uax}\qquad \xrightarrow{Le\text{-}intro\ \mathrm{AA}}\ \dfrac{Lab^{e}\quad bc^{a}}{Lac^{e}}$$

Darii LLL

$$Li\text{-}intro\ \mathrm{AA}\,\dfrac{La\text{-}elim\ \mathrm{AA}\,\dfrac{Lab^{a}\qquad L\text{-}subord\,\dfrac{Lbx}{bx}}{Lax}\qquad Lcx}{Lac^{i}}\qquad \xrightarrow{Li\text{-}elim\ \mathrm{AA}}\ \dfrac{Lab^{a}\quad Lbc^{i}}{Lac^{i}}$$

Ferio LLL

$$Lo\text{-}intro\ \mathrm{AA}\,\dfrac{Le\text{-}elim\ \mathrm{AA}\,\dfrac{Lab^{a}\qquad L\text{-}subord\,\dfrac{Lbx}{bx}}{Uax}\qquad Lcx}{Lac^{o}}\qquad \xrightarrow{Li\text{-}elim\ \mathrm{AA}}\ \dfrac{Lab^{e}\quad Lbc^{i}}{Lac^{o}}$$

Bocardo LLL

$$Lo\text{-}intro\ \mathrm{AA}\,\dfrac{Uax\qquad La\text{-}elim\ \mathrm{AA}\,\dfrac{Lbc^{a}\qquad L\text{-}subord\,\dfrac{Lcx}{cx}}{Lbx}}{Lab^{o}}\qquad \xrightarrow{Lo\text{-}elim\ \mathrm{AA}}\ \dfrac{Lac^{o}\quad Lbc^{a}}{Lab^{o}}$$

Barbara QXQ

$$Qa\text{-}elim\ \mathrm{AA}\,\dfrac{Qab^{a}\qquad a\text{-}elim\,\dfrac{bc^{a}\quad cx}{bx}}{Qax}\qquad \xrightarrow{Qa\text{-}intro\ \mathrm{AA}}\ \dfrac{Qab^{a}\quad bc^{a}}{Qac^{a}}$$

Darii QXQ

$$Qi-intro\ \mathrm{AA}\ \dfrac{Qa-elim\ \mathrm{AA}\ \dfrac{Qab^a \quad bx}{Qax} \qquad cx}{Qac^i} \qquad \xrightarrow{i-elim} \qquad \dfrac{Qab^a \quad bc^i}{Qac^i}$$

Cesare QLX

$$Darii\ QXQ\ \dfrac{Q-conv\ \dfrac{Qab^e}{Qab^a} \qquad bc^i}{Qac^i} \quad \xrightarrow[Qi\ \|\ La]{K} \quad \dfrac{Qab^e \quad Lac^a}{bc^e}$$

This last proof relies on the opposition $Qi \parallel La$, which is available in $AAEQ0$ but not in $AEQ0$. This style of proof is used by Stahl to prove Camestres QLX.[6]

$AAEQ1$-3,5 are obtained by adding U-intro 1-3,5 and U-elim 1-3,5 to $AAEQ0$. Their semantics are formed from $AAEQ0$'s by replacing **40.1.1-3** with **22.1.2-3** and then adding **25.1-3,5** respectively.

SS is valid in $AAEQ2$,3,5, and is provable in $AAEQ2$ as in $AEQ2$. Baroco XLL is provable in all three systems thus:

$$Lo-intro\ \mathrm{AA}\ \dfrac{SS\ \dfrac{ab^a \quad Uax}{Ubx} \qquad Lcx}{Lbc^o} \qquad \xrightarrow{Lo-elim\ \mathrm{AA}} \qquad \dfrac{ab^a \quad Lac^o}{Lbc^o}$$

In $AAEQ3$ Celarent XLL is valid, and provable as follows:

$$U-intro\ 3\ \dfrac{ab^e \qquad La-elim\ \mathrm{AA}\ \dfrac{Lbc^a \quad cx}{Lbx}}{Uax} \qquad \xrightarrow{Le-intro\ \mathrm{AA}} \qquad \dfrac{ab^e \quad Lbc^a}{Lac^e}$$

Various additions may be made to these systems, producing more or less interesting results. The axiom

Weak Campsall (WC)

$$\dfrac{Lab^a \quad Uax}{Ubx}$$

---

[6] STAHL 1977 pp.211-212.

- corresponding to our S2 (§25) - when added to *AAEQ0* validates Baroco LLL and Bocardo LLL, which are then provable as follows:[7]

$$\text{Lo-intro AA}\ \dfrac{\text{WC}\ \dfrac{Lab^a \quad Uax}{Ubx} \qquad Lcx}{Lbc^o} \quad \xrightarrow{\text{Lo-elim AA}} \quad \dfrac{Lab^a \quad Lac^o}{Lbc^o}$$

$$\text{Lo-intro AA}\ \dfrac{Uax \qquad \text{La-elim AA}\ \dfrac{Lbc^a \quad \text{L-subord}\ \dfrac{Lcx}{cx}}{Lbx}}{Lab^o} \quad \xrightarrow{\text{Lo-elim AA}} \quad \dfrac{Lac^o \quad Lbc^a}{Lab^o}$$

In §20 we claimed that, however Aristotle's proofs of these syllogisms are understood, they can be transformed into proofs of Bocardo LXL and Baroco XLL. The above demonstrations appear to refute this claim. True, the proof of Bocardo LLL could be turned into a proof of Bocardo LXL if we could use *Lo-intro A* instead of *Lo-intro AA*. And the proof of Baroco LLL could be turned into a proof of Baroco XLL if we could invoke SS in place of WC. Our claim about the two Barocos remains true within the context of SS - the Sun set Theorem. Our other claim, about the two Bocardos, remains true provided we assume that, in order that some *a* cannot be *b*, it is required that some individual *is a* and cannot be *b* (*A* semantics), not that some individual *must be a* and cannot be *b* (*AA* semantics). The issue is which of these requirements is more Aristotelian. We defer its discussion to Chapter Ten.

McCall's axiom $\dfrac{}{Laa^i}$ can be added to *AAEQ0*. The semantics of *AAEQ0* then needs to be augmented by E1 (§25.1). Partial La-conv is then valid and is provable thus:

$$\text{Li-intro AA}\ \dfrac{\text{i-intro}\ \dfrac{\text{L-subord}\ \dfrac{Lby}{by} \quad \text{L-subord}\ \dfrac{Lby}{by}}{bb^i} \qquad Lax \quad Lbx}{Lba^i} \quad \xrightarrow{\text{Li-elim AA}} \quad \dfrac{\dfrac{\dfrac{\dfrac{}{Lbb^i}\,\text{McCall}}{bb^i} \qquad Lab^a}{Lab^i}\,\text{Darii LXL}}{Lba^i}$$

The rule

---

[7] JOHNSON 1993 p.179. Johnson here uses a superstructure in one respect different from the *AA* superstructure which is used in JOHNSON 1989. In place of the *AA* truth-condition for Li-forms - with the subject restricted to the necessary - he uses the disjunctive truth-condition of *A* semantics.

i-elim T $$\frac{R \quad ax \quad Lbx}{p} \longrightarrow \frac{R \quad ab^i}{p}$$

can be added. Correspondingly, we may add base assumption E3 - Thomason's $(f \cap g \neq \varnothing) \supset (f \cap g^* \neq \varnothing)$[8] - to our semantics. Darii LXL and Ferio LXL are then provable:

$$La\text{-}elim\ \mathrm{AA}\ \frac{\dfrac{Lab^a \quad bx}{Lax}\ Li\text{-}intro\ \mathrm{AA} \quad Lcx}{Lac^i} \xrightarrow{i\text{-}elim\ T} \frac{Lab^a \quad bc^i}{Lac^i}$$

$$Le\text{-}elim\ \mathrm{AA}\ \frac{\dfrac{Lab^e \quad bx}{Uax}\ Lo\text{-}intro\ \mathrm{AA} \quad Lcx}{Lac^o} \xrightarrow{i\text{-}elim\ T} \frac{Lab^e \quad bc^i}{Lac^o}$$

Le-conv can be added, along with the semantic principle E4 $(f_2 \subseteq f_1^{\text{¤}} \supset f_1 \subseteq f_2^{\text{¤}})$.

Qi-conv can be added, corresponding to the semantic principle E7 $(f_1 \cap f_2^{\ddagger} \neq \varnothing \supset f_2 \cap f_1^{\ddagger} \neq \varnothing)$.

Q'i-conv can be added, corresponding to the semantic principle E8 $(f_1^{\dagger} \cap f_2^{\ddagger} \neq \varnothing \supset f_2^{\dagger} \cap f_1^{\ddagger} \neq \varnothing)$.

§40.3 C-*style singular modal syllogistic*[9]

*CEQ*0 is based on axioms rep, a-subalt, a-elim, e-elim 1, e-elim 2, L-rep, U-rep, Q-rep, Q-intro, Darii QXQ, and rules Substitution, Permutation, *C*, *K*, Cut, i-elim, o-elim, plus

La-intro C $$\frac{R \quad Mbx}{Lax} \longrightarrow \frac{R}{Lab^a}$$ *La-elim* C $$\frac{Lab^a \quad Mbx}{Lax}$$

---

[8] THOMASON Definition 2.

[9] On Buridan's modal syllogistic see the introduction and notes in BURIDAN.

Le-intro C $\quad \dfrac{R \quad Mbx}{Uax} \longrightarrow \dfrac{R}{Lab^e}$ $\qquad$ *Le-elim* C $\quad \dfrac{Lab^e \quad Mbx}{Uax}$

Li-intro C $\quad \dfrac{Lax \quad Mbx}{Lab^i}$ $\qquad$ *Li-elim* C $\quad \dfrac{R \quad Lax \quad Mbx}{p} \longrightarrow \dfrac{R \quad Lab^i}{p}$

Lo-intro C $\quad \dfrac{Uax \quad Mbx}{Lab^o}$ $\qquad$ *Lo-elim* C $\quad \dfrac{R \quad Uax \quad Mbx}{p} \longrightarrow \dfrac{R \quad Lab^o}{p}$

Qa-intro C $\quad \dfrac{R \quad Mbx}{Qax} \longrightarrow \dfrac{R}{Qab^a}$ $\qquad$ *Qa-elim* C $\quad \dfrac{Qab^a \quad Mbx}{Qax}$

Qi-intro C $\quad \dfrac{Qax \quad Mbx}{Qab^i}$ $\qquad$ *Qi-elim* C $\quad \dfrac{R \quad Qax \quad Mbx}{p} \longrightarrow \dfrac{R \quad Qab^i}{p}$

Barbara QMQ $\quad \dfrac{Qab^a \quad Mbc^a}{Qac^a}$ $\qquad$ *Cesare QLL* $\quad \dfrac{Qab^e \quad Lac^a}{Lbc^e}$

Noteworthy among *CEQ*0's theses is

WC

$$\cfrac{{\scriptstyle M-subord}\cfrac{{\scriptstyle L-subord}\cfrac{{\scriptstyle La-elim\ C}\cfrac{Lab^a \quad Mbx}{Lax}}{ax}}{Max}}{} \quad \xrightarrow{\ C\ } \quad \dfrac{Lab^a \quad Uax}{Ubx}$$

*CEQ*0's semantics is like *AEQ*0's except that **26.3.1-4** take the place of **22.1.5-8**, and **32.3.1-2** take the place of **28.1.2-3**.

We noted in §32.3 that a minimal system with *C* semantics would contain Le-subord, Le-conv, Barbara LLL, Celarent LLL, Darii LLL, Ferio LLL, Baroco LLL, Bocardo LLL, Darii LXL, Ferio LXL, along with Cesare QLL, Barbara QQQ, Darii QQQ, Barbara QLQ, Darii QLQ and Darii QXQ. All of these are provable in *CEQL*0:

Le-subord

$$\text{U-subord}\ \dfrac{\text{Le-elim C}\ \dfrac{Lab^e \qquad \text{M-subord}\ \dfrac{bx}{Mbx}}{Uax}}{\sim ax} \xrightarrow{\ e\text{-intro}\ } \frac{Lab^e}{ab^e}$$

Le-conv

$$\text{Le-elim C}\ \frac{Lab^e \quad Mbx}{Uax} \xrightarrow{\ C\ } \frac{Lab^e \quad Max}{Ubx} \xrightarrow{\ \text{Le-intro C}\ } \frac{Lab^e}{Lba^e}$$

Barbara LLL

$$\text{La-elim C}\ \dfrac{Lab^a \qquad \text{M-subord}\ \dfrac{\text{L-subord}\ \dfrac{\text{La-elim C}\ \dfrac{Lbc^a \quad Mcx}{Lbx}}{bx}}{Mbx}}{Lax} \xrightarrow{\ \text{La-intro C}\ } \frac{Lab^a \quad Lbc^a}{Lac^a}$$

Celarent LLL

$$\text{Le-elim C}\ \dfrac{Lab^e \qquad \text{M-subord}\ \dfrac{\text{L-subord}\ \dfrac{\text{La-elim C}\ \dfrac{Lbc^a \quad Mcx}{Lbx}}{bx}}{Mbx}}{Uax} \xrightarrow{\ \text{Le-intro C}\ } \frac{Lab^e \quad Lbc^a}{Lac^e}$$

Darii LLL

$$\text{Li-intro C}\ \dfrac{\text{La-elim C}\ \dfrac{Lab^a \qquad \text{M-subord}\ \dfrac{\text{L-subord}\ \dfrac{Lbx}{bx}}{Mbx}}{Lax} \qquad Mcx}{Lac^i} \xrightarrow{\ \text{Li-elim C}\ } \frac{Lab^a \quad Lbc^i}{Lac^i}$$

Ferio LLL

$$\text{Lo-intro C}\ \dfrac{\text{Le-elim C}\ \dfrac{Lab^e \qquad \text{M-subord}\ \dfrac{\text{L-subord}\ \dfrac{Lbx}{bx}}{Mbx}}{Uax} \qquad Mcx}{Lac^o} \xrightarrow{\ \text{Li-elim C}\ } \frac{Lab^e \quad Lbc^i}{Lac^o}$$

Baroco LLL

$$\text{Lo-intro C}\ \dfrac{\text{WC}\ \dfrac{Lab^a \quad Uax}{Ubx} \qquad Mcx}{Lbc^o} \xrightarrow{\ \text{Lo-elim C}\ } \frac{Lab^a \quad Lac^o}{Lbc^o}$$

Bocardo LLL

$$\text{Lo-intro C}\;\dfrac{Uax \qquad \text{M-subord}\,\dfrac{\text{L-subord}\,\dfrac{\text{La-elim C}\,\dfrac{Lbc^a \quad Mcx}{Lbx}}{bx}}{Mbx}}{Lab^o} \qquad \xrightarrow{\text{Lo-elim C}} \qquad \dfrac{Lac^o \quad Lbc^a}{Lab^o}$$

Darii LXL

$$\text{Li-intro C}\;\dfrac{\text{La-elim C}\,\dfrac{Lab^a \qquad \text{M-subord}\,\dfrac{bx}{Mbx}}{Lax} \qquad \text{M-subord}\,\dfrac{cx}{Mcx}}{Lac^i} \qquad \xrightarrow{\text{i-elim}} \qquad \dfrac{Lab^a \quad bc^i}{Lac^i}$$

Ferio LXL

$$\text{Lo-intro C}\;\dfrac{\text{Le-elim C}\,\dfrac{Lab^e \qquad \text{M-subord}\,\dfrac{bx}{Mbx}}{Uax} \qquad \text{M-subord}\,\dfrac{cx}{Mcx}}{Lac^o} \qquad \xrightarrow{\text{i-elim}} \qquad \dfrac{Lab^e \quad bc^i}{Lac^o}$$

Cesare QLL

$$\text{Qa-elim C}\,\dfrac{Qab^a \quad Mbx}{Qax} \xrightarrow{K} \dfrac{Qab^a \qquad \text{La-elim C}\,\dfrac{Lac^a \quad Mcx}{Lax}}{Ubx} \qquad \xrightarrow{\text{Le-intro C}} \qquad \dfrac{\dfrac{Qab^e}{Qab^a}\text{Q-conv} \qquad Lac^a}{Lbc^e}$$

Barbara QQQ

$$\text{Qa-elim C}\;\dfrac{Qab^a \qquad \text{Q/M-subord}\,\dfrac{\text{Qa-elim C}\,\dfrac{Qbc^a \quad Mcx}{Qbx}}{Mbx}}{Qax} \qquad \xrightarrow{\text{Qa-intro C}} \qquad \dfrac{Qab^a \quad Qbc^a}{Qac^a}$$

Darii QQQ

$$\text{Qi-intro C}\;\dfrac{\text{Qa-elim C}\,\dfrac{Qab^a \qquad \text{Q/M-subord}\,\dfrac{Qbx}{Mbx}}{Qax} \qquad Mcx}{Qac^i} \qquad \xrightarrow{\text{Qi-elim C}} \qquad \dfrac{Qab^a \quad Qbc^i}{Qac^i}$$

Barbara QLQ Proof analogous to that for Barbara QQQ

Darii QLQ Proof analogous to that for Darii QQQ

Darii QXQ Proof analogous to that for Darii QQQ

We can add U-intro and -elim 1-3 or 5 to $CEQ0$'s syntax, at the same time replacing **40.1.1-3** with **22.1.2-3** and adding **25.1-3** or **5** to its semantics. In these ways we construct systems $CEQ1$-3,5.

In $CEQ2$ Lo-subord, Barbara LXL, Celarent LXL, Baroco XLL, Bocardo LXL and Barbara QXQ are valid. They are provable as follows:

$$SS\frac{ab^a \quad Uax}{Ubx} \xrightarrow{C} \frac{Uax \quad Mbx}{ab^o} \xrightarrow{Lo-elim\ C} \frac{Lab^o}{ab^o}$$

$$SS\frac{bc^a \quad Ubx}{Ucx} \xrightarrow{C} La-elim\ C\frac{\dfrac{bc^a \quad Mcx}{Mbx} \qquad Lab^a}{Lax} \xrightarrow{La-intro\ C} \frac{Lab^a \quad bc^a}{Lac^a}$$

$$SS\frac{bc^a \quad Ubx}{Ucx} \xrightarrow{C} Le-elim\ C\frac{\dfrac{bc^a \quad Mcx}{Mbx} \qquad Lab^e}{Uax} \xrightarrow{Le-intro\ C} \frac{Lab^e \quad bc^a}{Lac^e}$$

$$Lo-intro\ C\frac{SS\dfrac{ab^a \quad Uax}{Ubx} \quad Mcx}{Lbc^o} \xrightarrow{Lo-elim\ C} \frac{ab^a \quad Lac^o}{Lbc^o}$$

$$SS\frac{bc^a \quad Ubx}{Ucx} \xrightarrow{C} Lo-intro\ C\frac{\dfrac{bc^a \quad Mcx}{Mbx} \qquad Uax}{Lab^o} \xrightarrow{Lo-elim\ C} \frac{Lac^o \quad bc^a}{Lab^o}$$

$$SS\frac{bc^a \quad Ubx}{Ucx} \xrightarrow{C} Qa-elim\ C\frac{\dfrac{bc^a \quad Mcx}{Mbx} \qquad Qab^a}{Qax} \xrightarrow{Qa-intro\ C} \frac{Qab^a \quad bc^a}{Qac^a}$$

Li-subord is invalid in $CEQ2$: see model 26.11.

Various additions can be made to $CEQ0$. Li-conv can be added to the axioms (and E3 and E4 to the semantics). La-subord can be added to the axioms (and E5 to the semantics). Qa-conv and Qi-conv can be added to the axioms (and E7 to the semantics.)

Buridan says ecthetic proofs exist for MMM-3, LLL-3, LML-3, MLM-3;[10] Darapti, Felapton, Datisi, Ferison LXL;[11] and QLQ-3, QQM-3.[12] The MMM syllogisms can be proved by a rule

$$\textit{Mi-elim}\ \mathrm{C} \qquad \frac{R \quad Max \quad Mbx}{p} \longrightarrow \frac{R \quad Mab^i}{p}$$

which is justified as follows:

$$\frac{R \quad Max \quad Mbx}{p} \xrightarrow{\ C\ } \frac{R \quad \sim p \quad Mbx}{Uax} \xrightarrow{\ Le-intro\ \mathrm{C}\ } \frac{R \quad \sim p}{Lab^e} \xrightarrow{\ C\ } \frac{R \quad Mab^i}{p}$$

Datisi MMM (for example) can then be proved:

$$Mi-intro\ \mathrm{C}\ \frac{Ma-elim\ \mathrm{C}\ \dfrac{Mac^a \quad Mcx}{Max} \qquad Mbx}{Mab^i} \qquad \xrightarrow{\ Mi-elim\ \mathrm{C}\ } \frac{Mac^a \quad Mbc^i}{Mab^i}$$

The QLQ syllogisms (such as Bocardo QLQ) are provable by Qi-elim *C*:

$$Qi-intro\ \mathrm{C}\ \frac{Qax \qquad M\text{-}subord\ \dfrac{L\text{-}subord\ \dfrac{La-elim\ \mathrm{C}\ \dfrac{Lbc^a \quad Mcx}{Lbx}}{bx}}{Mbx}}{Qab^i} \qquad \xrightarrow{\ Qi-elim\ \mathrm{C}\ } \frac{\dfrac{Qac^o}{Qac^i}\ Q-conv \qquad Lbc^a}{\dfrac{Qab^i}{Qab^o}\ Q-conv}$$

The LXL syllogisms he mentions can be proved by i-elim, for example:

$$Li-intro\ \mathrm{C}\ \frac{La-elim\ \mathrm{C}\ \dfrac{Lac^a \qquad M-subord\ \dfrac{cx}{Mcx}}{Lax} \qquad M-subord\ \dfrac{bx}{Mbx}}{Lab^i} \qquad \xrightarrow{\ i-elim\ } \frac{Lac^a \quad bc^i}{Lab^i}$$

§40.4 *CC-style singular modal syllogistic*[13]

CCEQ0's axioms and rules are rep, a-subalt, a-elim, e-elim 1, e-elim 2, L-rep, U-rep, Q-rep, La-subord, Le-subord, Li-subord, Lo-subord, Barbara LXL, Celarent

---

10 BURIDAN Theorem IV-6.

11 BURIDAN Theorem IV-18 (a).

12 BURIDAN Theorem IV-23.

13 On Ockham's modal syllogistic see STAHL 1976, esp. §11; STAHL 1977, esp. §7.

LXL, Darii LXL, Ferio LXL, Q-intro, Barbara QXQ, Darii QXQ, Cesare QLX, Substitution, Permutation, C, K, Cut, i-elim, o-elim, plus

La-intro CC $$\frac{R \quad bx}{Lax} \longrightarrow \frac{R}{Lab^a}$$ *La-elim* CC $$\frac{Lab^a \quad bx}{Lax}$$

Le-intro CC $$\frac{R \quad bx}{Uax} \longrightarrow \frac{R}{Lab^e}$$ *Le-elim* CC $$\frac{Lab^e \quad bx}{Uax}$$

Li-intro CC $$\frac{Lax \quad bx}{Lab^i}$$ *Li-elim* CC $$\frac{R \quad Lax \quad bx}{p} \longrightarrow \frac{R \quad Lab^i}{p}$$

Lo-intro CC $$\frac{Uax \quad bx}{Lab^o}$$ *Lo-elim* CC $$\frac{R \quad Uax \quad bx}{p} \longrightarrow \frac{R \quad Lab^o}{p}$$

Qa-intro CC $$\frac{R \quad bx}{Qax} \longrightarrow \frac{R}{Qab^a}$$ *Qa-elim* CC $$\frac{Qab^a \quad bx}{Qax}$$

Qi-intro CC $$\frac{Qax \quad bx}{Qab^i}$$ *Qi-elim* CC $$\frac{R \quad Qax \quad bx}{p} \longrightarrow \frac{R \quad Qab^i}{p}$$

$CCEQ0$'s semantics is as for $AEQ0$ except that **26.4.1-4** replace **22.1.5-8** and **32.2.1-2** replace **28.1.2-3**.

In §32.4 we formulated a minimal system with CC semantics, containing La-subord, Le-subord, Li-subord, Lo-subord, Barbara LXL, Celarent LXL, Darii LXL and Ferio LXL, along with Cesare QLX, Barbara QXQ and Darii QXQ. All of these are provable in $CCEQ0$. With the following exceptions, the proofs run parallel to the corresponding ones in $AEQ0$.

Li-subord

$$\text{i-intro 3}\,\frac{\text{L-subord}\,\dfrac{Lax}{ax} \quad bx}{ab^i} \quad \xrightarrow{\text{Li-elim CC}} \quad \frac{Lab^i}{ab^i}$$

Celarent LXL

$$\text{Le-elim CC}\,\frac{Lab^e \quad \text{a-elim}\,\dfrac{bc^a \quad cx}{bx}}{Uax} \quad \xrightarrow{\text{Le-intro CC}} \quad \frac{Lab^e \quad bc^a}{Lac^e}$$

Ferio LXL

$$\text{Lo-intro CC}\,\frac{Uax \qquad \text{Le-elim CC}\,\dfrac{Lab^e \quad bx}{cx}}{Lac^o} \quad \xrightarrow{\ i\text{-elim}\ } \quad \frac{Lab^e \quad bc^i}{Lac^o}$$

U-intro and -elim 1-3 and 5 can be added to *CCEQ*0's syntactic base, **25.1-3** or **5** being added to the semantics, along with **22.1.2-3** (while **40.1.1-3** are removed). Thus we get *CCEQ*1-3,5. In *CCEQ*3 Celarent XLL is provable as in *AAEQ*3.

*CCEQ*2 contains Baroco XLL, which is provable as in *AEQ*2.

*CCEQ*0 can be augmented by La-conv (and E1), by Li-conv (and E3), by Le-conv (and E4).

Ockham gives details of ecthetic proofs for several syllogisms. His ecthetic proof of Bocardo LXL is stated thus:

> But if the major is a particular negative, the discourse is valid, because it follows 'Some creator is of necessity not an ass, every creator is a person of the divinity, so some person of the divinity is of necessity not an ass'. And such a syllogism is proved through an expository syllogism. If some creator is of necessity not an ass, let it be *a*. Then *a* is of necessity not an ass. And *a* is a person of the divinity, by the minor proposition. So some person of the divinity of necessity is not an ass.[14]

Putting "*x*" for Ockham's "*a*", this can be represented:

$$\text{Lo-intro CC}\,\frac{Uax \qquad a\text{-elim}\,\dfrac{cx \quad bc^a}{bx}}{Lab^o} \quad \xrightarrow{\ \text{Lo-elim CC}\ } \quad \frac{Lac^o \quad bc^a}{Lab^o}$$

Ockham gives analogous ecthetic proofs for Disamis LLL,[15] Bocardo LLL,[16] and (?) Darapti QXM.[17]

---

[14] OCKHAM IIIa.33.54-60.

[15] OCKHAM IIIa.22.17-23.

[16] OCKHAM IIIa.22.24-32.

[17] OCKHAM IIIa.39.6-12.

§41. REDUCTION TO ASSERTORIC SYLLOGISTIC

The formulae of $AEQ0$ can be reduced to those of a system of assertoric singular syllogisms. All our other systems of singular modal syllogistic can also be reduced either to $E_Q 0$ or to modifications of it.

§41.1 *System* $E_Q 0$

Corresponding to any general term v of system AEQ there are terms v_L, v_M and v_Q. The resulting language contains, in addition to ordinary assertoric (singular or categorical) forms, assertoric forms with modalised terms, such as $a_M x$. $E_Q 0$ contains singular term variables

$$x, y, z, \ldots$$

plain general term variables

$$a, b, c, \ldots$$

bold general term variables

$$\boldsymbol{a}, \boldsymbol{b}, \boldsymbol{c}, \ldots$$

a bar

$$\bar{}$$

forming negated terms from plain terms, and subscripts $_L$, $_M$, $_Q$, forming modalised terms from plain terms. A *term* is then a variable or the result of applying a bar or a subscript to a term. (We adopt the convention that $\bar{\bar{\tau}}$ is interchangeable with τ.) A *propositional form* is a sequence $\tau\xi$, $\sim\tau\xi$, $\tau\upsilon^a$, $\tau\upsilon^e$, $\tau\upsilon^i$ or $\tau\upsilon^o$, where τ, υ are terms and ξ is an individual variable.

The contradictories are $\tau\xi \mid \sim\tau\xi$, $\tau\upsilon^a \mid \tau\upsilon^o$ and $\tau\upsilon^e \mid \tau\upsilon^i$; the contraries are $\tau\xi \parallel \bar{\tau}\xi$ and $\tau\upsilon^a \parallel \tau\upsilon^e$.

The axioms are

a-identity $\dfrac{}{\boldsymbol{aa}^a}$ *rep* $\dfrac{ax}{ax}$

a-subalt $\dfrac{ab^a}{ab^i}$ *a-elim* $\dfrac{ab^a \quad bx}{ax}$

e-elim 1 $\dfrac{ab^e \quad bx}{\bar{a}x}$ *e-elim 2* $\dfrac{ab^e \quad ax}{\bar{b}x}$

The rules are Substitution (of terms for plain variables and of bold variables for bold variables), Permutation, *C, K,* Cut, i-elim and o-elim (from §38.1) plus

R_{EL} $\dfrac{R \quad aa_L{}^a}{p} \longrightarrow \dfrac{R}{p}$

provided R has some member $va_L{}^a, va_L{}^i, a_Lv^a, a_Lv^i$ or $a_L\xi$

R_{EM} $\dfrac{R \quad a_Ma^a}{p} \longrightarrow \dfrac{R}{p}$

provided R has some member va^a, va^i, av^a, av^i or $a\xi$

R_{EQM} $\dfrac{R \quad a_Ma_Q{}^a}{p} \longrightarrow \dfrac{R}{p}$

provided R has some member $va_Q{}^a, va_Q{}^i, a_Qv^a, a_Qv^i$ or $a_Q\xi$

R_{QL} $\dfrac{R \quad a_La_Q{}^e}{p} \longrightarrow \dfrac{R}{p}$

Theses include $\dfrac{\bar{a}_Lx}{\sim ax}$, $\dfrac{ax}{a_Mx}$ and $\dfrac{a_Lx}{ax}$.

Proof

$$a-elim\,\dfrac{\bar{a}\bar{a}_L{}^a \quad \bar{a}_Lx}{\bar{a}x} \xrightarrow{\quad R_{EL} \quad} \dfrac{\bar{a}_Lx}{\bar{a}x}$$

$$rep\,\dfrac{ax}{ax} \xrightarrow{\quad K \quad} \dfrac{\bar{a}x}{\sim ax}$$

$$a-elim\frac{a_M a^a \quad ax}{a_M x}\xrightarrow{R_{EM}}\frac{ax}{a_M x}$$

$$a-elim\frac{aa_L{}^a \quad a_L x}{ax}\xrightarrow{R_{EL}}\frac{a_L x}{ax}$$

Semantics for E_Q0 is as follows.

An E_Q0 model is defined as an ordered triple $\langle D,F,V\rangle$ satisfying **21.1** (domain), **21.7-8** (validity), **22.1.1** (star sets), **24.1-5** (categorical forms), **24.6-7** (L and M terms), **28.1.1** (double dagger sets), **31.1** (Q terms), **38.2.2-3,6** (singular forms) and **40.1.1-3** (sun sets) plus

41.1.1 An assignment *V* of values to the variables in a wff associates with each individual variable in the wff either a member of *D* or the null set, and with each un-negated and un-subscripted term a non-null member of *F*.

System E_Q0 is sound with respect to this semantics. The axioms are valid. (Proofs as for *AEL5* of §38.) The rules preserve validity. (Proofs as for *AEL5* plus the following.)

R_{EL} If all members of *R* are true then all members of $R\cup\{aa_L{}^a\}$ are true provided *R* has some member $va_L{}^a, va_L{}^i, a_Lv^a, a_Lv^i$ or $a_L\xi$

R_{EM}, R_{EQM}, R_{QL} Similarly.

For every form *f* of *AEQ0*, there is a form *g* - or a pair of forms g_1, g_2 - of E_Q0 such that either (i) under any valuation *f* is true iff *g* is true, or (ii) under any valuation *f* is true iff both g_1 and g_2 are true, or (iii) under any valuation *f* is true iff either g_1 or g_2 is true. The correspondences are as in A_Q, with the following additions:

ax is true in *AEQ0* iff *ax* is true in E_Q0

$\sim ax$ is true in *AEQ0* iff *ax* is not true in E_Q0

Lax is true in *AEQ0* iff a_Lx is true in E_Q0

Uax is true in *AEQ0* iff $\sim a_Mx$ is true in E_Q0

Max is true in $AEQ0$ iff $a_M x$ is true in $E_Q 0$
$\sim Lax$ is true in $AEQ0$ iff $\sim a_L x$ is true in $E_Q 0$
Qax is true in $AEQ0$ iff $a_Q x$ is true in $E_Q 0$.

$AEQ0$ validity-statements correspond to truth-functions of $E_Q 0$ validity statements as AQ validity statements to truth-functions of A_Q validity statements, with obvious additions.

The decision-procedure for $E_Q 0$ is:

41.1.2 A wff of $E_Q 0$ is valid iff its antilogism sequence is extendable to an Aristotelian chain using the following extension rules.

Extension rule E- If a variable v in one member is matched by $\bar{v}$ in another, then add $v\bar{v}^e$ to the sequence.

Extension rule EL If a variable v in one member is matched by v_L in another, then add $vv_L{}^a$ to the sequence (provided there is some member $wv_L{}^a, wv_L{}^i, v_L w^a$ or $v_L w^i$).

Extension rule EM If a variable v in one member is matched by v_M in another, then add $v_M v^a$ to the sequence (provided there is some member wv^a, wv^i, vw^a, vw^i or $v\xi$).

Extension rule EQM If a term v_Q in one member is matched by v_M in another, then add $v_M v_Q{}^a$ to the sequence (provided there is some member $wv_Q{}^a, wv_Q{}^i, v_Q w^a, v_Q w^i$ or $v_Q\xi$).

Extension rule EQ If a term v_Q in one member is matched by v_L in another, then add $v_L v_Q{}^e$ to the sequence.

In what follows "extendable" will mean "extendable in accordance with these rules". We need to specify (i) that $\tau\xi$ is affirmative and $\sim\tau\xi$ negative (ii) that ξ

is distributed in $\tau\xi$ and undistributed in $\sim \tau\xi$ and (iii) that τ is undistributed in $\tau\xi$ and distributed in $\sim \tau\xi$.

Example:

$\dfrac{Qab^a \quad Mbx}{ax}$ is $AEQ0$-valid iff $\dfrac{Qab^a \quad Mbx}{ax}$ is E_Q0-valid, *i.e.* iff $\langle a_Q b^a, b_M x, \sim ax \rangle$ is an E_Q0-chain. In fact, this is not an E_Q0-chain since while it is extendable to $\langle a_Q b^a, b_M b^a, b_M x, \sim ax \rangle$, the latter is not an Aristotelian chain, a_Q and a occurring only once each, and b_M being twice undistributed. (And $\langle a_Q b^a, b_M x, \sim ax \rangle$ cannot be extended to include a_M because a's occurrence in $\sim ax$ is not in accord with Extension rule EQM.)

§41.2 Alternative superstructures

Formulae from systems *AAEQ0*, *CEQ0* and *CCEQ0* can also be translated into E_Q0, using the following correspondences for modal categorical forms.

AAEQ0

Lab^a is true in *AAEQ0* iff $a_L b^a$ is true in E_Q0
Lab^e is true in *AAEQ0* iff $a_M b^e$ is true in E_Q0
Lab^i is true in *AAEQ0* iff $a_L b_L{}^i$ is true in E_Q0
Lab^o is true in *AAEQ0* iff $a_M b_L{}^o$ is true in E_Q0
Qab^a is true in *AAEQ0* iff $a_Q b^a$ is true in E_Q0
Qab^i is true in *AAEQ0* iff $a_Q b^i$ is true in E_Q0.

CEQ*0*

Lab^a is true in *CEQ0* iff $a_L b_M{}^a$ is true in E_Q0
Lab^e is true in *CEQ0* iff $a_M b_M{}^e$ is true in E_Q0
Lab^i is true in *CEQ0* iff $a_L b_M{}^i$ is true in E_Q0
Lab^o is true in *CEQ0* iff $a_M b_M{}^o$ is true in E_Q0
Qab^a is true in *CEQ0* iff $a_Q b_M{}^a$ is true in E_Q0
Qab^i is true in *CEQ0* iff $a_Q b_M{}^i$ is true in E_Q0.

CCEQ0

Lab^a is true in *CCEQ*0 iff $a_L b^a$ is true in $E_Q 0$

Lab^e is true in *CCEQ*0 iff $a_M b^e$ is true in $E_Q 0$

Lab^i is true in *CCEQ*0 iff $a_L b^i$ is true in $E_Q 0$

Lab^o is true in *CCEQ*0 iff $a_M b^o$ is true in $E_Q 0$

Qab^a is true in *CCEQ*0 iff $a_Q b^a$ is true in $E_Q 0$

Qab^i is true in *CCEQ*0 iff $a_Q b^i$ is true in $E_Q 0$.

Validity-statements in *AAEQ*0, *CEQ*0 and *CCEQ*0 correspond to truth-functions of $E_Q 0$ validity-statements in the usual way.

As an example of how $E_Q 0$'s decision-procedure works for *CEQ*0, consider

$$\frac{Lab^a \quad Mbx}{Max}$$

Corresponding to this in $E_Q 0$ we have

$$\frac{a_L b_M{}^a \quad b_M x}{a_M x}$$

whose antilogism sequence is $\langle a_L b_M{}^a, b_M x, \sim a_M x \rangle$. This is extendable to $\langle a_L b_M{}^a, b_M x, \sim a_M x, a_M a^a, a a_L{}^a \rangle$, an Aristotelian chain.

For *CCEQ*0, consider $\dfrac{ab^a \quad Mbx}{Max}$ and its correlate $\dfrac{ab^a \quad b_M x}{a_M x}$ which is $E_Q 0$-invalid since while its antilogism sequence is extendable to $\langle ab^a, b_M b^a, b_M x, \sim a_M x, a_M a^a \rangle$ the latter is not an Aristotelian chain, b_M being twice undistributed.

41.3 Alternative bases

System $E_Q 0$ can be augmented by syntactic analogues of one or other of our 5 sun set definitions. We can generate systems $E_Q 1$, $E_Q 2$, $E_Q 3$, $E_Q 4$ and $E_Q 5$ by adding appropriate M-intro axioms or rules and M-elim axioms. The M-intro axioms or rules and M-elim axioms for the different systems are as follows:

$E_Q 1$

M-intro 1 $\dfrac{\sim \bar{a}_L x}{a_M x}$

M-elim 1 $\dfrac{a_M x}{\sim \bar{a}_L x}$

$E_Q 2$

M-intro 2 $\dfrac{R \quad \bar{d}_L a^a}{\sim d_L x} \longrightarrow \dfrac{R}{a_M x}$

M-elim 2 $\dfrac{\bar{d}_L a^a \quad a_M x}{\sim d_L x}$

$E_Q 3$

M-intro 3 $\dfrac{R \quad ad^e}{\sim d_L x} \longrightarrow \dfrac{R}{a_M x}$

M-elim 3 $\dfrac{ad^e \quad a_M x}{\sim d_L x}$

$E_Q 4$

M-intro 4 $\dfrac{R \quad ad_L{}^e}{\sim d_L x} \longrightarrow \dfrac{R}{a_M x}$

M-elim 4 $\dfrac{ad_L{}^e \quad a_M x}{\sim d_L x}$

$E_Q 5$

M-intro 5 $\dfrac{R \quad d_L a^a \quad de^e}{\sim e_L x} \longrightarrow \dfrac{R}{a_M x}$

M-elim 5 $\dfrac{d_L a^a \quad de^e \quad a_M x}{\sim e_L x}$

Semantics for these five systems can be obtained by replacing **40.1.1-3** with **22.1.2-3** in $E_Q 0$-semantics, and adding (respectively) **25.1-5**.

Each of the systems is sound with respect to its semantics. All that needs to be shown here is that the M-intro axioms are valid by the appropriate standard and that the M-elim rules preserve validity. Proofs are straightforward.

Formulae from $AEQ1$-3,5 (§40) can be translated into $E_Q 1-3,5$ using our customary correspondences.

$AEQ1$-3,5 validity-statements correspond to truth-functions of $E_Q 1-3,5$ validity statements as AQ validity statements to truth-functions of A_Q validity statements, with obvious additions.

Before giving decision-procedures for $E_Q 1-3,5$ we define $E_Q 1-3,5$-sequences. These are obtained by making the following replacements to any members $\sim v_M\xi$, $v_M\xi$ of a wff's antilogism sequence.

TABLE 41.1

SYSTEM	For $\sim v_M\xi$ put	For $v_M\xi$ put
$E_Q 1$	$\bar{v}_L\xi$	$\sim\bar{v}_L\xi$
$E_Q 2$	$\bar{w}_L v^a, \bar{w}_L\xi$ where w is any term variable	$\bar{u}_L v^o, \sim u_L\xi$ where u is a term variable occurring elsewhere in the sequence
$E_Q 3$	$vw^e, w_L\xi$ where w is any term variable	$vu^i, \sim u_L\xi$ where u is a term variable occurring elsewhere in the sequence
$E_Q 4$	$w_L v^e, w_L\xi$ where w is any term variable	$u_L v^i, \sim u_L\xi$ where u is a term variable occurring elsewhere in the sequence
$E_Q 5$	$(w_1)_L v^a, w_1 w_2{}^e, (w_2)_L\xi$ where w_1, w_2 are any term variables	$(u_1)_L v^o, u_1 u_2{}^i, \sim(u_2)_L\xi$ where u_1, u_2 are term variables occurring elsewhere in the sequence

Conjectural decision-procedures for $E_Q 1-5$ are:

41.3.1 A wff of $E_Q 1$ is valid iff it has an $E_Q 1$-sequence that is extendable to an Aristotelian chain.

41.3.2 A wff of $E_Q 2$ is valid iff it has an $E_Q 2$-sequence that is extendable to a sequence R such that

(i) if for some v, ξ: $\bar{u}_L v^o, \sim u_L\xi$ are in R then for some R', R'': both $\langle R', \bar{u}_L v^o\rangle$ and $\langle R'', \sim u_L\xi\rangle$ are Aristotelian chains, where R is a permutation of $\langle R', R'', \bar{u}_L v^a, \sim u_L\xi\rangle$;

(ii) if for no v,ξ: $\bar{u}_L v^o, \sim u_L\xi$ are in R then R is an Aristotelian chain.

41.3.3 A wff of E_Q3 is valid iff it has an E_Q3-sequence that is extendable to a sequence R such that

(i) if for some v,ξ: $uv^o, \sim u_L\xi$ are in R then for some R', R'' both $\langle R', uv^o\rangle$ and $\langle R'', \sim u_L\xi\rangle$ are Aristotelian chains, where R is a permutation of $\langle R', R'', uv^o, \sim u_L\xi\rangle$;

(ii) if for no v,ξ: $uv^o, \sim u_L\xi$ are in R then R is an Aristotelian chain.

41.3.4 A wff of E_Q4 is valid iff it has an E_Q4-sequence that is extendable to a sequence R such that

(i) if for some v,ξ: $u_L v^o, \sim u_L\xi$ are in R then for some R', R'': both $\langle R', u_L v^o\rangle$ and $\langle R'', \sim u_L\xi\rangle$ are Aristotelian chains, where R is a permutation of $\langle R', R'', u_L v^o, \sim u_L\xi\rangle$;

(ii) if for no v,ξ: $u_L v^o, \sim u_L\xi$ are in R then R is an Aristotelian chain.

41.3.5 A wff of E_Q5 is valid iff it has an E_Q5-sequence that is extendable to a sequence R such that

(i) if for some v,ξ: $(u_1)_L v^o, u_1u_2{}^i, \sim (u_2)_L\xi$ are in R then for some R', R'', R''': both $\langle R', (u_1)_L v^o\rangle$ and $\langle R'', u_1u_2{}^i\rangle$ and $\langle R''', \sim (u_2)_L\xi\rangle$ are Aristotelian chains, where R is a permutation of $\langle R', R'', R''', (u_1)_L u_2{}^o, u_1u_2{}^i, \sim (u_2)_L\xi\rangle$;

(ii) if for no v,ξ: $(u_1)_L v^o, u_1u_2{}^i, \sim (u_2)_L\xi$ are in R then R is an Aristotelian chain.

Here are some examples where the validity of formulae from $AEQ1$-5 is decided by reference to E_Q1-5:

E_Q1

$\dfrac{Lab^a \quad Uax}{Ubx}$ has an E_Q1-sequence $\langle a_L b^a, \bar{a}_L x, \sim \bar{b}_L x\rangle$, which can be extended to $\langle a_L b^a, aa_L{}^a, a\bar{a}^e, \bar{a}\bar{a}_L{}^a, \bar{a}_L x, \sim \bar{b}_L x\rangle$. This is not an Aristotelian chain since it contains two negatives; besides, b_L cannot be linked to b.

$E_Q 2$

$\frac{Uax}{L\bar{a}x}$ has an $E_Q 2$-sequence $\langle \bar{d}_L a^a, d_L x, \sim \bar{a}_L x \rangle$. This can be extended to $\langle \bar{d}_L a^a, \bar{d}_L \bar{d}^a, d\bar{d}^e, dd_L{}^a, d_L x, \sim \bar{a}_L x, a\bar{a}^e \rangle$ but no further: $\bar{a}_L$ cannot be linked to $\bar{a}$ because $\bar{a}_L$ does occur in the ways specified in R_{EL}. Thus we cannot form a single Aristotelian chain, let alone a pair of them.

$E_Q 3$

$\frac{ab^a \quad Uax}{Ubx}$ has an $E_Q 3$-chain $\langle ab^a, ad^e, d_L x, \sim d_L x, bd^i \rangle$, which splits into the pair of Aristotelian chains $\langle ab^a, ad^e, bd^i \rangle$, $\langle d_L x, \sim d_L x \rangle$.

$E_Q 4$

$\frac{Lda^e \quad de^e \quad Lex}{Uax}$ has an $E_Q 4$-chain $\langle d_L a^a, de^e, e_L x, \sim e_L x, e_L a^i \rangle$, which can be extended to $\langle d_L a^a, dd_L{}^a, de^e, ee_L{}^a, e_L x, \sim e_L x, e_L a^i \rangle$. This splits into the Aristotelian chains $\langle e_L x, \sim e_L x \rangle, \langle d_L a^a, dd_L{}^a, de^e, ee_L{}^a, e_L a^i \rangle$.

$E_Q 5$

To $\frac{L\bar{a}x}{Uax}$ corresponds $\langle \bar{a}_L x, \sim e_L x, de^i, d_L a^o \rangle$. This can be extended to $\langle a\bar{a}^a, \bar{a}\bar{a}_L{}^a, \bar{a}_L x, \sim e_L x, de^i, d_L a^o \rangle$ but no further: d_L cannot be linked to d nor e to e_L. Thus we cannot form a single Aristotelian chain, let alone two.

CHAPTER TEN. ESSENTIALISM

Our semantics for modal forms can be given a metaphysical interpretation which fits certain aspects of Aristotle's essentialism, including part of the doctrine of the predicables. In this Chapter we consider the nature of metaphysical theories in general and essentialism in particular; we show that Aristotle was an essentialist and that our semantics for his modal syllogistic can be given an essentialist interpretation; we identify which parts of Aristotle's essentialism are, or can by extension be, included in our semantics, and which cannot; we develop rudimentary formal analyses of the latter and we mention some non-Aristotelian essentialist theories that are expressible in the systems $E_Q 0-5$.

§42. VARIETIES OF ESSENTIALISM

Metaphysical theories rest on a concept of the real, as the way things *really* are. The implied contrast may be with mere *appearances*;[1] it may be between what exists *in nature* and what exists *by convention*,[2] or between the *furniture of the world* and our *projections*.[3]

Opposed to all metaphysical theories are philosophers like Nietzsche:

> We have suppressed the true world [*die wahre Welt*]: what world survives? the apparent world perhaps? ... certainly not! *In abolishing the true world we have also abolished the world of appearance!*[4]

Concerning such philosophers, Putnam comments:

> These philosophers have in common a rejection - a total root-and-branch rejection - of the enterprise mentioned, and a concern with the quotidian, with the *Lebenswelt*, with what a philosophy free of the search for a 'true world' (Nietzsche's phrase!) would look like. I myself see Husserl as such a philosopher .... Wittgenstein and Austin were such philosophers. Nelson Goodman is such a philosopher.[5]

Aristotle, needless to say, is not such a philosopher.

The metaphysical privileging of reality over mere fact requires that some terms be distinguished as specifying ways at least some things really are.[6] Let us call these *metaphysical terms*. Any metaphysical theory will contain some such terms, though the choice of metaphysical terms varies from one metaphysical theory to

---

1 This is Plato's way of drawing the contrast.

2 See the quote from Democritus below.

3 PUTNAM p.52.

4 NIETZSCHE p.25.

5 PUTNAM p.52.

6 Modern metaphysicians can do this within a possible-worlds semantics. '*a* belongs necessarily to *x*' can be defined as meaning '*a* belongs to *x* in all worlds in which *x* exists'. (Some metaphysicians will want to restrict the set of worlds to those that are possible in some narrower sense than logical possibility.) A term 'a' can then be defined as a metaphysical term iff in some world and for some *x* '*a* belongs necessarily to *x*' is true. But there is no sign of such a treatment in Aristotle.

another. Metaphysicians believe that all knowledge worth having is ultimately knowledge of reality.[7] It follows for them that all worthwhile knowledge must ultimately be expressible through metaphysical terms. Everything else will be classified as pseudo-knowledge expressed in metaphysically misleading language.

The existence of metaphysical as well as non-metaphysical terms requires a distinction between two types of metaphysical proposition, *i.e.* two types of proposition in which metaphysical terms are predicated. One type contain *only* such terms, the other relates metaphysical terms to other terms that inadequately represent reality. The former I shall call *pure* propositions, the latter *mixed*. Mixed propositions have the force of saying "Such-and-such is *really* so-and-so". A metaphysical theory needs to assert some such propositions if it is to offer any account of how ordinary ways of speaking relate to the metaphysically adequate representation of reality (supposing that the latter is achieved through terms that are not part of ordinary language).

Let us call those propositions whose assertion is part of a metaphysical theory *internal* to that theory. All other propositions, even if they contain metaphysical terms drawn from the theory we will call *external*. Internal propositions may thus be either pure or mixed. Equally, a (true) pure proposition may be either internal or external to the metaphysical theory from which its metaphysical terms come: if it happens that the only animals at a particular time are horses then *All animals are horses* is a true proposition containing only metaphysical terms of Aristotelian metaphysics, but it is not internal to Aristotelian metaphysics.

Metaphysical terms are thought to apply necessarily to those of their subjects to which they really apply. This is because things are thought to be necessarily the way they really are.[8] Thus if you hold that something is in reality H_2O then you will think that that stuff is H_2O necessarily; and if you hold that something is in reality a horse then you will think that that thing is a horse essentially and thus necessarily. On the first view, water, or the stuff in the beaker, is necessarily H_2O; and on the second, Phar Lap, or the animal approaching, is necessarily a horse. It does not follow, of course, that H_2O is necessarily in the beaker, or that a horse is necessarily approaching. Neither 'in the beaker' nor 'approaching' is a metaphysical term. Let us call the necessity with which metaphysical terms apply to their subjects *metaphysical necessity*.

---

7 For an eloquent statement of the connection between 'real' and 'cognitively reliable' see VLASTOS p.49ff.

8 For the possible-worlds theorist, it follows from the fact that 'H_2O' is a metaphysical term that it applies necessarily to at least some of its subjects.

§42.1 Essentialism

Essentialism is one kind of metaphysical theory. I take its characteristic tenet to be that things are, in reality, grouped into a plurality of Kinds; and, with Quine, I take Kinds to be sets:

> There is no call to reckon kinds as intensional. Kinds can be seen as sets, determined by their members. It is just that not all sets are kinds.[9]

But of course essentialist metaphysics cannot rest content with the thought that in reality things divide into Kinds; it must also identify certain terms as picking out the Kinds. Relative to a particular metaphysical theory, a *Kind term* is a metaphysical term 'a' such that according to that theory every *a* is in reality (and thus necessarily) an *a*. The remaining metaphysical terms (terms 'a' such that some but only some *a*s are in reality *a*s) I shall call *quasi-Kind terms*, and in that case the *a*s that are in reality *a*s I shall call a quasi-Kind.[10] Members of a Kind, as identified through their Kind term, I shall call *realities*.

Essentialism is usually defined as the belief that some individuals non-trivially have necessary attributes not had by all individuals.[11] The definition adopted here does indeed imply that some individuals non-trivially have necessary attributes not had by all individuals, given that belonging to a Kind is a non-trivial necessary attribute of an individual. Conversely, those who adopt the usual definition have focussed their attention on two types of necessary attribute - membership of a Kind, and possession of an individual essence. But we can treat having an individual essence as belonging to a singleton Kind, so the two definitions come down in practice to a consideration of the same cases.

Some metaphysicians postulate singleton Kinds. Plato is arguably one such (see §42.2 and §47). Some believe in individual essences, and thus in singleton Kinds. And some believe that origin determines Kind, and are thus committed to believing in singleton Kinds (if two individuals cannot share an origin).[12] If for whatever

---

9 QUINE p.118.

10 A possible-worlds account would be: A *kind-term* is a metaphysical term 'a' such that in every world 'for all *x*, *a* belongs necessarily to *x* if *a* belongs to *x*' is true. A *quasi-kind-term* is a metaphysical term that is not a kind-term, *i.e.* one such that in some world there is some individual *x* such that '*a* belongs contingently to *x*' is true.

11 *E.g.* PLANTINGA.

12 Essentialists who believe in singleton Kinds determined by a reality's origin (*e.g.* by the sperm and egg from which it developed) may also believe in quasi-Kinds formed by merging a number of such singletons. For example, suppose (with McCALL 1994) that animals of several different Kinds are less than seven feet tall as a consequence of their origins. Such animals would, in reality, be among the less-than-seven-feet-tall things. The subset of the less-than-seven-feet-tall things to which they belong has just the characteristics of a quasi-Kind.

reason you believe in singleton Kinds then I count you as an essentialist, since you hold that things really divide into Kinds.

For some metaphysicians, one Kind embraces all of reality. You might, for example, believe that it's all really water, or all really "hills in space-time".[13] You might believe that "there is just one substance which may be called either God or Nature, for it answers to the implications of either word."[14] If you believe any of these things, I do not call you an essentialist unless you also believe that in reality there are differences in Kind.

Essentialists believe that all talk about things that does not allocate them to their Kinds does not pick them out precisely and is therefore metaphysically misleading. So metaphysical knowledge must be formulated in language that picks out Kinds. Ultimately, what the essentialist metaphysician seeks is knowledge of the laws governing these Kinds - the laws of nature.

§42.2 Examples of essentialist theory

Any scientific theory that puts forward empirically testable laws of nature about distinct Kinds is essentialist by this account. Whether the posited Kinds are electrons or biological classes makes no difference to the general point. Terms not precisely picking out scientific Kinds are labelled 'pre-scientific' or 'folk' talk. As examples of internal essentialist propositions relating non-metaphysical to metaphysical terms, consider Xenophanes' statement that what are commonly called rainbows are really clouds, or Democritus's "By convention sweet and by convention bitter, by convention hot, by convention cold, by convention colour: in reality atoms and void".[15]

Equally essentialist are many non-empirical metaphysical theories. Plato's theory of Forms, if it maintains that the Forms are realities, each unique to its Kind, is a type of essentialism. Metaphysical knowledge, according to Plato, is knowledge of the Forms. All that is worth knowing about non-Forms (particulars) is knowledge about the Forms in which those particulars participate. Terms not designating Forms either cut reality up the wrong way or designate what is not fully real. Platonic laws of nature state relationships among the Forms. Platonic internal propositions also include the proposition that what is in ordinary parlance called "beautiful" is really just a *natural representation* of the reality, Beauty. Like a

---

[13] QUINE p.136.
[14] SPRIGGE p.856.
[15] BARNES pp.98, 252-253. Democritus has these words spoken by mind [*phren*] to the senses. They reply "Poor *phren*, do you take your evidence from us and then try to overthrow us? Our overthrow is your fall."

footprint, it both resembles it and is caused by what it represents.[16] And the entire material world is only a footprint of a higher reality.

§42.3 Aristotelian essentialism: generalities

What is of immediate interest to us is Aristotle's essentialism. The *Categories* posits as Kinds the Aristotelian species and genera, principally those in the category of Substance. These Kinds sharply contrast with Plato's, being this-worldly species whose natures are revealed by the way their members develop in time. Some terms (ones that are *per se* in the third sense of that expression distinguished at *Posterior Analytics* A4, 73b9-10) pick Kinds out precisely: these include 'man' and 'swan'. Others, lacking the appropriate unity, do not do so: these include paronyms like 'the just' and compound terms like 'white musician'. The items these terms pick out are in reality men, though they are not picked out as such. Quasi-Kind terms include 'white', a term which applies necessarily to swans but contingently to men.

Some Aristotelian internal propositions state class-relations between two Kinds, *e.g. No swans are horses.* Others state class-relations between a Kind and a Quasi-Kind, *e.g. All swans are white.* Aristotelian external propositions include propositions saying that something picked out by a non-metaphysical term really is so-and-so (*A musician is really a man*). Such propositions are included among the class of *per accidens* predications implicitly distinguished at *Posterior Analytics* A4, 73b6-8. The grammatical subject of such an external proposition is, as Aristotle puts it, "said of some other underlying subject", which subject is picked out by the proposition's grammatical predicate.

Within the Aristotelian framework, Richard Patterson draws a distinction between 'strong' and 'weak' universal affirmative apodeictics. He distinguishes[17] two kinds of case in which such propositions are true:

> In the first, illustrated by 'Animal necessarily applies to all human', the logical predicate is part of the essence of all things designated by the subject term, *and* animal is part of what-it-is-to-be a human. In the second - and here we may borrow Aristotle's example 'Animal necessarily applies to all white' - the predicate can (in any possible situation in which all white things are animals) again belong to the essence of everything designated by the subject term, but without animal being part of what-it-is-to-be white. Aristotle includes examples of both sorts in his treatment of syllogistic propositions.[18]

---

[16] Plato's theory of Forms supposes that the phenomenal world stands to reality simultaneously as a natural representation and an exemplification of it. This is absurd. The footprint does not exemplify the foot, but those properties which the foot transmits to it. Nor is the footprint a natural representation of footliness, but of the subject of that property. The theory thus confuses properties with the subjects that exemplify them.

[17] PATTERSON 1989 pp.3-4.

[18] At 30b5-6 Aristotle claims that even though to-move applies to all animal and animal applies necessarily to some white, to-move need not apply necessarily to some white. At 31a1 he claims that

He continues:

For the purposes of explicating this contention and of tracing the logical differences between these two readings, I shall describe them as 'weak *cop*' and 'strong *cop*', respectively ....[19]

The use of the terms 'weak' and 'strong' suggests that the strong sense is meant to imply the weak sense. That this is so is confirmed by Patterson's formal definitions. He defines the weak sense thus:[20]

For *some* definable essence C (a) A is part of what-it-is-to-be a C, and (b) All the Bs are essentially Cs.

And the strong sense:

(c) A is part of what-it-is-to-be a B, and (d) All the Bs are essentially B.

The entailment from strong to weak sense holds provided that the phrase 'what-it-is-to-be a B' implies that B is a definable essence.[21]

Patterson's strong apodeictics, it seems, are pure internal propositions (or at least internal propositions whose *subject* is pure). In weak apodeictics, the subject may be any term that is included within the subject-term of a true strong La-proposition.[22] Weak La-propositions are more general and thus in a sense primary: strong La-propositions are just a special case distinguished by the facts that the included set is a star set and that they are accepted by the metaphysical theory. In another sense, the strong propositions are primary: we can generate all the true weak La-propositions by applying Barbara LXL to conjugations of strong La-propositions with true a-propositions.

All of this I will call the *general* character of Aristotelian essentialism. I now turn to some more specific aspects of Aristotelian essentialism that will be of special interest for us later in this Chapter.

---

even though animal belongs necessarily to all man and belongs to no white, man need not necessarily belong to no white.

[19] PATTERSON 1989 *ibid.*

[20] PATTERSON 1989 pp.19-20. That the weak sense is implied by the strong is explicitly stated in PATTERSON 1990 p.166 note 31.

[21] Patterson is right about the logical relation between strong and weak La-propositions. But I do not agree with him in distinguishing two *forms* of La-proposition. There is only one La-form. There is indeed a question whether the terms of an La-proposition are metaphysical, pure or mixed. That is a metaphysical question. There is also a question whether the La-proposition is asserted by some given metaphysical theory. Patterson's own example (*All whites are animals*) has a metaphysical subject and a metaphysical predicate according to Aristotelian essentialism, but it is not internal to Aristotelian essentialism. The status of an La-proposition as strong or weak is not simply a question of logical form.

[22] PATTERSON 1989.

§42.4 Aristotelian essentialism: specifics

(1) Reality is not pervasive, *i.e.* some things are not realities. Can a metaphysics allow that there are pockets of unreality? The clear answer is Yes. By its nature, metaphysics privileges certain terms over others, as picking out the realities. The privileging of some terms as picking out realities, of course, does not imply that some individuals are unreal, since an individual not designated as real by one term may be so designated by another. But at least it can be said that the metaphysical spirit is *consistent* with the recognition of some individuals as unreal. More than that, it is easy to find examples of metaphysical systems which designate some individuals as unreal. If, for example, your metaphysics says that everything *emanates from* realities, then you can quite happily say that whole regions of the universe are reality-free, so long as the items there can be traced through emanation-relations back to realities. Similarly, a metaphysics that has everything *composed of* or *tending towards* realities allows unrealities to be traceable through relations of composition or teleology back to realities. How does one justify one's choice of metaphysical relations of derivation? In general one can say that these relations need to be chosen in ways that further the aim of essentialist theory, namely to know reality. So, one condition on them is that it turns out that the metaphysically derivative is such that one can know all there is to know about it by knowing the metaphysically primary.

Surprising as it may seem, Aristotle is committed to there being pockets of unreality. First of all, he is committed to holding that there are non-beings: *De Interpretatione* 3, 16b11-15 states that denials are about non-beings as well as beings, and *Topics* Δ4, 121a22-24 says that non-beings are thinkable.[23] Non-beings are of course not realities. Secondly, as suggested earlier, Aristotle may want to say that aggregates are not themselves realities even if their parts are, and that purely potential members of a Kind are not (yet) realities. The existence of such individuals poses no threat to the comprehensiveness of his metaphysical theory, provided that they can be accounted for in terms of ontological relations to individuals that do belong to Kinds. For instance, an unreal individual might be *composed of* real individuals or might be teleologically related to a Kind. Aristotle in *Metaphysics* Γ2, 1003a33-b10 details a number of such ontological relations.

(2) The Kinds are uniformly determinate. All individual members of a genus are members of the same number of its species. This appears to underlie the principle that if one species falls under two genera then one of those genera falls under the

---

[23] For a discussion of these passages see THOM 1991 p.536.

other, as stated at *Topics* Δ2, 121b24-122a2.[24] If one Kind were to fall under two Kinds which merely overlapped then individuals in the overlap would be in more Kinds than individuals not in the overlap.

In the *Topics* (Δ2, 122a26-27) Aristotle claims that each member of a genus is a member of some immediate species of that genus. More exactly, if he holds the uniform determinacy of Kinds, he must mean that a member of a genus that is not a member of a given species of that genus is a member of some distinct species of the same genus.

(3) There are no negative Kinds. The theory of categories implies that an unlimited negative term such as 'non-man' does not pick out a species, since there is no single species to which categorially diverse non-men such as a horse and a harmony both belong. The same applies to all unlimited terms.

(4) Aristotle recognises Quasi-Kind terms like 'white', and one feature of his treatment of them deserves notice. What characterises the term 'white' for Aristotle is that, though it applies with metaphysical contingency to some things, it applies with metaphysical necessity to all members of the Kind swans and all members of the Kind snow. Now, this treatment does not follow formally from the fact that 'white' is a quasi-Kind term, *i.e.* a term applying with metaphysical necessity to some and only some of its subjects. For, the formal possibility exists that it apply necessarily to some and only some of its subjects, but not Kind by Kind. It is therefore significant that Aristotle takes it to apply with metaphysical necessity Kind by Kind.

The term 'white' does not completely misrepresent reality, since it picks out two Kinds. However, the reason it picks them out is not that swans and snow together *are* a Kind. How then should we characterise the relation between 'white' and these two Kinds? One answer is that whiteness is a necessary accident of swans and snow. Another is that it is a differentia of swans and of snow. Either way, we have an explanation for Aristotle's treatment of it as applying with metaphysical necessity to whole Kinds. For, it is whole Kinds, and only they, to which Aristotelian differentiae or necessary accidents apply.

The metaphysical implications of treating whiteness as a necessary accident of swans are far-reaching, requiring both an account of the necessity involved and a theory of the relation between accidents and their substrata.[25] This latter in turn

---

[24] Aristotle considers a weaker formulation, according to which if one species falls under two genera then both genera fall under some one genus. He says that some people think that only the weaker version is correct, citing the example of Prudence [*phronesis*] which they say is a species of both Virtue and Knowledge, neither of these being a species of the other. But Aristotle seems to hold to the stronger version, at least in this text. Whether he ought to hold to it, and whether indeed he does so in his biological works, are matters of some interest. See FREELAND 1994 pp.163ff.

[25] For a historical survey of accounts of necessary accidents, see VAN RIJEN Chapter 7.

demands a theory of abstract entities. We shall outline one way of developing a theory of necessary accidents later in this section.

By contrast, if whiteness were a constitutive differentia of swans then that would imply that 'white' does not pick out anything in reality over and above what is picked out by the Kind-terms it qualifies, since that is Aristotle's view of differentiae.[26] The trouble is that whiteness is according to Aristotle not a differentia of swans or snow.[27] So we are left with the problem of characterising the relation between 'white' and the Kinds it includes.

What I propose is that, even though he was clear that 'white' is not a differentia-term, Aristotle in the *Prior Analytics* treated it as resembling a differentia-term in not picking out anything in reality over and above what is picked out by the Kind-terms it qualifies. The items in reality that are picked out by 'white' are picked out by the Kind-terms 'swan' and 'snow'; and these Kind-terms pick them out *precisely*.

(5) Finally, I mention two aspects of Aristotle's essentialism that cannot be expressed within the semantics we have been using. The first of these is a theory of abstract entities, required for the theory of categories. That theory affirms the existence of abstract realities: qualities, quantities and so on. Let us call all these abstract realities attributes. Aristotle holds that attributes, just like concrete things, are grouped in Kinds. Attributes include whiteness (or better, individual whitenesses - *Categories* 2, 1a27), and these all belong to the Kind colour. Thus abstract reality displays the same pattern of class relationships as is found in concrete reality.

If this were all there is to the theory of categories, our semantics would not be embarrassed by it: we could arbitrarily designate some of our star sets as abstract, others as concrete, and deal in our usual fashion with intra-categorial relations. However, Aristotle's theory requires relationships between, not just within, categories. He makes two types of statement about inter-categorial relations.

At *Categories* 1, 1a12-15 he says that the brave get their name from bravery. He thinks that to certain concrete general terms (and not others) there corresponds an abstract general term such that some reality falling under the abstract term inheres in any reality picked out by the concrete term. 'Brave' is such a term. 'Human' is not: he does not recognise 'humanity' as standing for a Kind of attribute. Nor is 'bravery' a term to which a further, second-order, abstract term corresponds, standing for a Kind of attribute common to all bravery. Our modal semantics lacks any device for marking the syntactic relation (which Aristotle calls 'paronymy') between 'brave' and 'bravery'.

---

26 See FEREJOHN p.98.

27 PATTERSON 1995 p.44 notes that "... letting White be ... a differentia of Swan would violate Aristotle's strictures on division, because White would not entail all the essential ancestors of Swan."

Aristotle's second type of inter-categorial statement is illustrated by *Categories* 2, 1a28 which says that colour exists only in a body. He thinks that certain abstract realities exist only by inhering in certain concrete realities. Our modal semantics lacks anything answering to this inter-categorial relation of inherence.

The existence of inherence and paronymy relationships gives rise to a class of metaphysical truths not dealt with in our semantics. These are truths like *Everything white is a body*. Aristotle has a metaphysical analysis of such truths. Analysing them so that we explicitly refer to the underlying realities, he can say *Everything in which some whiteness inheres is a body*. Analogously, the theory of categories has a metaphysical analysis of statements like *All snow is white*, namely: *All snow has some whiteness inhering in it*. This is rather more informative than saying that the Kind snow is included in a quasi-Kind comprising snow and swans.

(6) The second bit of Aristotelian metaphysics not expressible within our semantics is the theory of natural change presented in *Physics* B1, 192b21-23 where nature is said to be a principle of change. This bears on Aristotle's essentialism, since he links natural change with the processes whereby a thing 'attains its nature':

We also speak of a thing's nature as being exhibited in the process of growth by which its nature is attained. (*Physics* B1, 193b12-13)

And, as Waterlow points out,[28] at *Physics* B1, 193a9-10 he seems to use 'nature' and 'substance' as inter-changeable expressions. If this is right then Aristotle took the natural processes whereby a thing 'attains its nature' somehow to constitute that nature. This eludes our semantics because it has specifically to do with laws of natural development - whereas our semantics is atemporal. We shall offer a rudimentary formal analysis of the theory of attributes, and the theory of natural change, later in this Chapter.

Note

There is a differences between the way Aristotle formulates his essentialism and the form I have given it. He takes as basic a special kind of predication, in the *Categories* called 'said of' predication. A primary substance is then defined as whatever is not said of anything but has something said of it. A secondary substance is defined as what is said of a primary substance. Semantically, this amounts to taking as basic a strong membership relation which we could designate $\in^*$, rather than employing the device of star sets.

[28] WATERLOW pp.52-53.

The two approaches are inter-translatable. On the 'strong membership' approach, star sets can be defined as sets to which some individual stands in the relation $\in^*$. On the 'star sets' approach, strong membership can be defined as the relation holding between an individual and any star set to which it belongs. In the one case we have to specify which individuals are strong members of which sets; in the other case we have to specify which sets are star sets.

Does the Aristotelian approach make better philosophical sense than the one I have adopted? As Ferejohn points out, Aristotle's approach is linked to his understanding of the "What is it?" question.[29] This question has for Aristotle a sense exactly corresponding to strong membership. What an individual *is*, in this sense, is just the sets of which it is a strong member. What makes this philosophically interesting is the idea that this special sense of the question is pre-theoretically available. If it is, then Aristotle has a means of getting non-Aristotelians to accept his essentialism.

I do not find this argument convincing. One might just as well maintain that there is a pre-theoretically available sense of "kind" which Aristotle could have used for the same purpose. Non-Aristotelians, it can be claimed, understand the question "What *kinds* of thing are there?" exactly as well as they understand the question "What *is* it?". Their answers to both questions will produce the same results, given the inter-translatability noted above. In neither case can we be sure that their answers will coincide with Aristotle's. So neither question provides a sure foundation for convincing non-Aristotelians. But neither one is superior to the other.

Patterson makes much of the distinction between these two approaches;[30] and he is right to insist that Aristotle thinks in terms of strong membership, not star sets. But, because of the inter-translatability of the two approaches, we do not misrepresent Aristotle's system by expressing its semantics in terms of star sets. Indeed, Patterson's own formulations can be translated into ones having modalised predicates. Where he writes[31] that something (say, *c*) is in the essence of each *b*, I gloss this as "Every *b* is essentially (necessarily) *c*" and translate this as "Every *b* is an essential (a necessary) *c*".

---

29 FEREJOHN Chapter 4.

30 *E.g.* PATTERSON 1995 §2.3.

31 PATTERSON 1995 Appendix.

§43. THE ESSENTIALISM OF THE MODAL SYLLOGISTIC

Does the modal syllogistic expounded in the *Prior Analytics* depend on Aristotle's essentialist metaphysics? There are four different questions one might understand here. The first question is whether the text shows that Aristotle was an essentialist. An affirmative answer will enhance our understanding of Aristotle's philosophical development to the extent that it spells out the details of the text's essentialism. A second question asks whether the text presents its logical theory in essentialist terms. Part of the answer to this question is clear: at least one aspect of Aristotle's logical theory - his choice of examples - is determined in part by his commitment to essentialism. A third question is whether an adequate account of the syntactic and semantic theory of modal syllogistic needs to make use of concepts or propositions drawn from Aristotle's essentialism. In this connection I shall argue that a semantically adequate account can be formulated in the language of Aristotelian essentialism, incorporating the generalities of that essentialism. A fourth question asks whether Aristotle's motivation for constructing the modal syllogistic might derive from his essentialist metaphysics. Patterson's answer to this question, as I understand it, is affirmative and attributes to Aristotle the following train of thought. (i) Aristotle's primary cases of metaphysical necessity and impossibility are universal metaphysical laws stating inclusions or exclusions between Kinds (or quasi-Kinds); (ii) when terms which contingently fall under Kind or quasi-Kind-terms are substituted for them, the resulting propositions, *qua* applications of the metaphysical laws, inherit a derivative metaphysical necessity or impossibility; (iii) the truth-conditions of particular metaphysical necessities and impossibilities reduce to those for universals; (iv) having thus extended the ambit of metaphysical necessity and impossibility from the primary cases to the derivative, and from the universal to the particular, Aristotle thought it would be good to work out the logic of these modalities in their widened sense. This account of the modal syllogistic's genesis seems plausible.

§43.1 A metaphysical interpretation of the modal syllogistic

An essentialist metaphysic posits (i) a domain of individuals, (ii) a set of Kinds and quasi-Kinds whose members are members of the domain, (iii) a set of propositions concerning those Kinds and quasi-Kinds (the laws of nature), (iv) a set of propositions applying the laws of nature to subjects not identified in metaphysical terms. The models of our modal semantics contain analogues of all these elements. The star sets in a model correspond to the Kinds and quasi-Kinds. Some of the propositions true in a model which state class-relations between star sets correspond to the laws of nature, others to applications of such laws.

On the face of it, any given model under our formal semantics divides terms into two types, depending on whether their star sets are non-empty or empty. This distinction maps onto our distinction between metaphysical and non-metaphysical terms. There is a further semantic distinction, that between terms whose star set f^* satisfies the condition $f \subseteq f^*$ and those that have non-empty star sets which do not satisfy that condition. A metaphysical interpretation can be placed on this distinction too. Broadly, terms of the first type can be interpreted as differing from the others in standing for Kinds. This sort of interpretation is not specific to Aristotle's, or anyone else's, metaphysical theory. If we narrow our focus to the Aristotelian metaphysical context, then our metaphysical interpretation comes to this. The Kinds are what is definable; and what is definable according to Aristotle are species and genera. Thus the terms in any model for which $f \subseteq f^*$ are interpreted as standing for the species and genera of Aristotelian metaphysics.

I agree with Patterson's claim that "the terms of the premises and conclusions of a [modal] syllogism are related to one another as signifiers of genus, definition, proprium, or accident".[32] But unlike Patterson I take it that accidents may be either contingent or necessary to their subjects: horses move contingently, swans are white necessarily.[33]

Terms which in a given model have empty star sets are of two types. Some are true of at least one individual belonging to *some* star set (other than the term's own star set). Others are true only of individuals that belong to no star set. In the context of Aristotelian essentialism, the first type correspond to terms which though they themselves do not pick out a Kind nevertheless are true of some reality. The term 'approaching' is an example: things approaching do not as such form a Kind, but what is actually approaching may be of some Kind. The second type corresponds to terms true only of non-realities ('goat-stag' and, if I am right about aggregates and purely potential realities, 'white man on white horse' and 'seed').

§43.2 Interpreting the base

We have given a semantic interpretation of Aristotle's modal syllogistic which (i) is not committed to accepting any existential assumptions, and (ii) relies on definition **25.5** of sun sets. And we have constructed a metaphysical interpretation of that semantics in Aristotelian essentialist terms. The question that arises now is, Does

---

32 PATTERSON 1990 p.161.

33 PATTERSON 1990 p.162 says "accidents apply contingently or two-way possibly to things of which they are accidents". He seems (n.24) to take white as *genus* to swans and accident to humans. This makes it necessary for him to treat swans as belonging to two genera (animals and whites) neither of which is subordinate to the other.

the absence of existential assumptions in our semantics, along with the presence of definition **25.5**, fit Aristotelian essentialism?

With regard to existential assumptions, (1) E1 - the condition that every star-set is non-empty - is not Aristotelian, its translation being that every term determines a (non-empty) Kind or quasi-Kind. This is no part of Aristotle's essentialism, since there are terms like 'walking' which determine no Kind or quasi-Kind. (2) E2 - the condition that every term determining a non-empty star set is coextensive with that star set - has for its metaphysical translation the proposition that every term that determines a Kind or quasi-Kind is itself a Kind-term or a quasi-Kind-term. This is also unAristotelian, since there are terms like 'white' which, though determining a quasi-Kind, are not themselves Kind- or quasi-Kind-terms. (3) E3 - stating that if an individual satisfies two terms then it satisfies one them necessarily - has for its translation that if an individual satisfies two terms then it is a member of a Kind or quasi-Kind determined by one of them. This is also unAristotelian, since the terms 'walking' and 'musician' overlap in walking musicians but neither of these terms determines a Kind or quasi-Kind. Thus, our interpretation accords with Aristotelian essentialism in so far as it does *not* adopt the metaphysical translation of any of the three existential assumptions.

With regard to the definition of sun sets, **25.1** is ruled out for purely logical reasons relative to any of the superstructures *A*, *AA* or *C*. Given superstructure *A*, **25.1** does not validate the Sun set Theorem, which is required for the validity of Celarent LXL (see model 26.1). Relative to superstructure *AA*, **25.1** does not validate S4, without which Baroco LLL is invalid (see model 26.7). Lo-subord is invalid relative to superstructure *C* and **25.1** (see model 26.13). So logic alone excludes **25.1** (unless we adopt superstructure *CC*), but it does not force **25.5** on us.

There are metaphysical reasons, however, for excluding all definitions but **25.5**. **25.1** and **25.2** are not consonant with Aristotelian essentialism, since their metaphysical translations imply that if there are any metaphysical impossibilities then there are negative Kinds. For Aristotle, by contrast, there are metaphysical impossibilities but there are no negative Kinds (as will be shown in §44). **25.3** and **25.4**, on the other hand, are not Aristotelian since their metaphysical translations imply that a metaphysical impossibility arises from a merely factual exclusion between a class and a Kind. Such an exclusion might be underpinned by nothing in reality. This is not the way metaphysical impossibilities are for Aristotle. Just as metaphysical necessities reduce to inclusions between Kinds, so metaphysical impossibilities reduce to exclusions between Kinds. We shall see below (in discussing the semantics of Le- and Lo-propositions) that Aristotle's attributions of metaphysical impossibility all reduce to exclusions between Kinds (or quasi-Kinds). Thus, of our five definitions of sun sets only **25.5** - which does not require the positing of negative Kinds but does reduce all metaphysical impossibilities to Kind-exclusions - is, under metaphysical translation, consistent with Aristotelian essentialism.

§43.3 Interpreting the superstructure

Our semantics for the modal syllogistic uses an *A* style superstructure. I believe this superstructure is more Aristotelian than its rivals. Formal considerations lend some support to this conclusion. We saw in §32.5 that, granting each superstructure the assumptions most favourable to it, the *A* approach was marginally closer to Aristotle's formal results than any of its rivals. But that analysis took no account of the Aristotelicity of these extra assumptions. If that is done, the picture changes. *AA*, *C* and *CC* semantics are excluded since if they are to reach a competitive level of correspondence with Aristotle's logical theses they have to rely on the unAristotelian existential assumption E3 (and also on the unAristotelian E1 in the case of *AA* and *CC*). So, if we want the metaphysics implicit in our modal semantics to be Aristotelian, only the *A* superstructure will do. There is then only one defensible interpretation compatible with both the logical and metaphysical data, namely the one that reduces the modal syllogistic to *AEQ*5. Further, since **25.4** is ruled out, it follows that we cannot avail ourselves of the unified approach to contingency-forms which it affords, but will have to rest content with the Alexandrian ploy of one system for the pure moods and another for the mixed.

The best evidence supporting our use of the *A* superstructure comes, however, not from formal considerations but from the examples Aristotle uses to show the invalidity of modal formulae. With exceptions to be discussed presently, these examples support our interpretation.

43.3.1 La-propositions

On our interpretation we would expect Aristotle's examples of true universal affirmative apodeictic propositions to be such that their predicates stand for Aristotelian Kinds or quasi-Kinds. For on our semantics the predicates of universal affirmative apodeictic propositions stand for star sets, and these we interpret as Kinds or quasi-Kinds. In general this is what we do find. Aristotle's examples of true La-propositions have predicates like *animal* (which picks out a Kind) or *white* (which picks out a quasi-Kind). Figure 43.1 shows the relations Aristotle asserts to hold among the main terms he habitually uses in counter-examples. The figure shows some of the Kinds and quasi-Kinds to which he repeatedly refers in his examples.

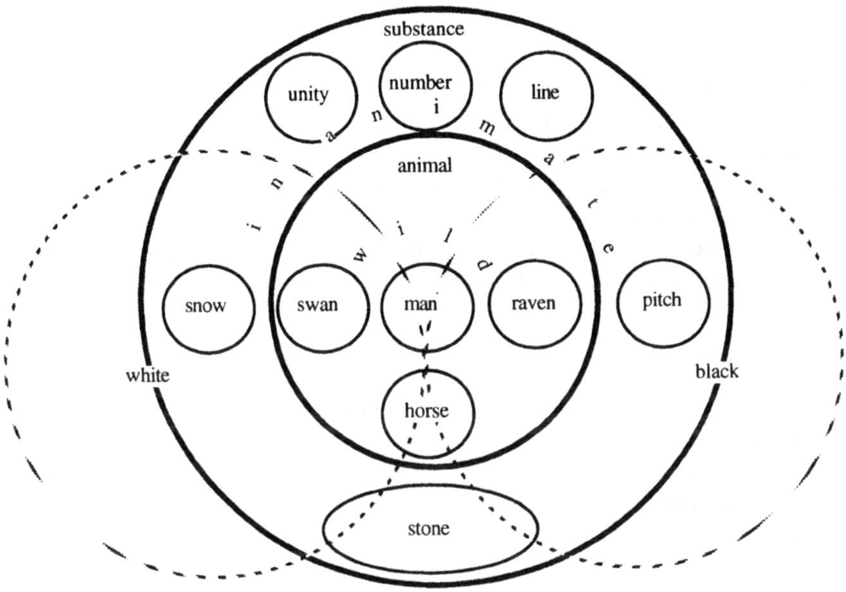

Fig. 43.1

AA and *CC* agree with *A*-semantics on the analysis of La-forms. All of them have the edge on the *C* analysis here. The *C* style superstructure faces two difficulties. First, it conflicts with Aristotle's assertion of *Every man is necessarily an animal.* This proposition is not true on the *C* analysis (given our account of impossibility), since it requires that everything for which being a man is not impossible be a necessary animal, and this is false since artefacts and other non-animals are such that being a man is not impossible for them, nothing being impossible for them since they are not necessarily anything. The second difficulty is that the *C* analysis conflicts with the validity of Barbara LXL: unless that mood's assertoric minor premiss is ampliated to the possible the ampliated apodeictic conclusion will not follow.

Patterson's account, with its distinction between strong and weak senses, its frank use of metaphysical language and its use of the modalised copula, seems to be in a class of its own; but the appearance is deceptive. As we shall see, Patterson's analysis is remarkably similar to Ockham's. First of all, we need only discuss the weak sense of modal propositions since it is the more general, strong modals being only a particular case of weak ones. According to Patterson, a weak La-proposition with predicate *a* and subject *b* is true iff for some *c*, *a* is in the definition of *c*, and every b is essentially c.[34] This is equivalent to the A account (and the AA and C accounts) on the following plausible assumptions:

43.3.1 If a is in the definition of c then whatever is essentially c is essentially a.

43.3.2 If every b is essentially a then for some c: a is in the definition of c, and every b is essentially c.

43.3.3 x is essentially a iff x is a necessary a.

There are two passages that do not fit our interpretation of La-propositions (or any of the other three interpretations). The first occurs in A19 where it is asserted that everything awake is necessarily moving. The problem is that the movers do not constitute an Aristotelian Kind. Independent reasons were given in §15.3.2 for finding this example odd.

The second passage is in A22 where Aristotle asserts that all sleeping horses are necessarily asleep. Our interpretation, applied to this example, delivers the consequence that for Aristotle the sleeping constitute either a Kind or a quasi-Kind. Now, it would be ludicrous to suggest that the sleeping make up an Aristotelian species. But there is room for confusion about whether sleep is a necessary accident. Sleep is indeed necessary for animals in one sense, as Aristotle points out in *On Sleep*:

> ... it is necessary that every creature which wakes must also be capable of sleeping, since it is impossible that it should always be actualizing its powers. (454b7-8. Beare/Barnes translation.)

The problem is that this is not the same sense as that in which whiteness is a necessary accident of snow. All animals must have the *capacity* for sleep, but need not *be* asleep. All snow, on the other hand, must *be* white. Perhaps this confusion infects the present passage. Whatever the truth of the matter, the example sits ill with the general character of Aristotle's examples. And it seems not to fit our formal semantics, since there is no plausible reason for designating the sleepers as a special sort of set. To the extent that this is so, the example poses a threat to the capacity of our formal semantics to apply to the concrete cases dealt with in the text. But, beyond that, it does not challenge our metaphysical interpretation; for this is in the first instance a metaphysical interpretation of that formal semantics.

---

[34] PATTERSON 1995 Appendix., where Patterson replaces his earlier (§2.7) talk of 'being in the definition' with talk of being entailed by. This is no insignificant change. I have used the earlier formulation.

43.3.2 Le-propositions

Le-propositions are of two types, according to our semantics. They either state that two star sets exclude one another, or that what is included in some star set excludes what is included in another star set which excludes the first. On our metaphysical reading, these propositions either state that two Kinds or quasi-Kinds exclude one another, or that what is included in one Kind or quasi-Kind excludes what is included in another. Aristotle's Le-assertions, in his demonstrations of invalidity for contingency-syllogisms, are typified by (i) $L(\text{man})(\text{horse})^e$, a proposition in which two Kinds are said to exclude each other necessarily, (ii) $L(\text{white})(\text{pitch})^e$, where a Kind and a quasi-Kind are said necessarily to exclude each other, and (iii) $L(\text{sleeping horse})(\text{man})^e$, where a Kind is said necessarily to exclude a sub-class of an exclusive Kind. Our interpretation fits all of these.

Le-propositions are taken as false in Aristotle's counter-examples to Celarent XLL and Camestres LXL. In the first case $L(\text{moving})(\text{man})^e$ is said to be false even if no animal is moving and every man is a necessary animal. Our semantics shows why this is so. The movers exclude the men in the example, but they are not a Kind or a quasi-Kind; nor are they included in any Kind or quasi-Kind. (But if we expand the example so that the movers *are* included in a Kind or quasi-Kind, *e.g.* by supposing that only rocks are moving, then $L(\text{moving})(\text{man})^e$ will become true. So Aristotle's example tacitly supposes that the movers are not all of one Kind.) In the second example, $L(\text{man})(\text{white})^e$ is said to be false even if no white thing is an animal and every man is a necessary animal. Again, our semantics shows why, and under what conditions, this is so. Assuming that the whites are not included in any one Kind or quasi-Kind, the Le-proposition is not true; if they are (*e.g.* if it happens that only snow is white) then $L(\text{man})(\text{white})^e$ becomes true.

The C analysis ("No possible b is a possible a") does not fit examples like $L(\text{man})(\text{horse})^e$ given our definition of impossibility. Suppose the universe consists of a man and a horse, the man being on the horse. Then the man and the horse are the only realities: the-man-on-the-horse is an unreal individual. The man is the sole member of its Kind and the horse the sole member of its Kind, and these are exclusive Kinds. Thus, the impossible men (the men's sun set) comprise just the horse. The-man-on-the-horse, and all other unreal individuals, are not among the impossible men. Hence they are possible men. Similarly these unreal individuals are possible horses. Thus, some possible men are possible horses, so $L(\text{man})(\text{horse})^e$ is false on C semantics.

The *AA* and *CC* analyses do not make it a requirement for the truth of an Le-proposition that its subject and predicate classes be included in two mutually exclusive Kinds or quasi-Kinds. They allow such a proposition to be true merely by virtue of a class's excluding some Kind. Patterson's account of weak Le-propositions ("*a* is incompatible with the essence of each *b*")[35] also has this feature, at least on the plausible assumption of **43.3.3** and:

43.3.4 If *a* is contrary to *c* then whatever is essentially *c* is not a possible *a*

43.3.5 If no *b* is a possible *a* then for some *c*, *a* is contrary to *c*, and every *b* is essentially *c*.

For on these assumptions, Patterson's truth-conditions for weak Le-propositions are the same as Ockham's *CC* conditions for divisive Le-propositions.

Patterson raises the following objection to my interpretation of Le-propositions. Consider the proposition *Nothing that is asleep is possibly in the Agora* in a situation where all sleeping things are horses and everything in the Agora is human. He takes this proposition to be "manifestly false", though it is true on my analysis.[36] But I say that this result ought not be surprising to someone who believes in a class of weak modal propositions which *apply* metaphysical laws to the present state of the world. It is a consequence of Aristotle's use of modal terms to cover such statements that what is necessary (or possible) at one time may not be so at another - not in the strong but in the weak sense. If we speak with Aristotle, it is not *eternally* impossible for sleeping things to be in the Agora but it is impossible under the stated conditions. All that this alleged manifest falsehood says is that the things that are in fact sleeping and the things that are in fact in the Agora differ essentially. In the imagined case this is true, though it would not be true if, say, the sole inhabitant of the Agora were a horse and the sole sleeper another horse.

43.3.3 Li-propositions

Our metaphysical interpretation treats an Li-proposition as true when it states that a Kind or quasi-Kind intersects some class. Aristotle's Li-assertions are typified by (i) $L(\text{biped})(\text{animal})^i$, where the subject and predicate pick out two Kinds having common members, (ii) $L(\text{animal})(\text{white})^i$, where a Kind picked out by the predicate intersects with a quasi-Kind picked out by the subject, and (iii)

[35] PATTERSON 1995 Appendix, where Patterson replaces his earlier talk of contrariety with talk of incompatibility. I have used the earlier formulation. A question not raised by Patterson is whether the 'incompatibility' mentioned here is metaphysical or logical.

[36] PATTERSON 1995 p.122.

$L(\text{white})(\text{man})^i$, where a Kind picked out by the subject intersects with a quasi-Kind picked out by the predicate. These all fit our interpretation.

The *AA* analysis conflicts with the validity of Darii LXL, given the truth of *Everything moving is an animal*. For, if that proposition is true then so is *Something moving is an animal*; and if in addition *Every animal is necessarily a body* is true then by Darii LXL so is *Something moving is necessarily a body*. But nothing is essentially moving; whereas *AA* semantics requires that an Li-proposition be true only when both subject and predicate stand for Kinds or Quasi-Kinds.

The *C* analysis of Li-propositions suffers from the defect that it makes Li-subordination invalid.

The *CC* analysis (like the *C*) makes Li-conversion invalid. Patterson's definition of weak Li-propositions ("For some *c*, *a* is in the definition of *c*, and some *b* is essentially *c*")[37] reduces to Ockham's *CC* analysis of divisives, given the plausible assumptions **43.3.1** and **43.3.3** plus

43.3.6 If some *b* is essentially *a* then for some *c*, *a* is in the definition of *c*, and some *b* is essentially *c*.

43.3.4 Lo-propositions

Aristotle's Lo-assertions are typified by (i) $L(\text{biped})(\text{animal})^o$ and (ii) $L(\text{animal})(\text{white})^o$. In the first case, the subject class is a Kind, in the second a quasi-Kind, which intersects with some Kind that excludes the predicate class - which in both cases is a Kind. Our interpretation (and the identical *CC* reading) fits both these cases (as do the *AA* and *C* analyses). But the validity of Ferio LXL, along with the fact that Le-propositions may be true even when the subject class is a non-Kind (provided it is included in a Kind that excludes a Kind including the subject class), implies that there is third type of true Lo-proposition. These are typified by (iii) $L(\text{sleeping horse})(\text{white})^o$, true by virtue of the fact that some white thing is a member of a Kind (snow) which excludes a Kind (horse) in which the subject class is included. Our interpretation also fits this case.

Patterson's definition of weak Lo-propositions ("*a* is contrary to the essence of some *b*")[38] is equivalent to the *A* and *CC* analysis, given earlier plausible assumptions plus

[37] PATTERSON 1995 Appendix.
[38] PATTERSON 1995 Appendix.

43.3.7 If some *b* is not possibly *a* then for some *c*, *a* is contrary to *c*, and some *b* is essentially *c*.

In §18.3 we proposed an interpretation of Aristotle's suggestion that there are circumstances under which some men are necessarily not white. We supposed that men existed but only snow was white. In that case, we said, some men would be of a nature that excludes the nature of all white things. It should now be clear that that example fits our proposed metaphysical interpretation of Lo-forms.

The *AA* analysis, however, does not, as is clear from a consideration of the way in which ephemeral are generated from non-ephemeral necessities. All snow is necessarily white, and this is not an ephemeral matter (given that snow exists). But if what is in my hand is snow then according to Aristotle what is in my hand is necessarily white, at least for the present. Notice that the generation of ephemeral from non-ephemeral necessities widens the class of subject-terms in true necessity-propositions but does not affect the class of predicates. Only a restricted class of terms can be predicates in true La-propositions; but any term at all could be subject. Because Ferio LXL is valid, the same is true of negative apodeictics. Eternally it is true that bears cannot be dogs; but if the animal approaching is a dog then according to Aristotle the animal approaching cannot be a bear, at least for the present. Any term at all could be subject of a true Lo-proposition. Thus the subjects of true Lo-propositions cannot be restricted to terms that can be predicate of true La-propositions. Yet this is what the *AA* analysis requires.

The *C* analysis of Lo-propositions ought not to be taken seriously as an analysis of Aristotelian usage, since it allows an Lo-proposition to be true while the corresponding assertoric is false.

43.3.5 Q-propositions

Universal contingency propositions asserted by Aristotle are typified by $Q(\text{white})(\text{animal})^a$ and $Q(\text{sleeping})(\text{sleeping horse})^a$. The predicate class is always a non-Kind (maybe a quasi-Kind), and the subject class may or may not be a Kind. Particular contingency assertions are typified by $Q(\text{white})(\text{man})^i$ and $Q(\text{animal})(\text{white})^i$. In these cases either the subject or the predicate is a Kind-term. But that cannot be required of true Qi-propositions if these are to be implied by Qa-propositions, since the latter may be true when the subject class is not a Kind. All this accords both with our interpretation of Q-propositions as unampliated and with our interpretation of them as ampliated to the possible. However, the data do not readily allow an interpretation according to which Q-propositions ampliate to the contingent:[39] the truth of Q(white)(animal)^a would then imply the existence of contingent animals.

§44. ARISTOTLE'S STATIC ESSENTIALISM

In §42.4 we mentioned six specifics of Aristotelian essentialism: (1) the non-pervasiveness of realities, (2) the uniform determinacy of Kinds, (3) the absence of negative Kinds, (4) the reducibility of quasi-Kinds to Kinds, (5) the theory of categories and (6) the theory of natural change. Let us now characterise these aspects of Aristotelian essentialism formally and see whether they are implicit in the modal syllogistic. This section deals with the first five points, the next with the sixth.

(1) The pervasiveness of realities can be stated thus:

44.1 $\forall x \exists g^* \left[x \in g^* \right]$

This is implied by the proposition that some term 'a' is what Angelelli calls an o-term,[40] *viz.* one satisfying the condition $\forall x[Lax \vee Uax]$. (At any rate, this implication holds under any of the definitions **25.1-5** of the impossible; for in each of those cases what belongs to a sun set belongs to some star set, so that if there is a term such that everything belongs either to its sun set or its star set then everything belongs to some star set.) Evidently, if there are any o-terms then realities are pervasive. Transposing, if realities are not pervasive then there are no o-terms. Interesting, therefore, as Angelelli's idea of an o-term is, it has no application to Aristotle's metaphysics if we are right in saying that for Aristotle the realities are not pervasive.

Our semantics is consistent with there being non-real individuals and thus with there being no o-terms. For each of our modal systems there are models like

44.1		*
f_1	d_1	–

in which every star-set is empty. But adding **44.1** to our semantics would make no difference to which formulae are valid. Only twice did we make use of models allowing for non-real individuals. This was in model 22.7 (Baroco LXL) and model 27.22 (LaQaLo-2). We also showed in both cases that a model in which the realities are pervasive can be used to refute those moods, namely models 22.8 and 27.23.

[39] See THOM 1994 §4.1.2.3.

[40] ANGELELLI p.203.

If Aristotle's essentialism rejects **44.1**, we can say that this aspect of his essentialism is consistent with, but not demanded by, the modal syllogistic as we have interpreted it.

Nino Cocchiarella is one philosopher who recognises that artefacts may be individuals belonging to no Kind:[41]

> ... it is quite inappropriate to assume ... that everything ... is of a natural kind .... In particular, most artifacts do not belong to a natural kind; for unlike the ring or pendant which is a bit of gold, the materials out of which most artifacts are made are so mixed and varied as to render the artifact as a whole as not being of a natural kind.[42]

Yet Cocchiarella also proposes a 'minimal logic of natural kinds' which includes the principle that

> ... an individual *can* belong to a natural kind only if being of that kind is essential to it, i.e. only if it *must* be of that kind whenever it exists ....[43]

We saw above that being in a given Kind may be possible for an individual which is not actually in that Kind. If Cocchiarella means that individuals not in a given set's sun set are in its star set if anything is, then he is mistaken. A given individual may be neither in the *a*s' sun set nor in the *a*s' star set even when that star set is non-empty. It only has to be an individual not belonging to any Kind.

(2) The uniform determinacy of Kinds can be expressed as follows:

44.2[44]
$$\left(x\in g^{*}\wedge g\subseteq g^{*}\wedge f^{*}\subset g^{*}\wedge f\subseteq f^{*}\wedge x\notin f^{*}\right)\supset$$
$$\exists h^{*}\left[h\subseteq h^{*}\wedge x\in h^{*}\wedge h^{*}\subset g^{*}\wedge h^{*}\cap f^{*}=\varnothing\right]$$

This is not a requirement of the modal syllogistic, as the following model shows.

---

[41] For textual evidence that Aristotle was of this view, see WATERLOW, ch.2 n.4. But see also her Appendix to ch.1.

[42] COCCHIARELLA 1977 p.456.

[43] COCCHIARELLA 1976 p.205.

[44] Cf. (K11) in COCCHIARELLA 1976, p.219. Brian Ellis rejects this principle, giving the example of cuprous oxide, copper compounds, and oxides, all of which he takes to be Kinds, the first being a species of overlapping genera. Concerning this example, one wonders whether all copper compounds (or all oxides) are similar enough to form a Kind. One also wonders whether cuprous oxide should be taken as a Kind. The alternative would be to restrict chemical Kinds to the elements, counting samples of cuprous oxide not as belonging to a Kind but as derived by a relation of composition from things that do belong to Kinds.

44.2		*
f	d_1	d_1
g	d_1, d_2	d_1, d_2
i	d_2	–

$d_2 \in g^*, f^* \subset g^*$ and $d_2 \notin f^*$ but $\sim \exists h^*\left[d_2 \in h^* \wedge h^* \subset g^* \wedge h^* \cap f^* = \varnothing\right]$.

A formal representation of Aristotelian essentialism would have to include this principle. The modal syllogistic would have to be extended to include it, perhaps in the shape of the rule

$$\frac{R \quad Lcx \quad Lbc^a \quad Lca^e}{p} \longrightarrow \frac{R \quad Lbx \quad \sim Lax \quad Lba^a}{p}$$

Adding this requirement of uniform determinacy of Kinds to our semantics would make no difference to which Aristotelian syllogisms are valid. Only twice did we make use of models allowing for non-uniform determinacy of Kinds. This was in model 22.8 (Baroco LXL) and model 27.23 (LaQaLo-2). We showed in both cases that a model in which the Kinds are uniformly determinate can be used to refute those moods, namely models 22.7 and 27.22.

In §42.4 we noted, as following from the uniform determinacy of Kinds, a law to the effect that if two genera are superordinate to a single Kind then one of them is subordinate to the other. This can be formulated:

44.3 $\left(f \subseteq f^* \wedge g \subseteq g^* \wedge h \subseteq h^* \wedge h \subseteq f \wedge h \subseteq g \wedge g \not\subset f\right) \supset \left(f \subseteq g\right)$

This is not true for the modal syllogistic, as the following model shows.

44.3		*
f	d_1, d_2	d_1, d_2
g	d_2, d_3	d_2, d_3
h	d_2	d_2

$h \subseteq f, h \subseteq g$ and $g \not\subset f$ but $f \not\subset g$.

It could be added in the shape of the axiom

$$\frac{a_L a^a \quad b_L b^a \quad c_L c^a \quad ac^a \quad bc^a \quad ab^o}{ba^a}$$

We have seen that Aristotle's modal syllogistic is compatible, on our metaphysical reading, with the realities being pervasive, and with the Kinds being uniformly determinate. Is it compatible with pervasiveness plus uniform determinacy? No. If every individual is required to belong to some star set, and in addition the depth of all star sets is uniform, then the white non-animal in Aristotle's counter-example to Baroco LXL must be in a star set that excludes the men. For if it is only in star sets that (i) include or (i) overlap the men then the realities are not uniform but are of greater depth (i) among the men or (i) in the overlap with the men. More generally, Baroco LXL will be valid given both pervasiveness and uniform determinacy: If $f_2 \subseteq f_1^*, f_3 \not\subset f_1$ then $f_3 \cap f_2^{¤} \neq \varnothing$

Proof By **44.1**, each member of f_3 is in some star set;
so take some member d_1 of f_3 and let it be in g^*.
By **44.2**, either (i) g^* excludes f_1^*
or (ii) some other star set h^* containing d_1 excludes f_1^*;
and in either case $f_3 \cap f_2^{¤} \neq \varnothing$.

Since Aristotle rejects Baroco LXL and accepts the uniform determinacy of Kinds, this constitutes indirect evidence that he did not the accept pervasiveness of realities.

With regard to LaQa-2, these premisses cannot both be true together, given uniformity and pervasiveness, if the middle term is pure (as it is in Aristotle's counter-example to Baroco LXL). But if the middle term is mixed, then given uniformity and pervasiveness, an Lo-conclusion will follow. This can be proved by an adaptation of our proof of Baroco LXL.

(3) Our treatment of sun sets as heterogeneous and mediated is consistent with, but does not require, the Aristotelian supposition that there are no 'negative species'. The validity of Celarent LXL requires that the Sun set Theorem (S3) holds. But S3 is logically independent of the existence of 'negative species'. S1.1, by contrast, while not being formally implied by our *AL* semantics, is implied by the principle that there are no negative realities:

Proof Suppose $\bar{f}^* = \varnothing$
then $\bar{f}^* \subseteq f^{¤}$.

(4) can be expressed as the **Quasi-Kinds Theorem**:

44.4 $$\left(f \not\subset f^* \wedge f^* \neq \varnothing\right) \supset \exists g_1, \ldots, g_n \left[g_1 \subseteq g_1^* \wedge \ldots \wedge g_n \subseteq g_n^* \wedge f^* = \bigcup_{i=1}^{n} g_i^* \right]$$

Quasi-Kinds are unions of Kinds. This is not a requirement of our semantics, as there are models in which a star set is not a union of pure star sets (even given uniform determinacy of Kinds).

44.4		*
f_1	d_1, d_2	d_1
f_2	d_2	d_2

The Quasi-Kinds Theorem is not to be confused with the unAristotelian idea that all unions of Kinds are themselves star sets:

$$\left(g_1 \subseteq g_1^* \wedge g_2 \subseteq g_2^*\right) \supset \exists f^*\left[f^* = \left(g_1^* \cup g_2^*\right)\right]$$

(5) We can sketch a theory of abstract entities and their relations with their *concreta* as follows. First, we expand the language of $E_Q 0-5$ to include an abstract term $\hat{v}$ for each general term v. Then we expand the semantics of §38-39, by defining a *Categories* model as an ordered septuple $\langle D, V, F, F^*, \hat{D}, \mu, \hat{F}\rangle$ where D, V, F, F^* are as before. D is intended to include only concrete individuals (*i.e.* substances). $\hat{D}$ is intended as a domain of abstract individuals (attributes). μ is a relation from members of D to members of $\hat{D}$, intended as the relation of *participation* linking concrete to abstract realities.[45] $\hat{F}$ is a non-empty set of possibly empty sets whose members are members of $\hat{D}$; it is intended as a set of Kinds of abstract realities.

An assignment of values to the variables in a wff associates with each such variable v an ordered pair $\langle \alpha_1, \alpha_2\rangle$ where $\alpha_1 \in F$, $\alpha_2 \in F^*$, and with $\hat{v}$ a possibly null set α_3 where $\alpha_3 \in \hat{F}$. In addition to **22.3** we adopt the constraint

44.5 If the pair $\langle \alpha_1, \alpha_2\rangle$ is assigned to v, and α_3 to $\hat{v}$ then $\forall d_1\left[d_1 \in \alpha_3 \supset \exists d_2, \alpha\left[d_2 \in \alpha \wedge \alpha \in F^* \wedge d_2 \mu d_1\right]\right]$

This formalises Aristotle's requirement that

All the other things are either said of the primary substances as subjects or are in them as subjects. (*Categories* 5, 2a34-35. Ackrill translation.)

In addition we may want to adopt one of the following:

44.6 $\alpha_1 = \alpha_2 \supset \alpha_3 = \varnothing$

44.7 $\alpha_1 = \alpha_2 \supset \alpha_3 \neq \varnothing$

44.8 $\forall d_1, d_2, d_3\left[\left(d_1 \mu d_3 \wedge d_2 \mu d_3\right) \supset d_1 = d_2\right]$.

[45] Aristotle uses this Platonic term *methexis* in his *Topics*. Participation is the converse of inherence.

The first of these reflects the view evident in the *Categories* that corresponding to a Kind of substance, such as horses, there is no abstract reality, equinity. The second reflects the view, seemingly implicit in *Metaphysics Z*, that there *are* such abstract realities as "what it was to be a horse". The third reflects the view, taken by some interpreters of the *Categories*, that abstract realities can be participated in by at most one concrete reality, *i.e.* that they are tropes.[46]

Further, we may wish to adopt the following:

44.9 If f is of the form g_L then $\hat{f} = \varnothing$

44.10 If f is of the form g_M then $\hat{f} = \varnothing$

44.11 If f is of the form g_Q then $\hat{f} = \varnothing$

These principles accord with Patterson's intuition that:[47]

> Aristotle ... simply does not, from a metaphysical point of view, believe in four different properties associated with, for example, the quality White ['being White', 'being (one-way) possibly White', etc.].

We then take it, on assigning $\langle f_1, f_1^* \rangle$ to a, $\hat{f}_1$ to $\hat{a}$, and d_1 to x, that

Lax is true iff $d_1 \in f_1^*$;

$L\hat{a}x$ is true iff $d_1 \in \hat{f}_1$;

Lab^a is true as before;

$L\hat{a}\hat{b}^a$ is true iff $\hat{f}_2 \subseteq \hat{f}_1$.

Frank A. Lewis makes the claim[48] that $\hat{f}_2 \subseteq \hat{f}_1$ iff

$$\exists d_1, g^*, h\left[d_1 \in g^* \wedge d_1 \mu h \wedge h \in f_2\right] \wedge$$
$$\forall d_2\left[\left\{\exists g^*\left[d_2 \in g^*\right] \wedge \exists d_3\left[d_3 \in f_2 \wedge d_2 \mu d_3\right]\right\} \supset \exists d_4\left[d_4 \in f_1 \wedge d_2 \mu d_4\right]\right]$$

This says that (i) if all whitenesses are colours then everything participating in a whiteness participates in a colour (which is formally true), and (ii) if everything

---

[46] This is the view of ACKRILL. It is discussed in LEWIS p.79 [*A3]. The contrasting position, which holds that though there are abstract individuals they may be participated in by a plurality of concrete individuals, is put by FURTH, and discussed by LEWIS pp.70-71, 82. Lewis himself does not think abstract individuals are recognised in the *Categories* ontology. Yet he wants there to be *relations* between the concrete and the abstract (including the converse of our relation μ). It seems to me that the best way to formalise such an ontology is to use a Leśniewskian, rather than a Fregean, logic, in the style of THOM 1986.

[47] PATTERSON 1995 p.32. Of course, Patterson's point does not count against our use, for semantic purposes, of *sets* of necessary-whites, possible-whites *etc.*

[48] LEWIS, p.75 (D2).

participating in a whiteness participates in a colour then all whitenesses are colours (which is not).

We can develop an essentialist metaphysics of substances and attributes one step further by letting singleton Kinds generate non-empty abstract realities. These will be haecceities (individual essences). A concrete Kind f^*, even if it has just one member, can generate an abstract Kind $\hat{f}$.

In order to construct an essentialist metaphysic from this semantic framework we would need to list the Kinds (both concrete and abstract) and the laws of nature, specifying (i) the extensional relations holding among the Kinds of substance, (ii) the extensional relations holding among the Kinds of attribute, (iii) the μ-relations holding between substances and attributes, Kind by Kind. For instance an Aristotelian essentialism might say that animals are a Kind of substance, and that sleep and waking are Kinds of attribute, and that attributes of these Kinds are peculiar to animals.

According to *Categories* semantics there are three metaphysically different cases in which ab^a is true.

(i) It might be that the *a*s and the *b*s are both Kinds or quasi-Kinds, the former including the latter (*All swans are animals, All whiteness is a colour*); in this case Lab^a is true.

(ii) It might be that the *a*s are a Kind or quasi-Kind of substance and the $\hat{b}$s a Kind of attribute belonging only to *a*s (*All sleepers are animals, All approaching are animals*); again Lab^a is true.

(iii) It might be that the $\hat{a}$s are a Kind of attribute and the *b*s a Kind or quasi-Kind of substance, all of which possess an attribute of the first Kind (*All swans are white, All swans are asleep*); Lab^a may or may not be true.

Categories semantics recognises laws of nature of three types, corresponding to these three groups of proposition; whereas Johnson-style semantics recognises only one type of natural law - that corresponding to group (i). But it is striking that both group (ii) and group (iii) here include laws of nature alongside accidental generalisations. No sufficient criterion, either semantic or syntactic, is given for distinguishing among propositions of group (ii) or (iii) those that are natural laws from those that are accidental generalisations. In particular, we still have no semantic or syntactic means of distinguishing necessary from contingent accidents: we are no better off than we were when we relied on the notion of a quasi-Kind. The upshot is that, though the *Categories* semantics is metaphysically more discriminating than our semantics of concrete realities and their Kinds, as a modal logic it comes to the exactly same thing.

§45. ARISTOTLE'S DYNAMIC ESSENTIALISM

(6) In §19.2 we introduced the idea of *dynamic* M- and Q-propositions, true by virtue of what the subject is or can become. We suggested that Aristotle's rejection of Barbara XQM because of the supposed truth of *All horses can move* results from a confusion between a dynamic and a static M-proposition, the latter being understood as contradictory to an L-proposition. What I want to do now is (i) give truth-conditions for dynamic M-forms, and (i) see whether *All horses can move* is true on those conditions. I approach these questions via an analysis of Aristotle's theory of natural change.

This theory takes some laws of nature to report natural changes. By a natural change I mean a change which is nomologically possible. When such a change occurs, an individual becomes something it previously was not. Basic natural changes are reported by laws of nature. There are also chains of basic changes. When an individual is in the state at the start of such a chain, it is *potentially* in the chain's end-state.

Relative to our system E_Q0, a *natural change* can be defined as an ordered quintuple $\langle m_1, m_2, d_1, f_1, f_2\rangle$, where
(i) m_1 and m_2 are E_Q0 models;
(i) $d_1 \in f_1, d_1 \notin f_2$ in m_1;[49]
(ii) $d_1 \in f_1, d_1 \in f_2$ in m_2.
Note that f_1 and f_2 may or may not be star sets.

Suppose there are two natural changes $\langle m_1, m_2, d_1, f_1, f_2\rangle$ and $\langle m_3, m_4, d_1, f_3, f_4\rangle$. These changes are *linked* iff $d_1 \in f_1, d_1 \in f_2$ in m_3. A sequence of such links forms a *chain*. A chain represents a possible pattern of development for members of the given set f_1. Then we could say that, relative to a given A_Q model m_1, an individual d_1 is a *potential* member of a set f_n iff there is a chain of natural changes, the first having the form $\langle m_1, m_2, d_1, f_1, f_2\rangle$ and the last having the form $\langle m_{n-1}, m_n, d_1, f_{n-1}, f_n\rangle$.[50]

A metaphysics might distinguish among natural changes a sub-class of *normal* changes. Such a distinction would generate two types of potentiality, broad and narrow. In the broad sense a potentiality exists if nature is such that a change of that type will occur given certain conditions, *i.e.* if a change of that type is merely

---

49 This, of course, is shorthand for saying that $d_1 \in f_1$ is true and $d_1 \in f_2$ false in m_1.

50 The models in the chain are analogous to (partial) possible worlds, and potentiality to truth in an accessible world; the accessibility relation is given by the constraints on linking.

possible. In the narrow sense a potentiality exists if nature is such that a change of this type is normal. Thus, relative to Aristotle's dynamic essentialism, humans are potentially grey-haired in the narrow sense; the *katamenia* is potentially human in the broad sense.[51]

In our definition of natural change, it may or may not be the case that $f_1 \subseteq f_1^*$ in both m_1 and m_2. If this is the case, the change represented is one in which a member of a Kind comes to be what it was not, while retaining its membership of that Kind. If the above condition is not met, but $f_2 \subseteq f_2^*$ in m_2, the change represented is one in which something not in a Kind comes to be so. The individual undergoing change is then potentially a member of that Kind. For Aristotle, the notion of being potentially a member of a Kind is the notion of broad potentiality. An egg is not a chicken but could develop into one. Such changes are nomologically possible rather than nomologically normal for members of the initial set. It would appear to be unnecessary to count chicken eggs themselves as a Kind. Thus a chicken egg is an unreal individual which is potentially (in the broad sense) a member of a Kind. This type of potentiality is characterised by the presence in a chain of at least one link of broad potentiality.

A13, 32b4-13 distinguishes two types of contingency-propositions - those stating what happens for the most part but not necessarily, and those expressing indeterminacy between being and not being. As examples of the first type Aristotle offers the statements that a man goes white, or grows, or gets smaller; these he calls 'natural' at 32b15-16. As examples of the second he offers the statement that an animal walks, or anything that happens by chance. Propositions of both kinds, he says (32b13-18) obey the laws of Q-conversion, but for different reasons. The natural obeys Q-conversion because it is not necessary, the indeterminate because there is no more reason why it should be one way rather than the other. He also says (32b18-22) that, though there is neither knowledge nor scientific syllogism regarding the indeterminate, there is regarding the natural. (This comment is pertinent, since the aim of the *Analytics* is to develop a logical theory for use in scientific demonstrations.) These two types of contingency-proposition correspond to a metaphysical difference in chains of natural change that depends on their end-states. Some end-states (going grey, in the case of humans) are such that chains of a certain temporal length, relating to a certain Kind, will have that end-state 'for the most part'. Other end-states (walking, for humans) are no more likely to occur than not to occur as the end-state of a chain of any given temporal length. It seems that Aristotle's 'natural contingencies' correspond to our narrow possibilities, and that his 'indeterminate contingencies' are included in our broad possibilities.

---

[51] For an illuminating discussion see FREELAND 1987 §2.

According to our definition, any horse is potentially in movement. Considering the initial state of any change a horse undergoes, there is a chain of natural changes leading from that initial state to the horse's being in movement. If there were not, the horse would be dead.

Potentiality in this sense - dynamic possibility - is distinct from the static possibility that is expressible in $E_Q 0-5$. The latter, though defined in terms of the static nature of Kinds, is an ephemeral matter, dependent on what is the case at a particular time. The potentiality of movement for a horse is not so dependent, but derives from the natural changes incident to horses in general (though other potentialities will vary with time). The difference is brought out by Johnson's example[52] of Max the non-chewing dog. Suppose that everything that is chewing is a bear and that Max the dog is necessarily not a bear; it does not follow, says Johnson, that Max is necessarily not chewing. We can certainly agree that it would be an unfortunate metaphysic which denied dogs the capacity for chewing at whatever times chewing happened to be confined to bears, on the basis that dogs and bears have incompatible natures. But notice that in saying this we assume (i) that there is an incompatibility between having the capacity to chew and being such as necessarily not to be chewing, and (ii) that such a capacity exists in dogs by virtue of their nature, just like the impossibility of their being bears. These assumptions can be sustained only by reading the predicate of *Dogs are possible chewers* intensionally so that the proposition asserts a compatibility between the canine and chewing natures. In other words, you need to assume that the possibility in question is a potentiality of dogs, attainable via a chain of natural changes. If you read it extensionally, then the possibility of dogs' being chewers depends on who is chewing. So long as the chewers include dogs, dogs are possible chewers. If the chewers include only bears then dogs are not possible chewers because they are not possible bears.

All this has been expressed within a concrete ontology of realities and Kinds. It would also be possible to construct a version of the theory of natural change based on an ontology of substances and attributes. This would be closer to Aristotle's own formulation in his *Physics*.

Clearly a theory of potentiality is sometimes suggested by Aristotle's use of examples in the modal syllogistic; but the theory is not well developed there and, to the extent that we supply that development, we come into conflict with the doctrine of modal equipollence.

---

52 JOHNSON 1993 p.179.

§46. THE METAPHYSICS OF UNQUALIFIED ASSERTORICS

The logic of unqualified propositions may, as we have argued, be captured by treating them as ampliated assertorics. Nonetheless, it cannot be maintained that the meaning of such propositions is fully reflected in this analysis. For, it is clearly part of that meaning (though not, as we have argued, the whole) that their truth-value does not fluctuate over time. But, the truth-value of an ampliated assertoric may vary over time, just as the membership of the class of possible *b*s may vary. It might be true of all currently possible men that they reside in our solar system, but not true of some future possible man.

Clearly, a full analysis of the meaning of unqualified propositions must have as a consequence that unqualifieds imply ampliated assertorics. Here is a suggestion: can pure apodeictics be identified with unqualifieds? No. Pure apodeictics are logically independent of ampliated assertorics.

(1) If every possible *b* is an *a*, it does not follow that every *b* is a necessary *a*. Given the antecedent, *nothing* may be a necessary *a*:

46.1		*	¤	†
f_1	d_1	–	–	d_1
f_2	d_1	–	–	d_1

$f_2^{\dagger} \subseteq f_1$ but $f_2 \not\subset f_1^{*}$.

(2) If every *b* is a necessary *a*, it does not follow that every possible *b* is an *a*. The consequent may be false simply because some non-*b* which is a possible *b* is unreal.

46.2		*	¤	†
f_1	d_1	d_1	–	d_1, d_2
f_2	d_1	–	–	d_1, d_2
f_3	d_2	–	–	d_1, d_2

$f_2 \subseteq f_1^{*}$ but $f_2^{\dagger} \not\subset f_1$.

(3) If every possible *b* is an *a*, it does not follow that every *a* is a necessary *a*. Consistent with the truth of the antecedent, *nothing* may be a necessary *a*. See model 46.1.

(4) If every possible *b* is an *a*, it does not follow that every *b* is a necessary *b*. Clearly, the antecedent would be true and the consequent false if the *b*s' star set were properly included among the *b*s while their sun set was identical with the non-*b*s and included in the *a*s' star set, as shown in Figure 46.1.

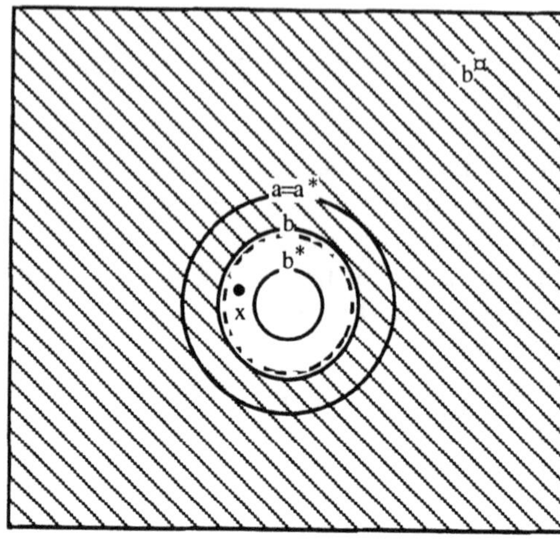

Fig. 46.1

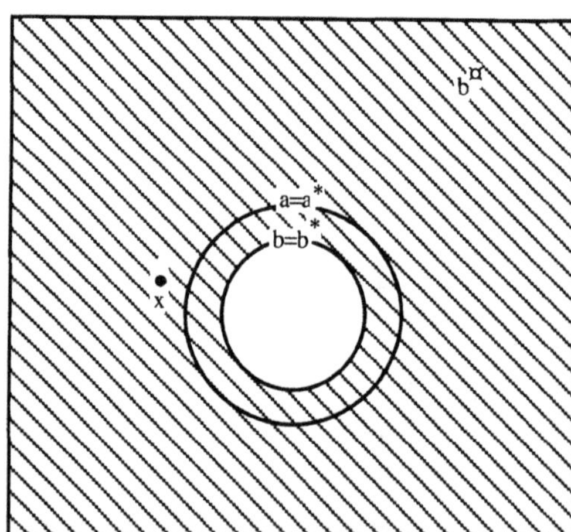

Figure 46.2

x in Figure 46.1 represents a b which is not a necessary b when all possible bs are bs (since all non-bs are impossible bs). Formally:

46.3		*	¤	†
f_1	d_1, d_2	d_1, d_2	d_3	d_1, d_2
f_2	d_1, d_2	d_2	d_3	d_1, d_2
f_3	d_3	d_3	d_1, d_2	d_3

$f_2^{\dagger} \subseteq f_1$ but $f_2 \not\subset f_2^{*}$.

(5) If every b is an a and every a is a necessary a and every b is a necessary b, it does not follow that every possible b is an a. Consistent with the truth of the antecedent, the consequent may be false because some individual (like x in Figure 46.2) is unreal. Formally:

46.4		*	¤	†
f_1	d_1	d_1	–	d_1, d_2
f_2	d_1	d_1	–	d_1, d_2
f_3	d_2	–	–	d_1, d_2

$f_2 \subseteq f_1, f_1 \subseteq f_1^{*}, f_2 \subseteq f_2^{*}$ but $f_2^{\dagger} \not\subset f_1$.

Could, then, the meaning of an unqualified proposition be captured by treating such a proposition as a conjunction of an ampliated assertoric and a pure apodeictic? No. Such a conjunction can vary its truth-value over time if one of its conjuncts can do so.

Something stronger is needed. We need to use not just pure, but logically pure, apodeictics. I mean propositions *Every (no) b is a* where it is not just true but logically true that all *b*s are (metaphysically) necessary *b*s and that all *a*s are metaphysically necessary *a*s. These propositions do not change their truth-value over time. And they seem to be the core propositions of Aristotelian essentialism, the non-derivative metaphysical necessities. But logically pure, like pure apodeictics in general, are logically independent of ampliated assertorics. So, ampliated assertorics have to be expressly written into the definition of unqualifieds. If Aristotle's aim is to develop a theory of unqualified assertorics which simultaneously gives the right logical results and captures the core class of internal apodeictics, then it seems that the only viable reading of those propositions is the conjunction of a logically pure apodeictic with an ampliated assertoric.

This reading works for negative unqualifieds. Such propositions will then be just those Le-propositions asserted by the metaphysical theory, both of whose terms are asserted by the theory to be pure. The second conjunct in our definition will be redundant because of the interchangeability of X'- and Le-forms. And Aristotle's example, *No raven is a reasoner*, is plausibly true if the reasoners are coextensive for Aristotle with the men.

But for affirmative unqualifieds, a serious problem arises from the fact that the required conjunctions will never be true. Consider the conjunction *All men are necessarily animals and all possible men are animals.* On Aristotelian principles, whereby some individuals are not necessarily anything, and all animals are necessary animals, the second conjunct will never be true. For, there will always be an unreal individual which, not being necessarily anything, is a possible man; and this individual will not be an animal. Thus, it seems that Aristotle's attempt to sketch a theory of unqualified assertorics is doomed to failure. When we combine the feature they must have to possess the logic he attributes to them with the feature they must have to possess the metaphysical significance he would like them to have, we are left with a class of propositions none of which are true. No wonder there are no examples in the text of affirmative unqualifieds.

§47. ARISTOTELIAN VS. PLATONIC ESSENTIALISM

The essentialism that is formally implicit in Aristotle's modal syllogistic is a static one. One might wonder, then, whether it is a metaphysical position more in tune with Plato's than with Aristotle's later (dynamic) essentialism. In this section, I put this doubt to rest.

Plato's essentialism agrees with Aristotle's in rejecting negative Kinds. According to Pelletier,[53] Plato rejected negative Kinds in the *Sophist*, taking the truth-makers of negative statements to include only positive realities and a relation of difference.

In terms of the formal relations between concrete and abstract realities, Platonism makes five claims: (i) that there are no concrete realities, (ii) that an abstract reality may have nothing participating in it, (iii) that an abstract reality may have several things participating in it, (iv) that all Kinds are singletons, and (v) that there are no quasi-Kinds. (The type of formal theory we are considering takes no account of the *content* of the participation relation. Platonism, in at least one of its versions, requires that that relation be a similarity relation. The beautiful things participate in Beauty because they are, more or less remotely, similar to it.)

The first three of these points are clearly Platonic doctrine. (i) is clear from the fact that for Plato the Forms are the sole realities and they are abstract. (ii) is clear. (iii) is clear from the fact that each Form is supposed to be a 'one over the many'. (iv) is Aristotle's pet accusation against the Forms that they are treated as 'this's: see for instance *Metaphysics*Z14, 1039b2-4, Z15, 1040a8-9, M9, 1086a32-34. That his point is well taken is suggested by a number of passages. Plato states (*Republic* 597c, *cf. Timaeus* 31a) that God could not have made more than one

[53] PELLETIER pp.50-52.

Form of bed. The 'third man argument' (*Parmenides* 132a) takes a Form to be one thing alongside the concrete particulars that participate in it. *Sophist* 255a argues that if Motion were Difference then since Rest participates in Difference, Rest would be participate in Motion (*i.e.* Rest would be moving); and this argument is valid only if it is assumed that at most one thing is Difference. Again, *Sophist* 256d argues that what is not is, since Motion both is and is not; and this argument is valid only if it is assumed that at most one thing is Motion. In general with regard to the 'interweaving of Forms' in the *Sophist*, it is noteworthy that Plato never suggests that there is a Form of Form, or of other Kinds we might suppose the Forms to belong to; he only mentions Forms in which Forms *participate*. This silence is consistent with a doctrine that there are no many-membered Kinds to which Forms belong. (v) is clear from the fact that Platonic essentialism is what we may call a technical metaphysics. We have already seen that Aristotelian essentialism countenances mixed terms, terms having a non-null star set which is properly included in their non-star set. By contrast some metaphysical theories require that each star set either is null or is identical with the corresponding non-star set. A *technical* metaphysical language divides into a non-metaphysical part (whose star sets are null) and a metaphysical part (whose star sets are identical with their corresponding non-star sets) in accordance with the principle:

47.1 $f \subseteq f^* \vee f^* = \varnothing$

If our theory acknowledges abstract realities, we will have to add

47.2 $f \subseteq \hat{f} \vee f = \varnothing$

The characteristic axiom for such a metaphysic is:

$$\frac{Lax \quad ay}{Lay}$$ [54]

This requirement is often satisfied when a metaphysical theory is entirely expressed in a technical metaphysical language specially devised to capture the realities, a 'perfect language'.

In terms of the theory of abstract realities outlined in §44, (i) means that α_2 is always null. (i) means that **44.5** - the Aristotelian requirement that each abstract reality be participated in by a concrete reality - is rejected. (ii) can be formulated:

47.3 $\left(x \in f^* \wedge y \in f^*\right) \supset x = y$

(iv) can be formulated as the axiom

[54] This is the thought expressed in THOM 1991, and incorporated in JOHNSON 1993. (It is not true for all terms, according to Aristotelian metaphysics, but the ones for which it is false may be regarded as secondary to those for which it is true.)

$$\frac{L\hat{a}b^a \quad L\hat{a}x}{bx}$$

or more generally

Plato $$\frac{Lab^a \quad Lax}{bx}$$

The Platonic realities, the Forms, are not only pure but are singletons.

The effect of adding this last axiom to our system is drastic. It introduces modal theses that have no assertoric counterpart, for instance LaLaXa-2 $\dfrac{Lab^a \quad Lac^a}{bc^a}$:

Proof

$$\text{Plato}\,\frac{Lab^a \qquad \text{La}-\text{elim}\,\dfrac{Lac^a \quad cx}{Lax}}{bx} \quad \xrightarrow{a-\text{intro}} \quad \frac{Lab^a \quad Lac^a}{bc^a}$$

Aristotelian essentialism differs from Platonism in (i) recognising concrete realities, (i) requiring that abstract realities have some concrete realitity participating in them, (ii) denying that an abstract reality may have several things participating in it, (iv) denying that all Kinds are singletons and (v) recognising quasi-Kinds. The essentialism that is implicit in the modal syllogistic differs importantly from Platonism in respect of (iv) and (v).

§48. CONCLUSIONS

The syllogistic logic of *de re* modalised propositions can be elaborated in a family of inter-related formal systems, all of them treatable in a semantically uniform manner, all of them reducible to singular modal systems, and all decidable by reference to non-modal syllogistic with modalised terms. The most important features differentiating one member of the family from another are (i) the assumptions made regarding the relation between impossibility and necessity (the sun set definitions), (i) the way in which modal categoricals are constructed from modal singulars (the superstructure), and (ii) the existential assumptions made by different members of the family.

Aristotle's modal syllogistic is a member of this family - its first but not its best-looking. Buridan's modal syllogistic has a stronger claim to the latter title. But, though the Aristotelian exposition has its imperfections, there is a small number of alternative ways in which one can make overall sense of it. I have argued that one of these interpretations emerges as the best, if one takes a broad view of the data for interpretation, so that they include the ecthetic proofs, the counter-examples, and the metaphysical context. The best interpretation defines the impossible as in **25.5**, uses superstructure *A*, and makes no existential assumptions.

There is an essentialism implicit in Aristotle's modal syllogistic on this interpretation, and this essentialism is what motivated the whole project. But it is not an essentialism that without qualification can be identified either as Aristotle's or as Plato's. It is a metaphysics that agrees with Plato's essentialism in its atemporality but diverges from Platonic thought on the issue of singleton Kinds. On the other hand it agrees with the general features of Aristotle's essentialism while not incorporating his doctrine of uniform determinacy, his doctrine of abstract entities, or his theory of natural change.

Io non posso ritrar di tutti appieno;
perocchè sì mi caccia il lungo tema ....[55]

[55] DANTE, *Inferno* 4.145-146.

REFERENCES

ACKRILL. J.L. Ackrill, *Aristotle's Categories and De Interpretatione* (Oxford 1963)

ALEXANDER. Alexander of Aphrodisias, *In Aristotelis Analyticorum Priorum Librum I Commentarium* ed. M. Wallies (Commentaria in Aristotelem Graeca; Berlin 1883)

ANGELELLI. Ignacio Angelelli, 'The Aristotelian modal syllogistic in modern modal logic', in *Konstruktionen versus Positionen, Beiträge zur Diskussion um die Konstruktive Wissenschaftstheorie herausgegeben von Kuno Lorenz* (Walter de Gruyter, Berlin & New York 1979)

ANSELM. Desmond Paul Henry, *Commentary on De Grammatico* (Dordrecht: Reidel 1974)

BARNES. Jonathan Barnes, *Early Greek Philosophy* (Penguin 1987)

BECKER. Albrecht Becker, *Die Aristotelische Theorie der Möglichkeitsschlüsse: eine logisch-philologische Untersuchung der Kapitel 13-22 von Aristoteles Analytica priora I* (Berlin: Junker und Dünnhaupt 1933)

BOCHEŃSKI. I.M. Bochenski, *La logique de Théophraste* (Librairie de l'université Fribourg en Suisse, 1947)

BRANDT. Reinhard Brandt, 'Zur aristotelischen Modallogik', *Hermes* **94** (1966) 113-115

BRENNER. Holger Brenner, 'Eine vollständige Formalisierung der Aristotelischen Notwendigkeitssyllogistik', *Arbeitsberichte des Arbeitskreises Mathematische Logik* **1** (1993) 1-19

BROGAN. A.P. Brogan, 'Aristotle's logic of statements about contingency', *Mind* **76** (1967) 49-61

BURIDAN. *Jean Buridan's Logic: the treatise on supposition, the treatise on consequences, translated with a philosophical introduction by Peter King* (Dordrecht: Reidel 1985)

BURKHARDT & SMITH. *Handbook of Metaphysics and Ontology* ed. Hans Burkhardt and Barry Smith 2 vols. (Munich: Philosophia 1991)

CAMPSALL. Richard of Campsall, *Questiones super librum Priorum Analeticorum* ed. Edward A. Synan (Toronto: Pontifical Institute of Mediaeval Studies 1968)

COCCHIARELLA 1976. Nino Cocchiarella, 'On the logic of natural kinds', *Philosophy of science* **43** (1976) 202-222

COCCHIARELLA 1977. Nino Cocchiarella, 'Sortals, natural kinds and re-identification', *Logique et analyse* **20** (1977) 439-474

FEREJOHN. Michael Ferejohn, *The origins of Aristotelian science* (New Haven: Yale University Press, 1991).

FLANNERY. K.L. Flannery, 'Alexander of Aphrodisias and others on a controversial demonstration in Aristotle's modal syllogistic', *History and Philosophy of Logic* **14** (1993) 201-214

FREDDOSO. *Ockham's Theory of Propositions: Part II of the Summa Logicae translated by Alfred J. Freddoso and Henry Schuurman, introduction by Alfred J. Freddoso* (University of Notre Dame Press 1980)

FREELAND 1987. Cynthia A. Freeland, 'Aristotle on bodies, matter, and potentiality', in Allan Gotthelf & James G. Lennox (eds.), *Philosophical issues in Aristotle's biology* (Cambridge University Press 1987)

FREELAND 1994. Cynthia A. Freeland, 'Nourishing speculation: a feminist reading of Aristotelian science', in Bat-Ami Bar On (ed.), *Engendering origins: critical feminist readings in Plato and Aristotle* (Albany: State University of New York Press) 145-187

FURTH. Montgomery Furth, *Substance, Form and Psyche: an Aristotelian metaphysics* (Cambridge 1988)

GEACH. P.T. Geach, Review of McCALL 1963, *Ratio* **6** (1964) 200-206

GERSONIDES. *The Logic of Gersonides: a translation of Sefer ha-Heqqesh ha-yashar (The Book of the correct syllogism) of Rabbi Levi ben Gershom with introduction, commentary, and analytical glossary by Charles H. Manekin* (Dordrecht: Kluwer 1992)

HENLE. Paul Henle, 'On the fourth figure of the syllogism', *Philosophy of Science* **16** (1949) 94-104

HINTIKKA. J. Hintikka, *Time and Necessity: studies in Aristotle's theory of modality* (Oxford 1973)

HUGHES & CRESSWELL. G.E. HUGHES, and M.J. CRESSWELL, *An Introduction to Modal Logic* (London: Methuen 1968)

JOHNSON 1989. Fred Johnson, 'Models for modal syllogisms', *Notre Dame Journal of Formal Logic* **30**, 271-284

JOHNSON 1993. Fred Johnson, 'Modal ecthesis', *History and Philosophy of Logic* **14**, 171-182

JOHNSON 1995. Fred Johnson, 'Apodeictic syllogisms: deductions and decision procedures', *History and Philosophy of Logic* **16**, 1-18

KNUUTTILA. Simo Knuuttila, *Modalities in Medieval Philosophy* (London: Routledge 1993)

KOSMAN. L.A. Kosman, 'Aristotle on inconvertible modal propositions', *Mind* **79** (1970) 254-258

LEWIS. Frank A. Lewis, *Substance and Predication in Aristotle* (Cambridge University Press 1991)

ŁUKASIEWICZ. Jan Łukasiewicz, *Aristotle's Syllogistic from the standpoint of modern formal logic* 2nd. ed. (Oxford: Clarendon 1957)

MARSILIUS. Marsilius of Inghen, *Quaestiones super libros Priorum Analyticorum* (Venice 1516)

McCALL 1963. Storrs McCall, *Aristotle's Modal Syllogisms* (Amsterdam: North-Holland)

McCALL 1994. Storrs McCall, *A Model of the Universe: space-time, probability and decision* (Oxford: Clarendon)

MIGNUCCI 1969. Mario Mignucci, *Aristotele: Gli analitici primi* (Naples)

MIGNUCCI 1991. Mario Mignucci, 'Expository proofs in Aristotle's syllogistic', *Oxford Studies in Ancient Philosophy Supplementary volume: Aristotle and the later tradition* ed. Henry Blumenthal and Howard Robinson (Oxford: Clarendon), 9-28

NIETZSCHE. Friedrich Nietzsche, *The Twilight of the Idols, or How to Philosophise with the Hammer* translated by Anthony M. Ludovici (Edinburgh: Foulis 1911)

NORTMANN 1990. Ulrich Nortmann, 'Über die Stärke der Aristotelischen Modallogik', *Erkenntnis* **32**, 61-82

NORTMANN 1994. Ulrich Nortmann, 'Does Aristotle's modal logic rest on metaphysical assumptions?', *Analyomen* **1**, Proceedings of the 1st conference 'Perspectives in Analytical Philosophy', ed. Georg Meggle and Ulla Wessels (Berlin: Walter de Gruyter 1994)

OCKHAM. William of Ockham, *Summa Logicae* ed. Philotheus Boehner, Gedeon Gál, Stephanus Brown (St Bonaventure NY 1974)

PATTERSON 1989. Richard Patterson, 'The case of the two Barbaras: basic approaches to Aristotle's modal logic', *Oxford Studies in Ancient Philosophy* **7**, 1-40

PATTERSON 1990. Richard Patterson, 'Conversion principles and the basis of Aristotle's modal logic', *History and Philosophy of Logic* **11**, 151-172

PATTERSON 1995. Richard Patterson, *Aristotle's Modal Logic: essence and entailment in the Organon* (Cambridge University Press 1995)

PATZIG. Günther Patzig, *Aristotle's Theory of the Syllogism: a logico-philological study of Book A of the Prior Analytics*, translated from the German by Jonathan Barnes (Dordrecht 1968)

PELLETIER. F.J. Pelletier, *Parmenides, Plato, and the Semantics of Not-Being* (University of Chicago Press 1990)

PLANTINGA. Alvin Plantinga, 'Essentialism', in BURKHARDT & SMITH vol.1, pp.252-255

PUTNAM. Hilary Putnam, *Realism with a Human Face*, edited by James Conant (Cambridge Massachusetts and London: Harvard University Press 1990)

QUINE. W.V.O. Quine, 'Natural kinds', in *Ontological relativity & other essays* (New York: Columbia University Press, 1969)

ROSS. *Aristotle's Prior and Posterior Analytics, a revised text with introduction and commentary by W.D. Ross* (Oxford: Clarendon 1949)

SCHMIDT. Klaus. J. Schmidt, 'Eine modal prädikatenlogische Interpretation der modalen Syllogistik des Aristoteles', *Phronesis* **34** (1989) 80-106

SCOTUS. (Pseudo) John Duns Scotus, *In libros Priorum Analyticorum Aristotelis quaestiones, in Ioannis Duns Scoti Doctoris subtilis ordinis Minorum opera omnia ed. L. Wadding* (Lyon 1639) vol.I

SMILEY. Timothy Smiley, 'What is a syllogism?', *Journal of Philosophical Logic* **2** (1973) 136-154

SMITH. Robin Smith, *Aristotle, Prior Analytics translated with introduction notes and commentary* (Indianapolis: Hackett 1989)

SPRIGGE. T.L.S. Sprigge, 'Spinoza, Benedict (Baruch)', in BURKHARDT & SMITH vol.2 p.856

STAHL 1976. Gérold Stahl, 'Une formalisation de quelques syllogismes modaux', *Logique et Analyse* **19**, 175-217

STAHL 1977. Gérold Stahl, 'Indications formelles sur les syllogismes avec «contingent»', *Logique et analyse* **20**, 199-220

STRIKER. Gisela Striker, 'Modal vs. assertoric syllogistic' in *Logic, Dialectic and Science in Aristotle, special issue of Ancient Philosophy ed. Robert Bolton and Robin Smith* (1994), 39-51.

THOM 1981. Paul Thom, *The Syllogism* (Munich: Philosophia)

THOM 1986. Paul Thom, 'A Lesniewskian Reading of Ancient Ontology: Parmenides to Democritus', *History and Philosophy of Logic* **7**, 155-166

THOM 1991. Paul Thom, 'The two Barbaras', *History and Philosophy of Logic* **12**, 135-149

THOM 1993. Paul Thom, 'Apodeictic ecthesis', *Notre Dame Journal of Formal Logic* **34**, 193-208

THOM 1994. Paul Thom, 'Interpreting Aristotle's contingency syllogistic', *Oxford Studies in Ancient Philosophy* **12**, 91-109

THOMASON. S.K. Thomason, 'Semantic analysis of the modal syllogistic', *Journal of Philosophical Logic* **22** (1993) 111-128

TREDENNICK. *Prior Analytics translated by Hugh Tredennick, in The Organon I: The Categories, On Interpretation, Prior Analytics* (Cambridge Mass.: Harvard University Press 1938)

TRICOT. J. Tricot, *Aristote, Les Premiers Analytiques, traduction nouvelle et notes* (Paris: Vrin 1966)

VAN RIJEN. Jeroen van Rijen, *Aspects of Aristotle's Logic of Modalities* (Dordrecht: Reidel 1989)

VLASTOS. Gregory Vlastos, 'A metaphysical paradox' in *Platonic studies* 2nd ed. (Princeton University Press 1981)

WAITZ. Theodor Waitz, *Aristoteles, Organon Graece novis codicum auxiliis adiutus recognovit, scholiis ineditis et commentario instruxit* 2 vols. (Leipzig 1844)

WATERLOW. Sarah Waterlow, *Nature, Change and Agency in Aristotle's Physics: a philosophical study* (Oxford: Clarendon Press 1982)

WIELAND 1966. Wolfgand Wieland, 'Die Aristotelische Theorie der Notwendigkeitsschlüsse', *Phronesis* **11**, 35-60

WIELAND 1975. Wolfgang Wieland, 'Die Aristotelische Theorie der Syllogismen mit modal gemischten Prämissen', *Phronesis* **20**, 77-92

WIELAND 1980. Wolfgang Wieland, 'Die aristotelische Theorie der Konversion von Modalaussagen', *Phronesis* **25**, 109-116.

INDEX LOCORUM

INDEX OF AXIOMS

INDEX OF RULES

NAME	*FORMULA*	*SECTION*
a-intro	$\frac{R \quad bx}{ax} \longrightarrow \frac{R}{ab^{a}}$ provided R has some member $(L)bv^{a(i)}$ or $(L)vb^{a(i)}$	38.1
a-upgrading 1	$\frac{R \quad X'ab^{a} \quad bc^{a}}{p} \longrightarrow \frac{R \quad X'ab^{a} \quad Qbc^{a}}{p}$	37
a-upgrading 2	$\frac{R \quad X'ab^{e} \quad bc^{a}}{p} \longrightarrow \frac{R \quad X'ab^{e} \quad Qbc^{a}}{p}$	37
C	$\frac{p \quad R}{q} \longrightarrow \frac{\sim q \quad R}{\sim p}$	3.3
C'	$\left.\begin{array}{l} \frac{\sim r \quad s}{\sim p} \\ NOT\, \frac{q}{\sim s} \end{array}\right\} \longrightarrow \frac{p \quad q}{r}$	5
Cont	$\frac{p \quad p \quad R}{q} \longrightarrow \frac{p \quad R}{q}$	4.1
e-intro	$\frac{R \quad bx}{\sim ax} \longrightarrow \frac{R}{ab^{e}}$	38.1
i-ecthesis	$\frac{R \quad ad^{a} \quad bd^{a}}{p} \longrightarrow \frac{R \quad ab^{i}}{p}$	4
i-elim	$\frac{R \quad ax \quad bx}{p} \longrightarrow \frac{R \quad ab^{i}}{p}$ provided R has some member $(L)bv^{a(i)}$ or $(L)vb^{a(i)}$	38.1
i-elim T	$\frac{R \quad ax \quad Lbx}{p} \longrightarrow \frac{R \quad ab^{i}}{p}$	40.2
K	$\frac{p \quad R}{q} \longrightarrow \frac{\neg q \quad R}{\sim p}$	3.3
K'	$\left.\begin{array}{l} \frac{\sim r \quad s}{\neg p} \\ NOT\, \frac{q}{\sim s} \end{array}\right\} \longrightarrow \frac{p \quad q}{r}$	5
L	$\frac{Lp \quad Lq}{Lr} \longrightarrow \frac{p \quad q}{r}$	14
L1	$\left.\begin{array}{l} \frac{p}{q} \\ \Box p \end{array}\right\} \longrightarrow \Box q$	5

o-elim	$\frac{R \quad {\sim}ax \quad bx}{p} \longrightarrow \frac{R \quad ab^{o}}{p}$ provided R has some member $(L)bv^{a(i)}$ or $(L)vb^{a(i)}$	38
Q'a-intro	$\frac{R \quad Mbx}{Qax} \longrightarrow \frac{R}{Q'ab^{a}}$	39, 40.1
Q'i-elim	$\left.\begin{array}{c}\frac{R \quad Qax \quad Mbx}{p} \\ \\ \frac{R \quad Max \quad Qbx}{p}\end{array}\right\} \longrightarrow \frac{R \quad Q'ab^{i}}{p}$	39, 40.1
Qa-intro A	$\frac{R \quad bx}{Qax} \longrightarrow \frac{R}{Qab^{a}}$ provided R has some member $(L)bv^{a(i)}$ or $(L)vb^{a(i)}$	39, 40.1
Qa-intro AA	= Qa-intro A	40.2
Qa-intro C	$\frac{R \quad Mbx}{Qax} \longrightarrow \frac{R}{Qab^{a}}$	40.3
Qa-intro CC	= Qa-intro A	40.4
Qi-ecthesis AA	$\frac{R \quad Qad^{a} \quad bd^{a}}{p} \longrightarrow \frac{R \quad Qab^{i}}{p}$	32.2
Q"i-ecthesis	$\frac{R \quad Q''ad^{a} \quad Q''bd^{a}}{p} \longrightarrow \frac{R \quad Q''ab^{i}}{p}$	32.4
Qi-elim A	$\left.\begin{array}{c}\frac{R \quad Qax \quad bx}{p} \\ \\ \frac{R \quad Qby \quad ay}{p}\end{array}\right\} \longrightarrow \frac{Q \quad Qab^{i}}{p}$	39, 40.1
Qi-elim AA	$\frac{R \quad Qax \quad bx}{p} \longrightarrow \frac{R \quad Qab^{i}}{p}$	40.2
Qi-elim C	$\frac{R \quad Qax \quad Mbx}{p} \longrightarrow \frac{R \quad Qab^{i}}{p}$	40.3
Qi-elim CC	= Qi-elim AA	40.4
R_{EL}	$\frac{R \quad aa_{L}{}^{a}}{p} \longrightarrow \frac{R}{p}$ provided R has some member $va_{L}{}^{a}, va_{L}{}^{i}, a_{L}v^{a}, a_{L}v^{i}$ or $a_{L}\xi$	41.1
R_{EM}	$\frac{R \quad a_{M}a^{a}}{p} \longrightarrow \frac{R}{p}$ provided R has some member $va^{a}, va^{i}, av^{a}, av^{i}$ or $a\xi$	41.1

R_{EQM}	$\frac{R \quad a_M a_Q{}^a}{p} \longrightarrow \frac{R}{p}$ provided R has some member $va_Q{}^a, va_Q{}^i, a_Q v^a, a_Q v^i$ or $a_Q\xi$	41.1
R_L	$\frac{R \quad aa_L{}^a}{p} \longrightarrow \frac{R}{p}$ provided R has some member $va_L{}^a, va_L{}^i, a_L v^a$ or $a_L v^i$	24
R_M	$\frac{R \quad a_M a^a}{p} \longrightarrow \frac{R}{p}$ provided R has some member va^a, va^i, av^a or av^i	24
R_{QL}	$\frac{R \quad a_L a_Q{}^e}{p} \longrightarrow \frac{R}{p}$	31, 41.1
R_{QM}	$\frac{R \quad a_M a_Q{}^a}{p} \longrightarrow \frac{R}{p}$ provided R has some member $va_Q{}^a, va_Q{}^i, a_Q v^a$ or $a_Q v^i$	31
T	$\left.\begin{array}{l}\frac{R}{q}\\ \\ \frac{q \quad S}{r}\end{array}\right\} \longrightarrow \frac{R \quad S}{r}$	23
T_1	$\left.\begin{array}{l}\frac{p \quad R}{q}\\ \\ \frac{q \quad S}{r}\end{array}\right\} \longrightarrow \frac{p \quad R \quad S}{r}$	3.4
U	If x is a thesis, and y is a proper or improper substitution in x, then y is a thesis	23
U*	If x is a thesis, and y is an improper substitution in x, then y is a thesis	3.1
U-elim 2	$\frac{R \quad L\bar{d}a^a \quad Ldx}{p} \longrightarrow \frac{R \quad Uax}{p}$	40
U-elim 3	$\frac{R \quad ad^e \quad Ldx}{p} \longrightarrow \frac{R \quad Uax}{p}$	40
U-elim 5	$\frac{R \quad Lda^a \quad de^e \quad Lex}{p} \longrightarrow \frac{R \quad Uax}{p}$	38.1
U_t	Substitution of any term t for a variable v in a formula's a- or i-conclusion only if after the substitution there is an a- or i-premiss containing t.	38

INDEX OF SYSTEMS

*CCEQ*0	rep, a-subalt, a-elim, e-elim 1, e-elim-2, L-rep, U-rep, Q-rep, La-subord, Le-subord, Li-subord, Lo-subord, La-elim *CC*, Le-elim *CC*, Li-intro *CC*, Lo-intro *CC*, Qa-elim *CC*, Qi-intro *CC*, Barbara LXL, Celarent LXL, Darii LXL, Ferio LXL, Q-intro, Barbara QXQ, Darii QXQ, Cesare QLX	U, M, C, K, T, i-elim, o-elim, La-intro *CC*, Le-intro *CC*, Li-elim *CC*, Lo-elim *CC*, Qa-intro *CC*, Qi-elim *CC*	40.4
*CCEQL*1	As for *CCEQL*0 plus U-intro 1	As for *CCEQL*0 plus U-elim 1	40.4
*CCEQL*2	As for *CCEQL*0 plus U-intro 2	As for *CCEQL*0 plus U-elim 2	40.4
*CCEQL*3	As for *CCEQL*0 plus U-intro 3	As for *CCEQL*0 plus U-elim 3	40.4
*CCEQL*5	As for *CCEQL*0 plus U-intro 5	As for *CCEQL*0 plus U-elim 5	40.4
*CEQ*0	rep, a-subalt, a-elim, e-elim1, e-elim-2, L-rep, U-rep, Q-rep, Q-intro, Darii QXQ, La-elim *C*, Le-elim *C*, Li-introC, Lo-intro *C*, Qa-elim *C*, Qi-intro *C* Barbara QMQ, Cesare QLL	U, M, C, K, T, i-elim, o-elim, La-intro *C*, Le-intro *C*, Li-elim *C*, Lo-elim *C* Qa-intro *C*, Qi-elim *C*	40.3
*CEQ*1	As for *CEQL*0 plus U-intro 1	As for *CEQ*0 plus U-elim 1	40.3
*CEQ*2	As for *CEQ*0 plus U-intro 2	As for *CEQ*0 plus U-elim 2	40.3
*CEQ*3	As for *CEQ*0 plus U-into 3	As for *CEQ*0 plus U-elim 3	40.3
*CEQ*5	As for *CEQ*0 plus U-intro 5	As for *CEQ*0 plus U-elim 5	40.3
E_Q0	a-identity, rep, a-subalt, a-elim, e-elim 1, e-elim 2	U_t, M, C, K, T, i-elim, o-elim, R_{EL}, R_{EM}, R_{EQM}, R_{QL}	41.1
E_Q1	As for E_Q*0* plus M-intro 1,M-elim 1	As for E_Q*0*	41.3
E_Q2	As for E_Q0 plus M-elim 2	As for E_Q0 plus M-intro 2	41.3
E_Q3	As for E_Q0 plus M-elim 3	As for E_Q0 plus M-intro 3	41.3
E_Q4	As for E_Q0 plus M-elim 4	As for E_Q0 plus M-intro 4	41.3
E_Q5	As for E_Q0 plus M-elim 5	As for E_Q0 plus M-intro 5	41.3

GENERAL INDEX

Synthese Historical Library

Texts and Studies in the History of Logic and Philosophy

Series Editor: Simo Knuuttila (*University of Helsinki*)

1. M.T. Beonio-Brocchieri Fumagalli: *The Logic of Abelard.* Translated from Italian by S. Pleasance. 1969 ISBN 90-277-0068-0
2. G. W. Leibniz: *Philosophical Papers and Letters.* A Selection, translated and edited, with an Introduction, by L. E. Loemker. 2nd ed., 2nd printing. 1976 ISBN 90-277-0008-8
3. E. Mally: *Logische Schriften.* Grosses Logikfragment – Grundgesetze des Sollens. Herausgegeben von K. Wolf und P. Weingartner. 1971 ISBN 90-277-0174-1
4. L. W. Beck (ed.): *Proceedings of the Third International Kant Congress.* 1972 ISBN 90-277-0188-1
5. B. Bolzano: *Theory of Science.* A Selection with an Introduction by J. Berg. Translated from German by B. Terrell. 1973 ISBN 90-277-0248-9
6. J. M. E. Moravcsik (ed.): *Patterns in Plato's Thought.* 1973 ISBN 90-277-0286-1
7. Avicenna: *The Propositional Logic.* A Translation from *Al-Shifā': al-Qiyās*, with Introduction, Commentary and Glossary by N. Shehaby. 1973 ISBN 90-277-0360-4
8. D. P. Henry: *Commentary on* De Grammatico. *The Historical-Logical Dimensions of a Dialogue of St. Anselms's.* 1974 ISBN 90-277-0382-5
9. J. Corcoran (ed.): *Ancient Logic and its Modern Interpretations.* 1974 ISBN 90-277-0395-7
10. E. M. Barth: *The Logic of the Articles in Traditional Philosophy.* A Contribution to the Study of Conceptual Structures. 1974 ISBN 90-277-0350-7
11. J. Hintikka: *Knowledge and the Known.* Historical Perspectives in Epistemology. 1974 ISBN 90-277-0455-4
12. E. J. Ashworth: *Language and Logic in the Post-Medieval Period.* 1974 ISBN 90-277-0464-3
13. Aristotle: *The Nicomachean Ethics.* Translation with Commentaries and Glossary by H. G. Apostle. 1974 ISBN 90-277-0569-0
14. R. M. Dancy: *Sense and Contradiction.* A Study in Aristotle. 1975 ISBN 90-277-0565-8
15. W. R. Knorr: *The Evolution of the Euclidean Elements.* A Study of the Theory of Incommensurable Magnitudes and its Significance for Early Greek Geometry. 1975 ISBN 90-277-0509-7
16. Augustine: *De Dialectica.* Translated with Introduction and Notes by B. D. Jackson from the Text newly edited by J. Pinborg. 1975 ISBN 90-277-0538-9
17. Á. Szabó: *The Beginnings of Greek Mathematics.* Translated from German. 1978 ISBN 90-277-0819-3
18. Juan Luis Vives: *Against the Pseudodialecticians.* A Humanist Attack on Medieval Logic. Texts (in Latin), with Translation, Introduction and Notes by R. Guerlac. 1979 ISBN 90-277-0900-9

Synthese Historical Library

19. Peter of Ailly: *Concepts and Insolubles.* An Annotated Translation (from Latin) by P. V. Spade. 1980 ISBN 90-277-1079-1
20. S. Knuuttila (ed.): *Reforging the Great Chain of Being.* Studies of the History of Modal Theories. 1981 ISBN 90-277-1125-9
21. J. V. Buroker: *Space and Incongruence.* The Origin of Kant's Idealism. 1981 ISBN 90-277-1203-4
22. Marsilius of Inghen: *Treatises on the Properties of Terms.* A First Critical Edition of the *Suppositiones, Ampliationes, Appellationes, Restrictiones* and *Alienationes* with Introduction, Translation, Notes and Appendices by E. P. Bos. 1983 ISBN 90-277-1343-X
23. W. R. de Jong: *The Semantics of John Stuart Mill.* 1982 ISBN 90-277-1408-8
24. René Descartes: *Principles of Philosophy.* Translation with Explanatory Notes by V. R. Miller and R. P. Miller. 1983 ISBN 90-277-1451-7
25. T. Rudavsky (ed.): *Divine Omniscience and Omnipotence in Medieval Philosophy.* Islamic, Jewish and Christian Perspectives. 1985 ISBN 90-277-1750-8
26. William Heytesbury: *On Maxima and Minima.* Chapter V of *Rules for Solving Sophismata*, with an Anonymous 14th-century Discussion. Translation from Latin with an Introduction and Study by J. Longeway. 1984 ISBN 90-277-1868-7
27. Jean Buridan's *Logic. The Treatise on Supposition. The Treatise on Consequences.* Translation from Latin with a Philosophical Introduction by P. King. 1985 ISBN 90-277-1918-7
28. S. Knuuttila and J. Hintikka (eds.): *The Logic of Being.* Historical Studies. 1986 ISBN 90-277-2019-3
29. E. Sosa (ed.): *Essays on the Philosophy of George Berkeley.* 1987 ISBN 90-277-2405-9
30. B. Brundell: *Pierre Gassendi: From Aristotelianism to a New Natural Philosophy.* 1987 ISBN 90-277-2428-8
31. Adam de Wodeham: *Tractatus de indivisibilibus.* A Critical Edition with Introduction, Translation, and Textual Notes by R. Wood. 1988 ISBN 90-277-2424-5
32. N. Kretzmann (ed.): *Meaning and Inference in Medieval Philosophy.* Studies in Memory of J. Pinborg (1937–1982). 1988 ISBN 90-277-2577-2
33. S. Knuuttila (ed.): *Modern Modalities.* Studies of the History of Modal Theories from Medieval Nominalism to Logical Positivism. 1988 ISBN 90-277-2678-7
34. G. F. Scarre: *Logic and Reality in the Philosophy of John Stuart Mill.* 1988 ISBN 90-277-2739-2
35. J. van Rijen: *Aspects of Aristotle's Logic of Modalities.* 1989 ISBN 0-7923-0048-3
36. L. Baudry: *The Quarrel over Future Contingents (Louvain 1465–1475).* Unpublished Latin Texts collected and translated in French by L. Baudry. Translated from French by R. Guerlac. 1989 ISBN 0-7923-0454-3

THE NEW SYNTHESE HISTORICAL LIBRARY
Texts and Studies in the History of Philosophy

37. S. Payne: *John of the Cross and the Cognitive Value of Mysticism.* An Analysis of Sanjuanist Teaching and its Philosophical Implications for Contemporary Discussions of Mystical Experience. 1990 ISBN 0-7923-0707-0
38. D.D. Merrill: *Augustus De Morgan and the Logic of Relations.* 1990 ISBN 0-7923-0758-5
39. H. T. Goldstein (ed.): *Averroes' Questions in Physics.* 1991 ISBN 0-7923-0997-9
40. C.H. Manekin: *The Logic of Gersonides.* A Translation of *Sefer ha-Heqqesh ha-Yashar* (The *Book of the Correct Syllogism*) of Rabbi Levi ben Gershom with Introduction, Commentary, and Analytical Glossary. 1992 ISBN 0-7923-1513-8
41. George Berkeley: *De Motu* and *The Analyst.* A Modern Edition with Introductions and Commentary, edited en translated by Douglas M. Jesseph. 1992 ISBN 0-7923-1520-0
42. John Duns Scotus: *Contingency and Freedom.* Lectura I 39. Introduction, Translation and Commentary by A. Vos Jaczn., H. Veldhuis, A.H. Looman-Graaskamp, E. Dekker and N.W. den Bok. 1994 ISBN 0-7923-2707-1
43. Paul Thom: *The Logic of Essentialism.* An Interpretation of Aristotle's Modal Syllogistic. 1996 ISBN 0-7923-3987-8

Kluwer Academic Publishers – Dordrecht / Boston / London

www.ingramcontent.com/pod-product-compliance
Lightning Source LLC
Chambersburg PA
CBHW060747150926
51501CB00003B/119
* 9 7 8 0 7 9 2 3 3 9 8 7 8 *